Mathematics for Artificial Intelligence • Volume 3

Stochastic Dynamics for AI

Continuous-Time Methods for
Learning, Sampling, Generation, and Control

GUANGYU WANG

Stochastic Dynamics for AI: Continuous-Time Methods for Learning, Sampling, Generation, and Control
Mathematics for Artificial Intelligence, Volume 3
First Edition

Disclaimer. This work is provided for educational and research purposes. Every result is stated with the hypotheses under which it holds, and the status of each claim — classical, established, derived, model-dependent, interpretive, conjectural, or open — is marked in the text. Readers should verify that a result's stated hypotheses are satisfied before relying on it; in particular, carrying a theorem to a learned system changes the object, the metric, the quantifier, the regime and often the reference measure, and this book's purpose is to make those changes visible rather than to license the transfer. Citations identify the primary sources; the statements printed here are governed by their displayed hypotheses and scope, and readers should consult the sources for variants, extensions and stronger forms. The author makes no warranty, express or implied, as to the accuracy or completeness of the contents, and assumes no liability for any loss arising from their use. Nothing here is engineering advice, a safety guarantee, or a certificate about any deployed system. Views expressed here are the author's own.

Typeset by the author in LaTeX.

Contents

How to Read the Apparatus

Every numbered claim in this volume carries a status *badge*, and several carry a second, independent marker. The system is explained in Section 0.3; this page is the card to keep open while reading.

Badge	What it asserts
CLASSICAL	A standard result of the classical literature, quoted without proof or with a sketch. Safe to use; look up any textbook in the area.
ESTABLISHED	A result of the modern research literature, quoted with its conditions. The volume's contract is one primary source per claim, with its theorem number and normalisation.
DERIVED	Proved in this volume. Not a claim of priority: several are classical facts whose proofs appear because a later chapter depends on their exact form.
MODEL-DEPENDENT	True inside a stated idealised model, and asserted only there. The model is named in the statement, and the mini-ledger says what it assumes.
INTERPRETIVE	A reading, a taxonomy or an organising picture. Nothing may be derived from it, and no proof in this volume cites one.
CONJECTURE	Believed, not proved here, and flagged as such.
OPEN PROBLEM	A well-posed question the framework raises and does not answer.
WITHDRAWN	*Disposition, not status.* Attached to a proposed identity in Chapter 11 when this volume argues that no admissible version of it is being asserted. Withdrawn rows are kept rather than deleted, so that a reader arriving with the belief finds it addressed. It is neither an eighth status nor a sixth relation type.
EMPIRICAL REGULARITY	*Modality, not status.* A measured regularity across systems rather than a proof. Such claims are numbered as *Empirical Regularities*, never as theorems.

A third axis, used only in Chapter 11, classifies *relations* rather than claims. Five types are used, and only five: EXACT, LIMIT, REPRESENTATION, VARIATIONAL and ANALOGY. A proposed identity may additionally carry the disposition WITHDRAWN, which is not a sixth type.

Most results carry their formal hypotheses in the statement itself. Selected results are followed by a short interpretation or connection note; a notes block records what a result assumes and where it stops; and a compact caution appears only where a transfer error is genuinely likely. Empirical, model-dependent, conjectural and source-sensitive claims may carry a separate scope or evidence note. A chapter summary closes each chapter.

Prerequisite Map

The chapters are not a linear sequence. The dependency spine is

$$\text{Chapter 1} \longrightarrow \text{Chapter 2} \longrightarrow \{\text{Chapter 3, Chapter 4, Chapter 7}\} \longrightarrow \text{Chapter 10},$$

with Chapters 5 and 6 depending on Chapter 2 alone, and Chapters 8 and 9 depending on Chapters 2 and 7. Chapter 11 depends on everything and proves nothing.

To read	You need	Load-bearing results
Chapter 1	Multivariable calculus; measure-theoretic probability	Theorem 1.2, Theorem 1.33, Theorem 1.41
Chapter 2	Chapter 1	Proposition 2.12, Theorem 2.13, Proposition 2.25
Chapter 3	Chapters 1 and 2	Theorem 3.1, Proposition 10.2
Chapter 4	Chapter 2	Proposition 2.25, Corollary 2.14
Chapter 5	Chapter 2	Proposition 5.3, Theorem 5.10
Chapter 6	Chapter 1; Chapter 2 for §6.3	Proposition 6.2, Proposition 6.6
Chapter 7	Chapter 1; Chapter 3 for the path-space section	Proposition 7.2, Proposition 7.5
Chapter 8	Chapter 2	Proposition 8.2, Proposition 8.3
Chapter 9	Chapters 2, 7 and 8	Proposition 9.4, Proposition 9.5
Chapter 10	Chapters 3 and 7	Proposition 10.2, Proposition 10.13
Chapter 11	Any subset of the above	None; the chapter states no claims

CHAPTER 0

Introduction

0.1 The Question

Learning, sampling, generation, control and collective behaviour are usually studied with different tools. This book studies the continuous-time evolutions behind them: of states, of probability laws, of value functions, and of measures on whole paths. The dynamics of a probability law is the common spine — it is what Chapter 2 is about and what most later chapters return to — but not every object below is itself a probability law, and saying so at the outset matters: Chapter 1 works with parametrised ordinary differential equations, Chapter 7 with a value function solving a Hamilton–Jacobi–Bellman equation, and Chapter 9 with a coupled forward–backward system in which one equation runs in each direction. What the book is about is which constructions transfer between these levels, and under exactly which conditions.

That framing is productive and it is dangerous, in the same way and for the same reason. It is productive because a construction proved in one setting often has an exact counterpart in another — the time reversal that makes diffusion models work is the same theorem that gives the reverse dynamics of a physical process, and a likelihood-weighted score objective admits, under Girsanov, a reading as a relative entropy between path measures — the same quantity that appears as the cost of a stochastic control problem. It is dangerous because a resemblance between two constructions is easy to state and hard to bound, and an unbounded resemblance turns into an assertion that two problems are the same when they are merely similar.

The organising question of this book is therefore not "what unifies these subjects?" but a narrower one:

> *For each bridge between two of them, exactly how much weight does it carry?*

Answering that question for thirty-odd bridges is the content of the volume, and Table 11.3 is the summary. The last column of that table — what goes wrong outside the stated conditions — is the part that took the work, and it is the part not assembled in this cross-domain, condition-and-failure form in the references surveyed in Section 0.6.

0.2 Four Structures, Not One

The tempting version of this book's thesis is that there is one equation. There is not, and Table 11.1 says why: the constructions treated here fall into four families that behave differently enough that no single variational principle covers them.

A *dissipative system* descends a functional in a geometry, and the Fokker–Planck equation of a reversible diffusion is the model case (Theorem 2.13). A *Hamiltonian or kinetic-hypocoercive system* does not: its conservative core dissipates nothing, and where friction and noise are present they act on the velocity alone, so the dissipation available is degenerate and the arguments of Chapter 2 fail in a way that requires different tools (Proposition 4.8). A *controlled system* is characterised by a backward equation and by no forward descent at all (Proposition 7.2). A *strategic system* has an equilibrium rather than an optimum, and only a restricted class of them has a variational structure (Proposition 9.5).

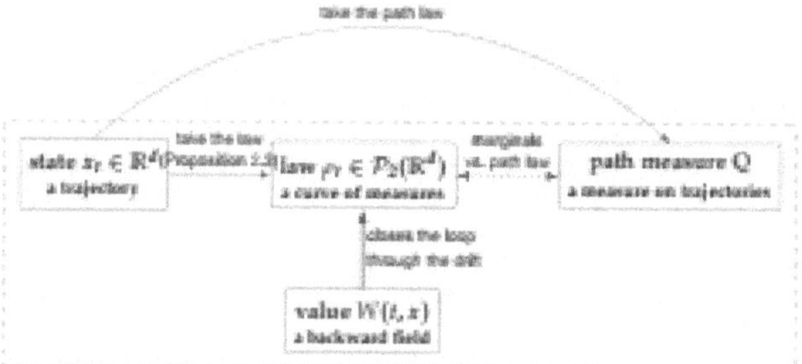

Figure 0.1: The four objects this volume moves, and the maps between them. Reading left to right loses information: a law does not determine a trajectory, and a family of marginals does not determine a path measure — the dotted arrow is the one most often crossed in the wrong direction. What this figure does not imply: that the four levels are related by a single construction, or that a statement proved at one level transfers to another. Which transfers are exact, and under what conditions, is the content of Chapter 11.

Each of the last three is separated from the first by a theorem in this volume, not by an assertion. That is the pattern the book follows throughout: where a general claim would be too strong, the correction is a construction that bounds it. Proposition 2.16 exhibits a diffusion that dissipates the free energy at exactly the rate of the W_2 gradient flow and is not one; Proposition 5.15 computes the curl of a velocity field that is widely described as a gradient; Proposition 6.6 gives the exact condition under which stochastic gradient descent has a temperature; Proposition 10.13 separates three constructions usually written as one. A caveat costs a sentence and is forgotten; a counterexample with a computation attached is harder to lose.

Principle 11.1 states the reading that organises all of this, and it is tagged INTERPRETIVE: it is a way of arranging the material and not a theorem, and no proof in this volume depends on it.

0.3 How to Read a Claim in This Book

Every numbered claim carries a badge saying how strongly it is being asserted. There are seven, and only seven.

- CLASSICAL Established in the standard mathematical literature. Hypotheses are restated in full rather than gestured at; Theorem 1.33 is the model.
- ESTABLISHED Established in the modern research literature and attributed to exactly one primary source, with its conditions — whether or not a proof is reproduced here. The badge records provenance, not whether this volume supplies the argument.
- DERIVED Established *in* this volume from preceding material, rather than imported as a quoted theorem, under conditions stated here. Most of the corrections in this book are of this kind.
- MODEL-DEPENDENT True inside an idealisation defined in the text. Every such claim carries a three-field ledger naming the model, what is exact inside it, and what does not transfer; Proposition 3.23 and Theorem 6.12 are the two in this volume.
- INTERPRETIVE A way of organising a phenomenon. Not a theorem, and no proof anywhere in the book may cite one.
- CONJECTURE Believed plausible and not proved. Each carries a test that would refute it; Conjecture 6.14 is the one in this volume.
- OPEN PROBLEM A well-posed question the framework raises and does not answer.

A second, independent axis records the *kind* of evidence. A claim that is a measured regularity across systems rather than a proof carries the badge EMPIRICAL REGULARITY in addition to its status, and is numbered as an *Empirical Regularity* rather than as a theorem or a proposition — a measured pattern is not a theorem, and calling it one while denying it in an accompanying note would be exactly the kind of label–content mismatch this apparatus exists to prevent. There are two: Empirical Regularity 6.10 and Empirical Regularity 6.15. They still carry a status badge, because a published measurement is attributable in the same way a theorem is; ESTABLISHED there means "reported in one primary source with its conditions", not "proved". This is not an eighth status: status says how firmly, modality says of what kind, and the environment name says which.

Formal hypotheses live in the theorem statement, and the proof shows where each is used. Around that, a few devices do different jobs.

- An interpretation note explains what a result says once the symbols are stripped away: the mechanism behind it, the object it moves, and why it is worth remembering. A connection note ties it to a construction in another chapter. These are the notes to read first.
- A notes block records what a result assumes and where it stops. It is a reading aid, not a second statement of the hypotheses.
- A caution appears only where the same formula is routinely misread across objects, metrics or regimes — rarely, and never as a matter of routine.
- A convention box fixes a normalisation once for the whole volume. Table 0.2 lists them.
- An operational reading says what a result means for practice, and goes exactly one step beyond the theorem. The step after that is always named and always deferred.

- A **what does not follow** box collects the inferences the surrounding material invites and does not support.

Each chapter closes with a **boundary statement**: what it established, what it proposed, what is interpretive, the common pitfalls, what is not established, and what is not addressed. Read backwards, those six fields are a map of the chapter's limits, and they are the most compressed statement of what the book does not do.

0.3.1 The transfer rule

A result proved about one object is used about another only if four things match: the *object* (a trajectory, a law, a path measure, an empirical measure), the *metric* in which the conclusion is stated, the *quantifier* (for all, for some, generically), and the *regime* (continuous time or discretised, finite or infinite particle number, fixed or vanishing step). Where a transfer changes one of the four, this volume says so and names what was changed. Where it cannot be made, the result stays where it was proved.

0.4 How Claims Fail

Almost every error found while writing this book was one of six substitutions. They are catalogued with their instances in Table 11.2; in brief:

- **Pathwise versus in law**: a statement about trajectories used as one about their law, or the reverse.
- **Metric substitution**: a rate proved in one divergence quoted in another.
- **Regime substitution**: a result proved at vanishing step size applied at a fixed one.
- **Geometry substitution**: "gradient flow" without a metric.
- **Constraint substitution**: a soft penalty read as a hard constraint.
- **Normalisation**: two conventions used inside one argument.

All six are invisible to ordinary proofreading, because both sides of the substitution are true statements about objects carrying the same symbols and the same units. They are the reason this volume states its conventions in boxes and its hypotheses in the first column of every table.

0.5 What This Volume Does Not Attempt

- **No code and no benchmarks.** Computational exercises describe a task and its acceptance criteria; they contain no implementation, and no number in this book is a measurement of a system.
- **No architecture survey.** Networks appear as objects with named properties — a bounded feature map, a mean-field scaling — and never as designs to be compared.

- **No complete course in stochastic analysis.** Chapter 1 is a working foundation with the hypotheses stated in full, not a substitute for the books in Section 0.6.
- **No guarantee about a deployed system.** Every rate in this book is a rate for an identified idealisation. Where a result would have to be composed across idealisations to say something about a program, the composition is described and not performed; Table 3.1 is the model case.
- **No empirical claims beyond two.** The two carrying the empirical-regularity badge are Empirical Regularity 6.10 and Empirical Regularity 6.15, and their scope is the settings in which they have been measured.

0.6 Where This Sits Among Other Books

This volume is mathematically narrower than each of the following and broader in the range of constructions it touches. It replaces none of them, and a reader who wants the theory rather than the map should go to them.

- For stochastic analysis — the Itô integral, stochastic differential equations, martingale theory — the standard texts of Øksendal and of Karatzas and Shreve. Chapter 1 states what this book needs and proves what it uses inside its own proofs.
- For diffusion processes and their applications, Pavliotis; for Markov diffusion operators, functional inequalities and curvature, Bakry, Gentil and Ledoux. Section 2.3 and Section 2.4 are a working subset of the latter, organised around which hypothesis buys which rate.
- For gradient flows in metric spaces, Ambrosio, Gigli and Savaré; for optimal transport, Villani and Santambrogio. Section 2.2 quotes the metric-space theory and proves the two identities it uses downstream.
- For log-concave sampling, Chewi. Chapter 4 is narrower and is organised around what a lower bound can and cannot say.
- For stochastic control, Fleming and Soner and Pham; for mean-field games, Carmona and Delarue. Chapters 7 and 9 take the verification direction and the two systems, and leave existence and regularity to those references.

The delta is the last column of Table 11.3, and the six substitutions of Table 11.2. The individual ingredients belong to established mathematical literatures, and a specialist in any one of the areas above will recognise most of them. What is new here is a transfer calculus: a systematic record of which cross-domain identifications are exact, which require a limit, a representation change or a variational reformulation, and which fail — and, for each, what it costs to use outside its hypotheses.

0.7 Two Paths Through the Book

Table 0.1: Two routes. The right-hand column is not a compressed version of the left: it covers one or two load-bearing results per chapter and omits the rest, and a reader following it will not have seen the examples, proofs and cross-connections that make the load-bearing results intuitive.

Chapter	Full path (no stochastic analysis assumed)	Short path (stochastic analysis assumed)
Chapter 1	All of it. Section 1.5.4 sets the proof register for the volume	Section 1.5.4's convention box; Section 1.5.5; Section 1.6.4
Chapter 2	All of it	Section 2.1.2; Section 2.2; Table 2.2
Chapter 3	All of it	Section 3.1; Proposition 3.17; Theorem 3.20; Table 3.1
Chapter 4	All of it	Proposition 4.2; Table 4.1; Proposition 4.17
Chapter 5	All of it	Proposition 5.3; Table 5.1; Table 5.2
Chapter 6	All of it	Proposition 6.4; Proposition 6.6; Table 6.1
Chapter 7	All of it	Proposition 7.4; Remark 7.8; Proposition 7.12
Chapter 8	All of it	Table 8.1; Proposition 8.8
Chapter 9	All of it	Proposition 9.4; Proposition 9.5
Chapter 10	All of it	Proposition 10.2; Proposition 10.13; Proposition 10.15
Chapter 11	All of it, last	All of it, first — it is the shortest description of the book

Exercises are in four tiers throughout: direct application, completing a derivation, probing the boundary, and computational. The third tier is the one this book is for. Its questions ask where a statement stops being true, and they can be answered by a reader who understands what the assumptions do, which object the statement concerns, and how the result connects to the constructions around it.

0.8 Conventions Fixed Once

Each of the following is fixed in one place and used everywhere without further comment. A reader who finds a second convention anywhere in the book has found a defect.

Table 0.2: The volume's conventions, and where each is fixed. The last column is the failure the convention prevents; every one of them has been observed in the literature.

Convention	Fixed in	What it prevents
Itô unless a circle appears; generator carries $\frac{1}{2}\operatorname{tr}(a\nabla^2)$	Section 1.5.4, with the conversion in Section 1.5.5	A drift differing by $\frac{1}{2}\sigma'\sigma$, invisible in the notation
$t = 0$ carries the data, $t = 1$ the reference law	Section 3.4.1	A sign flip on every velocity field, invisible until samples are drawn

continued overleaf

Table 0.2: (continued)

Convention	Fixed in	What it prevents
Reverse-time equations are members of a one-parameter family with λ named	Section 3.1.2	Reading $\frac{1}{2}g^2\nabla \log p$ and $g^2\nabla \log p$ as two conventions rather than two equations
LSI(κ) means $\mathrm{KL}(\|) \leq \mathcal{I}(\|)/(2\kappa)$	Section 2.2.1	A rate quoted from a source using the competing normalisation
Relative Fisher information $\mathcal{I}(\cdot \| \pi)$ is never the Fisher information matrix $G(\varphi)$	Section 2.2.1, discharged in Section 5.4	Two unrelated objects sharing a name
Every occurrence of "gradient flow" names its metric	Section 2.2.1	The geometry substitution; this one is mechanically checked

0.9 What Is New Here

Most of the results in this book are not. The classical material is classical, the modern results are attributed, and a specialist will recognise almost every theorem. Three things are the volume's own.

The first is the sorting: which claim belongs to which of the seven statuses, and the discipline that a claim's badge is chosen before its prose. The second is the set of constructions built to bound claims that would otherwise be stated too strongly — among them Proposition 2.16, Proposition 5.15, Proposition 6.4, Proposition 6.6, Proposition 6.13, Proposition 8.3, Proposition 9.5 and Proposition 10.13. Several of these are elementary; that is the point, since an elementary obstruction is one a reader can check.

The third is the Atlas of Chapter 11: thirty-odd relations between constructions in different areas, each with its type, its exact conditions, and what goes wrong outside them. Relations that this volume argues against are kept in that table rather than deleted from it, so that a reader who arrives holding one finds it addressed.

This chapter states no claim of its own and therefore closes without a boundary statement. Its limits are the limits of the chapters it describes, and they are stated there.

CHAPTER 1

Deterministic and Stochastic Dynamics

Many of the constructions in this volume are differential equations, but they move different objects: states, probability laws, value functions, and coupled forward–backward fields. This chapter supplies the state-space dynamics and the stochastic calculus on which the later law-level constructions depend. Before any of them can be written down, two ingredients have to be in place: the deterministic theory that says when a differential equation has a solution at all, and the stochastic calculus that says what it means to differentiate along a path that is nowhere differentiable. This chapter supplies both, and it is the only chapter of the book with no machine learning in it beyond Section 1.3.

That is deliberate, and it is worth saying why. The results here are old and their hypotheses are well understood, which makes them the right place to fix the habits the rest of the volume depends on: stating what a theorem assumes rather than what it feels like, naming the metric before using the word distance, and distinguishing a statement about a trajectory from a statement about the law of a trajectory. Every one of those habits will be load-bearing by Chapter 2, and each of them is cheapest to acquire on material where nothing is at stake.

The chapter has three movements. Sections 1.1 to 1.3 are deterministic: existence and uniqueness, the Euclidean gradient flow and its convergence, and the continuum limit of a residual network. Sections 1.4 and 1.5 construct Brownian motion and the Itô integral. Section 1.6 puts the two together, and Section 1.7 states what the chapter does and does not settle.

1.1 Ordinary Differential Equations

1.1.1 Existence, uniqueness, and the hypotheses that buy them

Definition 1.1 (Initial value problem). Given $b : [0, T] \times \mathbb{R}^d \to \mathbb{R}^d$ and $x_0 \in \mathbb{R}^d$, the *initial value problem* is to find $x \in C^1([0, T]; \mathbb{R}^d)$ with

$$\dot{x}(t) = b(t, x(t)) \quad \text{for } t \in [0, T], \qquad x(0) = x_0. \tag{1.1}$$

The field b is *locally Lipschitz in x, uniformly in t* if for every $R > 0$ there is $L_R < \infty$ with $|b(t, x) - b(t, y)| \le L_R |x - y|$ for all $t \in [0, T]$ and all $|x|, |y| \le R$; it is *globally Lipschitz* if L_R may be taken independent of R.

Continuity of b alone gives existence and not uniqueness. The standard example is $\dot{x} = |x|^{1/2}$ from $x_0 = 0$, which is solved by $x \equiv 0$ and, for every $s \ge 0$, by $x(t) = \frac{1}{4}(t - s)^2 1_{\{t > s\}}$: a continuum of solutions leaving the same initial point at different times. The square root is continuous at 0 and not Lipschitz there, and that single failure is what the uniqueness half of the next theorem is paying for.

Theorem 1.2 (Picard–Lindelöf) CLASSICAL. *Let b be continuous on $[0, T] \times \mathbb{R}^d$ and locally Lipschitz in x, uniformly in t. Then for every $x_0 \in \mathbb{R}^d$ there exist $\tau \in (0, T]$ and a unique $x \in C^1([0, \tau]; \mathbb{R}^d)$ solving (1.1) on $[0, \tau]$. If b is globally Lipschitz, then $\tau = T$ and the solution is unique on all of $[0, T]$.*

Proof. Fix $R = |x_0| + 1$, let $L = L_R$ be the Lipschitz constant on the ball B_R, and let $M = \sup\{|b(t, y)| : t \in [0, T],\ |y| \le R\}$, finite by continuity. Choose $\tau \le \min\{T, 1/M, 1/(2L)\}$ and let $\mathcal{X}$ be the set of continuous $x : [0, \tau] \to \overline{B_R}$ with $x(0) = x_0$, a closed subset of $C([0, \tau]; \mathbb{R}^d)$ with the supremum norm, hence a complete metric space.

A continuous x solves (1.1) on $[0, \tau]$ if and only if it solves the integral equation $x(t) = x_0 + \int_0^t b(s, x(s))\, ds$; the forward implication is the fundamental theorem of calculus and the converse follows because the right-hand side is C^1 whenever x is continuous. Define Φ on $\mathcal{X}$ by that right-hand side. Then Φ maps $\mathcal{X}$ into itself, since $|\Phi x(t) - x_0| \le \tau M \le 1$, and for $x, y \in \mathcal{X}$,

$$|\Phi x(t) - \Phi y(t)| \le \int_0^t |b(s, x(s)) - b(s, y(s))|\, ds \le L\tau \sup_{[0,\tau]} |x - y| \le \tfrac{1}{2} \sup_{[0,\tau]} |x - y|.$$

So Φ is a contraction, and Banach's fixed point theorem gives a unique fixed point in $\mathcal{X}$. Uniqueness among *all* continuous solutions, not only those staying in B_R, follows because any solution is continuous and therefore remains in B_R for a possibly shorter time, on which the contraction argument applies.

If b is globally Lipschitz with constant L, the same estimate holds on any subinterval of length $1/(2L)$ with $\mathcal{X}$ replaced by all of $C([0, \tau]; \mathbb{R}^d)$, and the solution is continued from τ to 2τ and so on, reaching T in finitely many steps. ∎

Remark 1.3 (Local is not global). Without a growth condition the solution can leave every compact set in finite time. For $\dot{x} = x^2$ in $\mathbb{R}$ with $x_0 > 0$ the solution is $x(t) = x_0/(1 - x_0 t)$, which blows up at $t = 1/x_0$; the field is locally Lipschitz everywhere and globally Lipschitz nowhere. Linear growth, $|b(t, x)| \le C(1 + |x|)$, is the usual sufficient condition for global existence, and it is exactly what Section 1.1.2 will show to be enough.

1.1.2 Grönwall's inequality, and the price of a Lipschitz constant

Proposition 1.4 (Grönwall's inequality, integral form) CLASSICAL. *Let $u : [0, T] \to [0, \infty)$ be continuous and suppose that for some constants $a \ge 0$, $L \ge 0$,*

$$u(t) \;\le\; a + L \int_0^t u(s)\, ds \qquad \text{for all } t \in [0, T]. \tag{1.2}$$

Then $u(t) \le a e^{Lt}$ on $[0, T]$.

Proof. Let $v(t) = a + L \int_0^t u$, so v is C^1, $u \le v$, and $v' = Lu \le Lv$. Hence $(e^{-Lt} v)' = e^{-Lt}(v' - Lv) \le 0$, so $e^{-Lt} v(t) \le v(0) = a$, and $u(t) \le v(t) \le a e^{Lt}$. ∎

Corollary 1.5 (Continuous dependence on the initial condition) DERIVED. *Let b be globally Lipschitz in x with constant L, uniformly in t, and let x, y solve (1.1) on $[0, T]$ from x_0 and y_0. Then*

$$|x(t) - y(t)| \leq |x_0 - y_0|\, e^{Lt} \qquad \text{for all } t \in [0, T], \tag{1.3}$$

and the exponent cannot be improved: for $b(x) = Lx$ in $\mathbb{R}$, equality holds for every t.

Proof. Subtracting the integral equations, $|x(t) - y(t)| \leq |x_0 - y_0| + L \int_0^t |x(s) - y(s)|\, ds$, which is (1.2) with $u = |x - y|$ and $a = |x_0 - y_0|$; Proposition 1.4 gives (1.3). For $b(x) = Lx$ the solutions are $x_0 e^{Lt}$ and $y_0 e^{Lt}$, so the difference is exactly $|x_0 - y_0| e^{Lt}$. ■

Interpretation

Equation (1.3) is an amplification law for uncertainty: whatever you did not know about the initial state, you may know e^{Lt} times less about the state at time t. The Lipschitz constant is the exchange rate, and the exponential is the worst case rather than the typical one — it is attained by a linear expanding field and is wildly pessimistic for a contractive one.

Two features recur throughout the volume. The estimate compares two solutions of the *same* equation from different data; comparing solutions of *different* equations, which is what every numerical scheme and every learned model does, needs a second term (Exercise 1.3). And the mechanism — differentiate a distance, bound its derivative, integrate — is the same one that produces exponential convergence in Chapter 2 once the distance is a divergence between probability laws instead of a distance between points.

Notes on Corollary 1.5

Where it is sharp and where it is useless. The bound is attained by a linear expanding field, so no better exponent holds under these hypotheses alone. Under stronger hypotheses it is enormously pessimistic: if b is one-sided Lipschitz with a negative constant — $\langle b(x) - b(y), x - y\rangle \leq -\mu\,|x - y|^2$, as holds for $b = -\nabla f$ with f μ-strongly convex — then the same argument with $\frac{d}{dt}|x - y|^2$ gives $|x(t) - y(t)| \leq |x_0 - y_0|\, e^{-\mu t}$, an exponential contraction rather than an expansion. Which of the two applies is not a matter of degree: they differ in the sign of a scalar, and no interpolation between them is available from the Lipschitz constant alone.

What it does not say. It is a statement about two solutions of the *same* equation from different starting points. It says nothing about two solutions of *different* equations, which is the situation in every numerical and every learned setting; that estimate carries an extra term $\int_0^t |b - \tilde{b}|$ inside the Grönwall argument and is Exercise 1.3. Confusing the two is the reason error bounds for learned dynamics are sometimes quoted without the approximation error in them.

Operational reading

Read (1.3) as a horizon. If the field is L-Lipschitz and the initial state is known to precision ε, the state is known to precision 1 only for times up to about $L^{-1} \log(1/\varepsilon)$: accuracy buys time logarithmically, which is why doubling the precision of a measurement extends a useful forecast by a fixed additive amount rather than a multiplicative one.

That is one step beyond the corollary, and this box stops there. Whether a particular trained dynamical model is usable over a particular horizon depends on its approximation error as well as its Lipschitz constant, and on whether the true dynamics contracts; both enter Exercise 1.3 and neither is supplied here.

1.1.3 Linear systems

The linear case is the one place in this chapter where everything can be computed, and it is worth doing carefully because it is the local model for everything else.

Proposition 1.6 (Linear systems and the matrix exponential) CLASSICAL. *Let $A \in \mathbb{R}^{d\times d}$. The unique solution of $\dot{x} = Ax$, $x(0) = x_0$, is $x(t) = e^{tA}x_0$ with $e^{tA} := \sum_{k\geq 0} t^k A^k / k!$, the series converging absolutely in any submultiplicative norm. Writing $\alpha(A) := \max\{\Re\lambda : \lambda \in \operatorname{spec}(A)\}$ for the spectral abscissa,*

$$\lim_{t\to\infty} \frac{1}{t} \log \left\| e^{tA} \right\| = \alpha(A), \tag{1.4}$$

so every solution converges to zero, for every initial condition, if and only if $\alpha(A) < 0$. The quantifier is over all initial conditions, and it has to be: the existence of a decaying solution says nothing about $\alpha(A)$, since the zero solution decays whatever A is, and an unstable A still has decaying solutions along the stable eigenspace.

Proof. Absolute convergence gives term-by-term differentiation, $\frac{d}{dt}e^{tA} = Ae^{tA}$, so $e^{tA}x_0$ solves the equation; uniqueness is Theorem 1.2, A being globally Lipschitz. For (1.4), put A in Jordan form $A = S(D+N)S^{-1}$ with D diagonal and N nilpotent commuting with D; then $e^{tA} = Se^{tD}e^{tN}S^{-1}$ with e^{tN} polynomial in t of degree less than d, so $\left\|e^{tA}\right\| = e^{\alpha(A)t}\,p(t)$ up to constants for a polynomial p of degree less than d, and taking $t^{-1}\log$ kills the polynomial factor. ■

What does not follow

The spectrum does not control the transient. Equation (1.4) is a statement about $t \to \infty$. At finite t, $\|e^{tA}\|$ can be enormous even when $\alpha(A) < 0$, and the gap grows with the non-normality of A. For $A = \begin{pmatrix} -1 & M \\ 0 & -2 \end{pmatrix}$ every eigenvalue is negative, yet $\|e^{tA}\|$ reaches order M before it decays. The quantity that detects the onset of the transient is the numerical abscissa $\frac{1}{2}\lambda_{\max}(A + A^{\mathsf{T}})$ — *for the Euclidean operator norm*, in which it equals $\frac{d}{dt}\big|_{t=0}\|e^{tA}\|$, so its sign decides whether the norm grows initially, and for this example it is of order $M/2$ and positive. The norm matters: (1.4) holds in any submultiplicative norm, whereas the initial derivative and hence the numerical abscissa do not, and a transient visible in one norm can be invisible in another. It is a diagnostic at $t = 0$ and not a description of the whole transient: how large the excursion becomes, and for how long, is governed by the non-normality of A through its pseudospectrum, and neither abscissa determines it.

This matters here because linearisation is how every local analysis in this volume is set up, and because the same confusion recurs in the discrete setting: a Jacobian with spectral radius below one does not imply that the iterates decrease in norm, and a recurrent map can be spectrally stable and still amplify a perturbation by orders of magnitude over the length of a sequence. Whenever a claim in this book is about behaviour at finite time, the spectrum is not the right object, and Section 2.4 says what is.

1.2 Gradient Flows in Euclidean Space

1.2.1 The flow and what it dissipates

Definition 1.7 (Euclidean gradient flow). Let $f \in C^1(\mathbb{R}^p)$. The *Euclidean gradient flow* of f is the solution of

$$\dot{\theta}(t) = -\nabla f(\theta(t)), \qquad \theta(0) = \theta_0. \tag{1.5}$$

The adjective is not decoration. Equation (1.5) is steepest descent with respect to the Euclidean norm on $\mathbb{R}^p$: among all unit-norm directions v, the one minimising $\frac{d}{ds} f(\theta + sv)|_{s=0} = \langle \nabla f, v \rangle$ is $-\nabla f / |\nabla f|$, and *unit-norm* refers to a choice of norm. Change the norm and the flow changes: with a positive definite G and the norm $|v|_G^2 = \langle v, Gv \rangle$, steepest descent is $\dot{\theta} = -G^{-1}\nabla f$, the natural gradient flow of Section 5.4. Chapter 2 replaces $\mathbb{R}^p$ by a space of probability measures and the Euclidean norm by the Wasserstein metric, and obtains a different equation again — from the same functional. This volume therefore names the metric at every occurrence of the phrase; an unnamed one would violate the convention of Section 0.8.

Proposition 1.8 (Energy dissipation) DERIVED. *Let $f \in C^1$ with ∇f locally Lipschitz, and let θ solve (1.5) on $[0, T]$. Then*

$$\frac{d}{dt} f(\theta(t)) = -|\nabla f(\theta(t))|^2 \le 0, \tag{1.6}$$

so $f(\theta(t))$ is non-increasing, and

$$f(\theta_0) - f(\theta(T)) = \int_0^T |\nabla f(\theta(t))|^2 \, dt. \tag{1.7}$$

If in addition f is bounded below, say $f \ge f_*$, then the solution is global: it extends to $[0, \infty)$, and

$$\int_0^\infty |\nabla f(\theta(t))|^2 \, dt \le f(\theta_0) - f_* < \infty, \qquad \textit{hence} \qquad \liminf_{t \to \infty} |\nabla f(\theta(t))| = 0. \tag{1.8}$$

Proof. The chain rule gives $\frac{d}{dt} f(\theta) = \langle \nabla f(\theta), \dot{\theta} \rangle = -|\nabla f(\theta)|^2$; integrating over $[0, T]$ gives (1.7).

For the second part, first note that the solution cannot escape in finite time. Let $[0, T_{\max})$ be its maximal interval of existence. By (1.7) and $f \ge f_*$, Cauchy–Schwarz bounds the path length on any $[0, T] \subset [0, T_{\max})$ by

$$\int_0^T |\dot{\theta}| \, dt \le \sqrt{T} \Big(\int_0^T |\nabla f(\theta)|^2 \, dt \Big)^{1/2} \le \sqrt{T} \, (f(\theta_0) - f_*)^{1/2},$$

which is finite as $T \uparrow T_{\max}$ whenever $T_{\max} < \infty$. A curve of finite length has a limit point, so $\theta(t)$ stays in a compact set and converges as $t \uparrow T_{\max}$; the local existence theorem (Theorem 1.2) then continues the solution past $T_{\max}$, contradicting maximality. Hence $T_{\max} = \infty$.

With the solution global, the left side of (1.7) is bounded by $f(\theta_0) - f_*$ uniformly in T, so the integral converges; a non-negative function with convergent integral has liminf zero along $t \to \infty$. ∎

Notes on Proposition 1.8

What the conclusion is, exactly. That $|\nabla f|$ has liminf zero. Neither of the two natural strengthenings follows.

Convergence of $\theta(t)$ does not follow, even on a bounded trajectory. The counterexample is the C^{∞} "Mexican hat" on $\mathbb{R}^2$: in polar coordinates a radial profile with a circular trough at $r = 1$, modulated in the angle so that the trough's depth rotates slowly, with all derivatives matched to zero at $r = 1$ so that the function is smooth there. The gradient flow then spirals inwards without settling, the trajectory is bounded, and its ω-limit set is the whole circle. The construction is delicate — the angular modulation must decay fast enough for smoothness and slowly enough to prevent convergence — and it is quoted rather than carried out here. Łojasiewicz-type inequality, which forces the trajectory to have finite length; that is a hypothesis about f and not a consequence of (1.5).

Limit zero of $|\nabla f|$ does not follow either — but only for unbounded trajectories. If the orbit is precompact, then $\dot\theta$ is bounded and ∇f is uniformly continuous on the closure of the orbit, so $t \mapsto |\nabla f(\theta(t))|^2$ is uniformly continuous with a convergent integral, and Barbălat's lemma upgrades the liminf to a limit. A trajectory along which $|\nabla f|$ returns to a fixed positive value infinitely often must therefore leave every compact set, and any example exhibiting one is using escape to infinity rather than oscillation in a bounded region. This is worth stating because the two failures are routinely quoted as one.

What it does not say. It says nothing about a discretisation. The monotonicity in (1.6) is exact and unconditional for the flow; for gradient descent it holds only for step sizes below a threshold set by the smoothness of f (Proposition 1.11 and Section 6.4), and in the regime where trained networks are usually run it fails. Reading (1.6) as a statement about training is the regime substitution of Table 11.2.

1.2.2 Rates, and the hypotheses that produce them

Dissipation gives monotonicity and no rate. A rate requires an inequality relating the objective to its gradient — exactly the pattern that Chapter 2 repeats with the free energy and the Fisher information, and Chapter 5 with the free energy and the kernel Stein discrepancy. It is worth seeing the pattern first in the case where it is elementary.

Theorem 1.9 (Convergence under strong convexity) CLASSICAL. *Let $f \in C^1(\mathbb{R}^p)$ be μ-strongly convex for some $\mu > 0$, that is $f(y) \ge f(x) + \langle \nabla f(x), y - x\rangle + \frac{\mu}{2}|y - x|^2$ for all x, y. Let θ_* be its unique minimiser and $f_* = f(\theta_*)$. Then along (1.5), for all $t \ge 0$,*

$$f(\theta(t)) - f_* \le e^{-2\mu t}(f(\theta_0) - f_*), \qquad |\theta(t) - \theta_*| \le e^{-\mu t}|\theta_0 - \theta_*|. \tag{1.9}$$

Proof. Minimising both sides of the strong convexity inequality over y gives the Polyak–Łojasiewicz inequality $|\nabla f(x)|^2 \ge 2\mu(f(x) - f_*)$: the left side of the inequality is minimised at $y = \theta_*$, giving $f_* \ge f(x) - \frac{1}{2\mu}|\nabla f(x)|^2$. With $h(t) = f(\theta(t)) - f_*$, Proposition 1.8 gives $h' = -|\nabla f|^2 \le -2\mu h$, and Grönwall gives the first bound. For the second, strong convexity implies $\langle \nabla f(x) - \nabla f(y), x - y\rangle \ge \mu|x - y|^2$, so with $g(t) = |\theta(t) - \theta_*|^2$ and $\nabla f(\theta_*) = 0$, $g' = 2\langle\dot\theta, \theta - \theta_*\rangle = -2\langle\nabla f(\theta) - \nabla f(\theta_*), \theta - \theta_*\rangle \le -2\mu g$, and Grönwall again. ∎

Theorem 1.10 (Convergence under the Polyak–Łojasiewicz inequality) CLASSICAL. *Suppose $f \in C^1$ is bounded below by $f_* = \inf f$ and satisfies*

$$\tfrac{1}{2}|\nabla f(x)|^2 \ge \mu(f(x) - f_*) \qquad \textit{for all } x \in \mathbb{R}^p. \tag{1.10}$$

Then $f(\theta(t)) - f_* \le e^{-2\mu t}(f(\theta_0) - f_*)$ along (1.5). The minimiser need not be unique, and no bound on $|\theta(t) - \theta_*|$ follows.

Proof. Identical to the first half of the proof of Theorem 1.9, with (1.10) in place of the consequence of strong convexity. For the second sentence, $f(x_1, x_2) = x_1^2$ on $\mathbb{R}^2$ satisfies (1.10) with $\mu = 1$ and has a line of minimisers; the flow converges to the point of that line with the same second coordinate as θ_0, so the limit depends on the initialisation and no bound in terms of $f(\theta_0) - f_*$ alone can hold. ∎

The two theorems have the same conclusion about function values and different hypotheses, and the difference is not cosmetic. The following table is the first instance in this volume of a construction used in three later chapters: a small comparison of hypotheses that are routinely conflated, each with what it does and does not buy. Its later relatives are the functional-inequality table of Section 2.4 and the regime table of Table 6.1.

Table 1.1: Four hypotheses on f, in decreasing strength. Only the first gives uniqueness and a rate for the distance to *the* minimiser; the Polyak–Łojasiewicz row gives a value rate and no distance statement at all; the Łojasiewicz row gives convergence to *a* point and, once its local constants and the entry time into the region are fixed, a tail-distance rate that follows from finite length. The last column is the one that is usually assumed rather than checked.

Hypothesis	**Rate on $f - f_*$**	**Rate on $\lvert\theta - \theta_*\rvert$**	**Survives explicit discretisation?**
μ-strong convexity	$e^{-2\mu t}$ (Theorem 1.9)	$e^{-\mu t}$, minimiser unique	Yes, for $\eta < 2/\beta$ with β the smoothness constant
Polyak–Łojasiewicz (1.10)	$e^{-2\mu t}$ (Theorem 1.10)	None; minimisers form a set	Yes, same threshold; the same proof applies
Łojasiewicz with exponent $\vartheta \in (0, \frac{1}{2})$	Polynomial, $t^{-1/(1-2\vartheta)}$	Convergence to a single point θ_∞, with the tail bound $\lvert\theta(t) - \theta_\infty\rvert \le (\vartheta c)^{-1}(f(\theta(t)) - f_*)^\vartheta$ of Exercise 1.10, hence $O(t^{-\vartheta/(1-2\vartheta)})$ — but only once c, ϑ and the entry time into the region where the inequality holds are fixed; no rate uniform in the initialisation	Yes, with the same exponent, *provided* ∇f is β-Lipschitz and $\eta < 2/\beta$; the inequality alone does not make the explicit scheme stable

continued overleaf

Table 1.1: (continued)

Hypothesis	Rate on $f - f_*$	Rate on $\|\theta - \theta_*\|$	Survives explicit discretisation?
Convexity alone	$O(1/t)$ for the *flow*, and no better in general	None	Not from convexity alone. The explicit scheme need not even be stable: a discrete $O(1/k)$ rate requires additional smoothness — a β-Lipschitz gradient — together with η below the corresponding threshold, and then the constant involves $\|\theta_0 - \theta_*\|$, which is not observable

Notes on Theorem 1.9 and Theorem 1.10

Assumed. A global inequality, on all of $\mathbb{R}^p$. Both hypotheses are routinely asserted locally — near a minimiser, or on a sublevel set — and a local version gives a local conclusion only, valid until the trajectory leaves the region. The proofs above use the inequality at $\theta(t)$ for every t and nowhere else, so they localise cleanly, which is worth knowing precisely because the localisation is so often left implicit.

Where the hypotheses fail. Standard multilayer parameterisations are generally non-convex in the weights, and many common architectures violate global Polyak–Łojasiewicz because they possess stationary points with $\nabla f = 0$ and $f > f_*$, which contradicts (1.10). It is worth being precise about what the usual symmetry argument does and does not give: permutation and rescaling symmetries produce non-identifiability and flat directions, and they imply that minimisers are non-unique, but flatness alone does not manufacture a *non-optimal* stationary point — that requires an argument about the architecture and the data, and for particular degenerate cases the conclusion can fail. The universal reading, "no multilayer loss is ever PL", is stronger than anything proved here or needed below. What is sometimes available is (1.10) on a neighbourhood of the trajectory in an overparameterised regime, which is a hypothesis about the initialisation and the width, and is treated as such in Chapter 8.

What they do not say. Neither theorem says anything about the minimiser reached when there are many, nor about generalisation, nor about any discretisation beyond what the last column of Table 1.1 records. The step from "the flow converges quickly" to "training converges quickly" crosses two regimes and is taken nowhere in this volume.

1.2.3 Gradient descent as a discretisation

Proposition 1.11 (First-order accuracy over a bounded horizon) [illegible]. *Let ∇f be globally β-Lipschitz, let θ solve (1.5) on $[0, T]$, and let θ_k be generated by explicit Euler, $\theta_{k+1} = \theta_k - \eta \nabla f(\theta_k)$ with $\theta_0 = \theta(0)$ and $\eta \leq 1/\beta$. Then, with $N = \lfloor T/\eta \rfloor$ — so that $N\eta \leq T$ and the exact flow is* *evaluated only where it is defined — and* $M = \sup_{[0,T]} |\nabla f(\theta(t))|$.

$$\max_{0 \le k \le N} \left|\theta_k - \theta(k\eta)\right| \le \frac{\beta M}{2} \frac{e^{\beta T} - 1}{\beta} \eta = O(\eta) \quad \text{as } \eta \downarrow 0,\ T \text{ fixed.} \tag{1.11}$$

Proof. Write $e_k = \theta_k - \theta(k\eta)$. Over one step, $\theta((k+1)\eta) = \theta(k\eta) - \int_{k\eta}^{(k+1)\eta} \nabla f(\theta(s))\mathrm{d}s$, so

$$e_{k+1} = e_k - \eta\big(\nabla f(\theta_k) - \nabla f(\theta(k\eta))\big) + \int_{k\eta}^{(k+1)\eta} \big(\nabla f(\theta(s)) - \nabla f(\theta(k\eta))\big)\mathrm{d}s.$$

The middle term has norm at most $\eta\beta\,|e_k|$. In the integral, $|\nabla f(\theta(s)) - \nabla f(\theta(k\eta))| \le \beta\,|\theta(s) - \theta(k\eta)| \le \beta M(s - k\eta)$, so the integral has norm at most $\beta M\eta^2/2$. Hence $|e_{k+1}| \le (1 + \eta\beta)\,|e_k| + \beta M\eta^2/2$, and with $e_0 = 0$ the discrete Grönwall argument gives $|e_k| \le \frac{\beta M\eta}{2} \cdot \frac{(1+\eta\beta)^k - 1}{\beta}$, which is bounded by (1.11) using $(1 + \eta\beta)^k \le e^{\beta k\eta} \le e^{\beta T}$. ■

What the limit is in. The statement is $\eta \downarrow 0$ with T held fixed. The constant $e^{\beta T}$ is exponential in the horizon, so the bound is useless for T large compared with $1/\beta$ — which is to say, for any training run. This is not a weakness of the proof: it is Corollary 1.5 appearing again, and no better constant is available from these hypotheses.

What it does not say. It does not say that gradient descent with a fixed η tracks the flow, and it does not say anything about the stationary behaviour of either. Two further approximations are needed before the discrete iteration can be discussed at fixed step size, and they are separate constructions with separate hypotheses: the modified equation of Section 6.2, valid to a stated weak order at small η, and the fixed-step analysis of Section 6.4, which is where trained networks actually live. Table 6.1 sorts the three, and no claim in this volume is quoted across its rows.

What does not follow

The Euclidean gradient flow is not gradient descent with a small step size. It is a different object with different properties, and three of the differences are structural rather than quantitative. The flow has no step size and therefore no stability threshold; its energy decreases unconditionally, whereas the iteration's does so only below a threshold. The flow's *geometric orbit* is unchanged by a positive constant rescaling of the vector field, which merely rescales the clock, while its *space parameterisation* changes the orbit itself; the iteration is invariant under neither. (The orbit statement is about a constant rescaling: a general $t \mapsto \varphi(t)$ multiplies the field by $\varphi'(t)$ and, if the field is also changed non-uniformly across components, moves the orbit — which is exactly what Example 5.4 exploits.) And the flow's trajectory is determined by θ_0, whereas the iteration's depends on η in a way that does not vanish as the number of steps grows.

A rate for the flow is not a rate for an algorithm. Theorem 1.9 gives a rate in continuous time. The number of gradient evaluations needed to reach accuracy ε is a different quantity, obtained by combining a rate with a step size and a stability constraint; the composition is elementary in the strongly convex case and unavailable in general. Quoting a continuous-time exponent as an iteration complexity is the regime substitution, and Section 11.6 gives the accounting template that prevents it.

1.3 Neural Differential Equations

This section is the only place in the chapter where a machine learning object appears, and it is here because it is the cleanest example of a continuum limit whose hypotheses are usually left out. The construction is genuinely useful; the claim that a residual network is a differential equation is not, and the difference between the two is a hypothesis about how the weights vary with depth.

1.3.1 Residual networks and their continuum limit

Let $F : \mathbb{R}^d \times \mathbb{R}^m \to \mathbb{R}^d$ be a layer map and consider the residual iteration

$$x_{\ell+1} = x_\ell + \tfrac{1}{L} F(x_\ell, \theta_\ell), \qquad \ell = 0, \ldots, L-1, \tag{1.12}$$

with L layers, parameters $\theta_0, \ldots, \theta_{L-1} \in \mathbb{R}^m$, and the factor $1/L$ written explicitly. Compare with the ordinary differential equation

$$\dot{x}(s) = F\big(x(s), \theta(s)\big), \qquad s \in [0,1], \qquad x(0) = x_0. \tag{1.13}$$

Proposition 1.12 (Continuum limit of a residual network) DERIVED. *Assume*

(H1) F is globally Lipschitz in x with constant L_x and in θ with constant L_θ, uniformly on the relevant range, and bounded by M there;

(H2) there is a Lipschitz $\theta : [0,1] \to \mathbb{R}^m$ with constant L'_θ such that $\theta_\ell = \theta(\ell/L)$ for every ℓ.

Let x_ℓ solve (1.12) and $x(\cdot)$ solve (1.13) from the same x_0. Then

$$\max_{0 \le \ell \le L} \big|x_\ell - x(\ell/L)\big| \le \frac{C}{L}, \qquad C = \frac{e^{L_x} - 1}{L_x}\Big(\frac{L_x M}{2} + L_\theta L'_\theta\Big). \tag{1.14}$$

Proof. The iteration (1.12) is explicit Euler for (1.13) with step $1/L$, except that the field is evaluated with the parameter frozen at the left endpoint of each step. Write $e_\ell = x_\ell - x(\ell/L)$ and split the one-step error as in the proof of Proposition 1.11, with one extra term. Over $[\ell/L, (\ell+1)/L]$,

$$e_{\ell+1} = e_\ell + \tfrac{1}{L}\big(F(x_\ell, \theta_\ell) - F(x(\ell/L), \theta_\ell)\big) + \int_{\ell/L}^{(\ell+1)/L} \big(F(x(\ell/L), \theta_\ell) - F(x(s), \theta(s))\big)\, ds.$$

The first bracket has norm at most $L_x |e_\ell|$. In the integral, split the difference into a change in x and a change in θ: since $|\dot{x}| \le M$ we have $|x(s) - x(\ell/L)| \le M(s - \ell/L)$, and by (H2) $|\theta(s) - \theta_\ell| \le L'_\theta(s - \ell/L)$, so the integrand is at most $(L_x M + L_\theta L'_\theta)(s - \ell/L)$ and the integral at most $(L_x M + L_\theta L'_\theta)/(2L^2)$. Hence $|e_{\ell+1}| \le (1 + L_x/L)|e_\ell| + C'/L^2$ with $C' = (L_x M + L_\theta L'_\theta)/2$, and the discrete Grönwall estimate of Exercise 1.16 with $e_0 = 0$ gives $|e_\ell| \le C' L^{-1}(e^{L_x} - 1)/L_x$, which is (1.14) after enlarging the constant to the stated form. ■

Where it fails. A residual network trained without any depth-coupling of the parameters does not satisfy (H2), and there is no reason it should: each layer's weights are free variables, and adjacent layers can differ by $O(1)$. Along such a family the right-hand side of (1.14) does not tend to zero as L grows, because L_θ' grows with L; in the extreme case θ_ℓ alternating between two values, L_θ' is of order L and the bound is vacuous. Empirically, weights of trained residual networks are found to vary smoothly with depth in some architectures and regimes and not in others, which is an *empirical* question about which this proposition says nothing.

What it does not say. It does not say that a residual network is a differential equation, nor that the two have the same expressive power, nor that training one approximates training the other. It says that under (H1) and (H2) the forward maps agree to $O(1/L)$. Every further statement — about memory, about generalisation, about invertibility — needs its own argument.

1.3.2 The adjoint equation

The practical reason to write (1.13) rather than (1.12) is that the gradient of a terminal loss can be computed by solving a second differential equation backwards, without storing the forward trajectory.

Proposition 1.13 (Adjoint equations) CLASSICAL. *Let F be C^1 in both arguments, let x solve (1.13) on $[0,1]$, and let $\mathcal{L}(x(1))$ be a C^1 terminal loss. Define the adjoint $a:[0,1]\to\mathbb{R}^d$ as the solution of the backward problem*

$$\dot{a}(s) = -\left(\partial_x F(x(s),\theta(s))\right)^{\top} a(s), \qquad a(1) = \nabla\mathcal{L}(x(1)). \tag{1.15}$$

Then $a(s) = \left(\partial x(1)/\partial x(s)\right)^{\top}\nabla\mathcal{L}(x(1))$, and for a parameter $\theta(\cdot)$ perturbed by $\delta\theta(\cdot)$,

$$\delta\mathcal{L} = \int_0^1 \left\langle a(s),\ \partial_\theta F(x(s),\theta(s))\,\delta\theta(s)\right\rangle \mathrm{d}s. \tag{1.16}$$

Proof. Let $J(s) = \partial x(1)/\partial x(s)$, the derivative of the flow map from time s to time 1. Differentiating the flow property $x(1) = \Psi_{s\to 1}(x(s))$ in s and using $\frac{d}{ds}\Psi_{s\to1}(y) = -\partial_y\Psi_{s\to1}(y)F(y,\theta(s))$ gives $\dot{J} = -J\,\partial_x F$, so $J^{\top}$ satisfies (1.15); at $s=1$, $J(1) = I$, so $J^{\top}\nabla\mathcal{L}$ has the stated terminal value and, by uniqueness of solutions of the linear system, $a = J^{\top}\nabla\mathcal{L}$.

For (1.16), let δx be the first variation of the trajectory, solving the linearised equation $\dot{\delta x} = \partial_x F\,\delta x + \partial_\theta F\,\delta\theta$ with $\delta x(0) = 0$. Then $\frac{d}{ds}\langle a,\delta x\rangle = \langle \dot{a},\delta x\rangle + \langle a,\dot{\delta x}\rangle = \langle a,\partial_\theta F\,\delta\theta\rangle$, the terms in $\partial_x F$ cancelling by (1.15). Integrating from 0 to 1 and using $\delta x(0) = 0$ gives $\langle a(1),\delta x(1)\rangle = \int_0^1\langle a,\partial_\theta F\delta\theta\rangle$, and the left side is $\delta\mathcal{L}$ by the chain rule. ∎

Constant memory is not free. The adjoint method avoids storing the forward trajectory by recomputing it backwards from $x(1)$. That recomputation is itself a numerical integration, and it inherits Corollary 1.5: an error made at the end of the backward pass is amplified by e^{L_F} by the time it reaches $s=0$. Whether the saving is worth the error depends on the Lipschitz constant of the trained field, which is not controlled by the training objective unless it is put there. Reported comparisons that hold the solver tolerance fixed while changing the method are not comparing like with like.

Invertibility of the forward map is not invertibility of the network. Equation (1.13) has a unique solution forwards and backwards under Theorem 1.2, so the continuum model is invertible. The residual iteration (1.12) is invertible only for L large enough that $I + L^{-1}\partial_x F$ is nonsingular, which is a condition on L_F/L and is not implied by Proposition 1.12. The two statements are about different objects, and the second is the one an implementation obeys.

Depth is not time. The variable s in (1.13) indexes layers and has no dynamical meaning; nothing in this section makes a residual network a physical or a probabilistic process. When Chapter 3 introduces a time variable that does carry such meaning, it will be a different construction with a different justification, and the notational similarity is a coincidence of the differential equation, not evidence of a connection.

1.4 Brownian Motion

1.4.1 Definition and existence

Definition 1.14 (Brownian motion). A process $W = (W_t)_{t\geq 0}$ with values in $\mathbb{R}^d$ on a probability space $(\Omega, \mathcal{F}, \mathbb{P})$ is a *standard d-dimensional Brownian motion* if

(i) $W_0 = 0$ almost surely;

(ii) for all $0 \leq s < t$ the increment $W_t - W_s$ is independent of $\sigma(W_r : r \leq s)$ and has law $\mathcal{N}(0, (t-s)I)$;

(iii) almost every path $t \mapsto W_t(\omega)$ is continuous.

Conditions (i) and (ii) determine all finite-dimensional distributions: for $0 = t_0 < t_1 < \cdots < t_n$ the increments $W_{t_i} - W_{t_{i-1}}$ are independent centred Gaussians with covariances $(t_i - t_{i-1})I$. Condition (iii) is a genuine additional requirement, because the finite-dimensional distributions do not see a null set of paths, and the process obtained from Kolmogorov's extension theorem alone need not have a continuous version. That Brownian motion exists is the content of the following.

Theorem 1.15 (Wiener) CLASSICAL. *A standard d-dimensional Brownian motion exists.*

Proof roadmap. Take $d = 1$; the general case is d independent copies. The Lévy–Ciesielski construction sets $W_t = \sum_{n\geq 0} \xi_n \int_0^t h_n(s)\,ds$ with (ξ_n) independent standard Gaussians and (h_n) the Haar basis of $L^2([0,1])$; the finite-dimensional distributions are Gaussian with the required covariance by Parseval, and the series converges uniformly almost surely because the Haar functions at level k have disjoint supports and the maximum of 2^k independent Gaussians is $O(\sqrt{k})$, which is summable against the $2^{-k/2}$ from the integration. Continuity is therefore inherited from the partial sums. The interval $[0,1]$ is extended to $[0,\infty)$ by concatenating independent copies.

This is a roadmap and not a proof: the summability estimate and the extension are each a page, and neither is used again in this volume. What is used is the statement, together with the three consequences isolated in Sections 1.4.2 and 1.4.3, all of which are proved in full. □

Proposition 1.16 (Invariances) CLASSICAL. *Let W be a standard one-dimensional Brownian motion. Then each of the following is again a standard Brownian motion:*

(i) (scaling) $t \mapsto c^{-1/2}W_{ct}$ *for any* $c > 0$;

(ii) (symmetry) $t \mapsto -W_t$;

(iii) (time inversion) $t \mapsto tW_{1/t}$ *for* $t > 0$, *extended by* 0 *at* $t = 0$;

(iv) (simple Markov property) $t \mapsto W_{t+s} - W_s$ *for any fixed* $s \geq 0$, *and it is independent of* $\sigma(W_r : r \leq s)$.

Proof. In each case (i)–(iv) the process is centred Gaussian with independent increments, so it suffices to check the covariance and, for (iii), continuity at 0. For (i), $\mathbb{E}[c^{-1/2}W_{cs}\,c^{-1/2}W_{ct}] = c^{-1}(cs \wedge ct) = s \wedge t$. (ii) is immediate. For (iii), $\mathbb{E}[sW_{1/s}\,tW_{1/t}] = st\,(1/s \wedge 1/t) = s \wedge t$; continuity on $(0,\infty)$ is clear, and continuity at 0 follows because the two processes have the same finite-dimensional distributions and hence the same law on paths restricted to rationals, so $\limsup_{t\downarrow 0}|tW_{1/t}| = 0$ almost surely by the strong law applied to W_n/n. (iv) is Definition 1.14(ii) with s fixed. ■

The scaling invariance (i) is the reason the exponent $\frac{1}{2}$ appears everywhere below: it forces $|W_{t+h} - W_t|$ to be of order $h^{1/2}$, and therefore $(W_{t+h} - W_t)^2$ to be of order h — the same order as h itself. That coincidence of orders is the entire content of Itô calculus.

1.4.2 Path properties

Theorem 1.17 (Regularity and irregularity of the paths) CLASSICAL. *Let W be a standard Brownian motion.*

(i) For every $\gamma < \frac{1}{2}$, almost every path is locally γ-Hölder continuous.

(ii) For every $\gamma > \frac{1}{2}$, almost surely no path is γ-Hölder at any point.

(iii) Almost surely, $t \mapsto W_t$ is differentiable at no point.

Proof sketch. (i) By scaling, $\mathbb{E}\,|W_t - W_s|^{2m} = C_m\,|t-s|^m$ for every m, so the Kolmogorov–Chentsov criterion applies with any $\gamma < \frac{m-1}{2m}$, and $m \to \infty$ gives every $\gamma < \frac{1}{2}$.

(ii) If a path were γ-Hölder at t with $\gamma > \frac{1}{2}$ then the sums $\sum_k (W_{t_{k+1}} - W_{t_k})^2$ over partitions of a neighbourhood would be $O(\|\Pi\|^{2\gamma-1}) \to 0$, contradicting Theorem 1.25 below, which gives the limit t.

(iii) Differentiability at t implies γ-Hölder at t for $\gamma = 1$, so (iii) follows from (ii). The classical proof of (iii) does not go through (ii) and gives more — an exact modulus of continuity — and is not reproduced; what is used later is only (iii) as stated, and (ii) with $\gamma = 1$. ■

Part (iii) is the reason the chain rule of Section 1.1.1 is unavailable and Itô's formula is needed. Part (ii) is the reason it is needed in the specific form it takes: the second-order term does not vanish because the paths are exactly rough enough for its order to match the order of dt.

1.4.3 Filtrations, martingales and stopping times

Definition 1.18 (Filtration and the usual conditions). A *filtration* is an increasing family $(\mathcal{F}_t)_{t\geq 0}$ of sub-σ-algebras of $\mathcal{F}$. It satisfies the *usual conditions* if it is right-continuous, $\mathcal{F}_t = \bigcap_{u>t} \mathcal{F}_u$, and complete, in the sense that $\mathcal{F}_0$ contains every $\mathbb{P}$-null set. A process X is *adapted* if X_t is $\mathcal{F}_t$-measurable for every t, and *progressively measurable* if $(s,\omega) \mapsto X_s(\omega)$ restricted to $[0,t] \times \Omega$ is $\mathcal{B}([0,t]) \otimes \mathcal{F}_t$-measurable for every t.

The usual conditions are assumed throughout the volume and are not mentioned again. They are what makes hitting times of open sets stopping times and allows every adapted right-continuous process to be modified on a null set to be progressively measurable; both are used silently in Section 1.5. A *martingale* with respect to $(\mathcal{F}_t)$ is an adapted integrable process M with $\mathbb{E}[M_t \mid \mathcal{F}_s] = M_s$ for $s \leq t$.

Proposition 1.19 (Two martingales) CLASSICAL. *Let W be a one-dimensional Brownian motion and $\mathcal{F}_t = \sigma(W_r : r \leq t)$, completed. Then W and $t \mapsto W_t^2 - t$ are martingales, and for $\lambda \in \mathbb{R}$ so is $t \mapsto \exp(\lambda W_t - \frac{1}{2}\lambda^2 t)$.*

Proof. Write $\Delta = W_t - W_s$, independent of $\mathcal{F}_s$ with law $\mathcal{N}(0, t-s)$. Then $\mathbb{E}[W_t \mid \mathcal{F}_s] = W_s + \mathbb{E}\Delta = W_s$. Next $W_t^2 = W_s^2 + 2W_s\Delta + \Delta^2$, so $\mathbb{E}[W_t^2 \mid \mathcal{F}_s] = W_s^2 + 0 + (t-s)$, which gives $\mathbb{E}[W_t^2 - t \mid \mathcal{F}_s] = W_s^2 - s$. Finally $\mathbb{E}[e^{\lambda\Delta}] = e^{\lambda^2(t-s)/2}$, so $\mathbb{E}[e^{\lambda W_t} \mid \mathcal{F}_s] = e^{\lambda W_s} e^{\lambda^2(t-s)/2}$, and multiplying by $e^{-\lambda^2 t/2}$ gives the martingale property. Integrability holds in each case because Gaussian moments and exponential moments are finite. ■

The second of these is the martingale whose predictable part is exactly the quadratic variation of Theorem 1.25; the third is the special case $\theta \equiv \lambda$ of the exponential martingale Corollary 1.34 that Chapter 10 uses to change measure.

Definition 1.20 (Stopping time). A random time $\tau : \Omega \to [0,\infty]$ is a *stopping time* if $\{\tau \leq t\} \in \mathcal{F}_t$ for every t. For a closed set A and a continuous adapted process X, the hitting time $\tau_A = \inf\{t \geq 0 : X_t \in A\}$ is a stopping time.

Theorem 1.21 (Optional stopping and Doob's inequality) CLASSICAL. *Let M be a continuous martingale and τ a stopping time.*

(i) $M_t^\tau := M_{t\wedge\tau}$ is a martingale; in particular $\mathbb{E}M_{t\wedge\tau} = \mathbb{E}M_0$ for every t.

(ii) If $\tau < \infty$ almost surely and $(M_{t\wedge\tau})_{t\geq 0}$ is uniformly integrable — for instance if it is bounded — then $\mathbb{E}M_\tau = \mathbb{E}M_0$.

(iii) For $p > 1$, $\mathbb{E}\big[\sup_{s\leq T} |M_s|^p\big] \leq \left(\frac{p}{p-1}\right)^p \mathbb{E}|M_T|^p$.

Proof roadmap. (i) is the discrete optional stopping theorem applied along dyadic times and passed to the limit by continuity and dominated convergence. (ii) adds L^1 convergence $M_{t\wedge\tau} \to M_\tau$, which is what uniform integrability supplies and which fails without it — see Proposition 1.22 and the notes there. (iii) is Doob's L^p maximal inequality, proved from the weak maximal inequality by integrating in p. These are standard and are quoted; the volume uses (i) and (ii) constantly and (iii) only in the construction of Section 1.5.3. □

The following is the smallest computation that exercises all of the above, and its hypotheses are the point.

Proposition 1.22 (Exit time from an interval) DERIVED. *Let W be a one-dimensional Brownian motion, let $a, b > 0$, and let*

$$\tau = \inf\{t \geq 0 : W_t \notin (-b, a)\}.$$

Then $\tau < \infty$ almost surely, $\mathbb{E}\tau = ab$, and

$$\mathbb{P}(W_\tau = a) = \frac{b}{a+b}, \qquad \mathbb{P}(W_\tau = -b) = \frac{a}{a+b}. \tag{1.17}$$

Proof. Apply Theorem 1.21(i) to the martingale $W_t^2 - t$ at the stopping time $\tau \wedge n$: $\mathbb{E}[W_{\tau\wedge n}^2] = \mathbb{E}[\tau \wedge n]$. The left side is at most $\max(a,b)^2$ because W stays in $[-b, a]$ up to τ, so $\mathbb{E}[\tau \wedge n] \leq \max(a,b)^2$ for every n, and monotone convergence gives $\mathbb{E}\tau \leq \max(a,b)^2 < \infty$; in particular $\tau < \infty$ almost surely. Now $W_{\tau\wedge n}$ is bounded, hence uniformly integrable, so Theorem 1.21(ii) applies to W and gives $\mathbb{E}W_\tau = 0$; writing $p = \mathbb{P}(W_\tau = a)$ this is $pa - (1-p)b = 0$, which is (1.17). Applying (ii) to $W_t^2 - t$, legitimate for the same reason together with $\mathbb{E}\tau < \infty$, gives $\mathbb{E}\tau = \mathbb{E}W_\tau^2 = pa^2 + (1-p)b^2 = \frac{b}{a+b}a^2 + \frac{a}{a+b}b^2 = ab$. ■

Where it fails. With one barrier, $\tau_a = \inf\{t : W_t = a\}$ is still finite almost surely, but $\mathbb{E}\tau_a = \infty$: letting $b \to \infty$ in $\mathbb{E}\tau = ab$ shows this, and it is confirmed by the density of τ_a, which has tails of order $t^{-3/2}$. Applying Theorem 1.21(ii) to W at τ_a regardless would give $0 = \mathbb{E}W_0 = \mathbb{E}W_{\tau_a} = a$, which is false. Nothing is wrong with the martingale or with the stopping time; the hypothesis that fails is uniform integrability, and its failure is invisible in the statement of the conclusion.

What it does not say. It is a statement about a single trajectory's exit, in expectation. It says nothing about the law of τ beyond its mean, and nothing about exit times for diffusions with drift, for which the corresponding identity involves the generator and is Section 2.1.

Operational reading

Read $\mathbb{E}\tau = ab$ as a statement about how expensive it is to see a rare excursion. Doubling one barrier doubles the expected waiting time; removing one barrier altogether makes the expected waiting time infinite even though the event still happens with probability one. Almost-sure occurrence and finite expected waiting time are different properties, and in this example they come apart at the simplest possible barrier.

That is one step beyond the proposition, and this box stops there. What it costs to sample a rare event, as opposed to wait for one, is the subject of Chapter 4 and Chapter 10, and depends on constructions not available here.

1.5 Brownian Motion and the Itô Integral

1.5.1 Why the classical integral will not do

Let g be a function of bounded variation on $[0, T]$ and f continuous. The Riemann–Stieltjes integral $\int_0^T f \, dg$ exists as a limit of sums $\sum_k f(\xi_k)(g(t_{k+1}) - g(t_k))$ and does not depend on where in $[t_k, t_{k+1}]$ the evaluation point ξ_k is chosen. Both halves of that sentence fail for Brownian paths, and the second fails first.

Proposition 1.23 (The evaluation point is not a detail) DERIVED. *Let W be a one-dimensional Brownian motion, $T > 0$, and let $\Pi = \{0 = t_0 < \cdots < t_N = T\}$ be a partition with mesh $|\Pi|$. Write*

$$L_\Pi = \sum_k W_{t_k}(W_{t_{k+1}} - W_{t_k}), \qquad R_\Pi = \sum_k W_{t_{k+1}}(W_{t_{k+1}} - W_{t_k}), \qquad C_\Pi = \tfrac{1}{2}(L_\Pi + R_\Pi).$$

Then, as $|\Pi| \to 0$, in $L^2(\mathbb{P})$,

$$R_\Pi - L_\Pi \longrightarrow T, \qquad L_\Pi \longrightarrow \tfrac{1}{2}W_T^2 - \tfrac{1}{2}T, \qquad C_\Pi \longrightarrow \tfrac{1}{2}W_T^2. \tag{1.18}$$

In particular the three limits are three different random variables, and no choice of evaluation point is distinguished by the analysis alone.

Proof. Write $\Delta_k = W_{t_{k+1}} - W_{t_k}$. Then $R_\Pi - L_\Pi = \sum_k \Delta_k^2$, which converges to T in L^2 by Theorem 1.25. Also $L_\Pi + R_\Pi = \sum_k (W_{t_{k+1}} + W_{t_k})\Delta_k = \sum_k (W_{t_{k+1}}^2 - W_{t_k}^2) = W_T^2$, exactly, for every partition. Adding and subtracting the two displays gives $2L_\Pi = W_T^2 - \sum_k \Delta_k^2$ and $2C_\Pi = W_T^2$, and the stated limits follow. ■

What it shows and what it does not. It shows that the limit depends on the evaluation point, and that the gap is exactly the quadratic variation. It does *not* show that one choice is correct. The left-endpoint choice is selected in Section 1.5.3 for a reason external to the analysis: it makes the integral a martingale, because W_{t_k} is known before Δ_k occurs. The midpoint choice gives the Stratonovich integral, whose limit $\frac{1}{2}W_T^2$ obeys the classical chain rule and is not a martingale. Both are legitimate; Proposition 1.37 states the conversion and Section 1.5.4 fixes the convention for the volume.

1.5.2 Quadratic variation

Definition 1.24 (Quadratic variation along a partition). For a process X and a partition Π of $[0,t]$ set $V_\Pi^{(2)}(X) = \sum_k (X_{t_{k+1}} - X_{t_k})^2$ and $V_\Pi^{(1)}(X) = \sum_k |X_{t_{k+1}} - X_{t_k}|$. The *quadratic variation* $\langle X \rangle_t$ is the limit of $V_\Pi^{(2)}(X)$ as $|\Pi| \to 0$, when it exists in probability, and the *total variation* is the supremum of $V_\Pi^{(1)}(X)$ over all partitions.

Theorem 1.25 (Quadratic variation of Brownian motion) CLASSICAL. *Let W be a one-dimensional Brownian motion and $t > 0$. Then for every sequence of partitions of $[0,t]$ with $|\Pi| \to 0$,*

$$V_\Pi^{(2)}(W) \longrightarrow t \qquad \text{in } L^2(\mathbb{P}), \tag{1.19}$$

so $\langle W \rangle_t = t$. For a d-dimensional Brownian motion, $\langle W^i, W^j \rangle_t = \delta_{ij} t$, where the cross-variation is defined by polarisation.

Proof. With $\Delta_k = W_{t_{k+1}} - W_{t_k}$ and $\delta_k = t_{k+1} - t_k$, the increments are independent with $\mathbb{E}\Delta_k^2 = \delta_k$ and $\mathbb{E}\Delta_k^4 = 3\delta_k^2$, so $\mathrm{Var}(\Delta_k^2) = 3\delta_k^2 - \delta_k^2 = 2\delta_k^2$. Hence $\mathbb{E}V_\Pi^{(2)} = \sum_k \delta_k = t$ and, by independence,

$$\mathbb{E}(V_{\Pi}^{(2)} - t)^2 = \operatorname{Var} V_{\Pi}^{(2)} = \sum_k 2\delta_k^2 \le 2\|\Pi\| \sum_k \delta_k = 2\|\Pi\|\, t \longrightarrow 0.$$

For the cross-variation, polarisation gives $\langle W^i, W^j \rangle = \frac{1}{2}(\langle W^i + W^j \rangle - \langle W^i \rangle - \langle W^j \rangle)$, and $W^i + W^j$ is $\sqrt{2}$ times a Brownian motion for $i \neq j$, so its quadratic variation is $2t$ and the bracket vanishes; for $i = j$ it is t. ■

Proposition 1.26 (Unbounded variation) DERIVED. *Almost surely, W has infinite total variation on every interval $[s, t]$ with $s < t$. Consequently $\int_s^t f\,dW$ cannot be defined pathwise as a Riemann–Stieltjes integral for all continuous f.*

Proof. On a fixed interval, $V_{\Pi}^{(2)} \le \max_k |\Delta_k| \cdot V_{\Pi}^{(1)}$. If the total variation were finite on an event of positive probability, then on that event, along a sequence of partitions with $\|\Pi\| \to 0$, the first factor tends to 0 by uniform continuity of the path while the second stays bounded, so $V_{\Pi}^{(2)} \to 0$. But $V_{\Pi}^{(2)} \to t - s > 0$ in probability by Theorem 1.25, and a subsequence converging to 0 on a set of positive probability contradicts convergence in probability to a positive constant. Taking a countable union over rational endpoints and using monotonicity of total variation in the interval gives the statement for every interval simultaneously. The final sentence is the classical fact that Riemann–Stieltjes integration against g requires g of bounded variation if it is to be defined for every continuous integrand. ■

The proof of Itô's formula needs Theorem 1.25 in a weighted form, with the weight adapted and evaluated at the left endpoint. It is convenient to state it for the class of integrators that appear there.

Lemma 1.27 (Weighted quadratic variation) DERIVED. *Let σ be progressively measurable with $|\sigma_s| \le K$ for all $s \le t$, let $M_r = \int_0^r \sigma_s\,dW_s$, and let H be adapted, continuous and bounded by K'. Then, as $\|\Pi\| \to 0$,*

$$\sum_k H_{t_k}(M_{t_{k+1}} - M_{t_k})^2 \longrightarrow \int_0^t H_s \sigma_s^2\,ds \qquad \text{in } L^1(\mathbb{P}). \tag{1.20}$$

Proof. Write $\Delta_k M = M_{t_{k+1}} - M_{t_k}$ and $I_k = \int_{t_k}^{t_{k+1}} \sigma_s^2 ds$, and split (1.20) as $\sum_k H_{t_k}(\Delta_k M^2 - I_k) + \sum_k H_{t_k} I_k$.

For the second sum, $s \mapsto H_s$ is continuous and bounded and $\sum_k H_{t_k} I_k$ is a Riemann sum for $\int_0^t H\sigma^2 ds$ with weights $I_k \le K^2\delta_k$; it converges almost surely, hence in L^1 by boundedness.

For the first, by Theorem 1.30 the process $M_r^2 - \int_0^r \sigma^2$ is a martingale, so $\mathbb{E}[\Delta_k M^2 - I_k \mid \mathcal{F}_{t_k}] = 0$ and the terms $H_{t_k}(\Delta_k M^2 - I_k)$ are orthogonal in L^2: for $j < k$, $\mathbb{E}[H_{t_j}(\cdot)_j H_{t_k}(\cdot)_k] = \mathbb{E}\big[H_{t_j}(\cdot)_j H_{t_k} \mathbb{E}[(\cdot)_k \mid \mathcal{F}_{t_k}]\big] = 0$. Hence

$$\mathbb{E}\Big(\sum_k H_{t_k}(\Delta_k M^2 - I_k)\Big)^2 \le K'^2 \sum_k \mathbb{E}(\Delta_k M^2 - I_k)^2 \le CK'^2K^4 \sum_k \delta_k^2 \le CK'^2K^4 \|\Pi\|\, t,$$

where the middle inequality uses the Burkholder–Davis–Gundy bound $\mathbb{E}\Delta_k M^4 \le C\,\mathbb{E} I_k^2 \le CK^4\delta_k^2$. The right-hand side tends to 0, giving convergence in L^2 and hence in L^1. ■

Why the boundedness is there and how it is removed. Both bounds are used only to make the fourth-moment estimate finite. In the proof of Itô's formula they are supplied by the localisation of Step 1, which is why that step comes first; without localisation Lemma 1.27 would have to be stated for local martingales and the convergence would be in probability rather than in L^1. Nothing else changes.

What they do not say. Lemma 1.27 evaluates the weight at the *left* endpoint. With the weight at the right endpoint the limit acquires an extra term, by exactly the mechanism of Proposition 1.23, and the resulting object is not the one appearing in Itô's formula. Adaptedness of H is what makes the left-endpoint choice work, and it is used in the orthogonality step and nowhere else.

1.5.3 Construction of the Itô integral

Definition 1.28 (Elementary processes). A process H is *elementary* on $[0,T]$ if $H_s = \sum_{i=0}^{n-1} H_i \mathbf{1}_{(t_i,t_{i+1}]}(s)$ for a partition $0 = t_0 < \cdots < t_n = T$ and random variables H_i that are $\mathcal{F}_{t_i}$-measurable and square integrable. For such H define

$$\int_0^T H_s\,dW_s := \sum_{i=0}^{n-1} H_i\big(W_{t_{i+1}} - W_{t_i}\big). \tag{1.21}$$

The definition is forced once the left-endpoint choice of Proposition 1.23 is made, and everything below is a completion.

Proposition 1.29 (Itô isometry on elementary processes) CLASSICAL. *For H elementary, $\mathbb{E}\int_0^T H_s dW_s = 0$ and*

$$\mathbb{E}\Big(\int_0^T H_s\,dW_s\Big)^2 = \mathbb{E}\int_0^T H_s^2\,ds. \tag{1.22}$$

Proof. Write $\Delta_i = W_{t_{i+1}} - W_{t_i}$. For $i < j$, $\mathbb{E}[H_i\Delta_i H_j\Delta_j] = \mathbb{E}\big[H_i\Delta_i H_j\,\mathbb{E}[\Delta_j \mid \mathcal{F}_{t_j}]\big] = 0$, since $H_i\Delta_i H_j$ is $\mathcal{F}_{t_j}$-measurable and Δ_j is independent of $\mathcal{F}_{t_j}$ with mean zero. The same computation with a single factor gives $\mathbb{E}[H_i\Delta_i] = 0$. For the diagonal terms, $\mathbb{E}[H_i^2\Delta_i^2] = \mathbb{E}\big[H_i^2\,\mathbb{E}[\Delta_i^2 \mid \mathcal{F}_{t_i}]\big] = \mathbb{E}[H_i^2](t_{i+1} - t_i)$, and summing gives (1.22). ■

Theorem 1.30 (The Itô integral) CLASSICAL. *Let $\mathcal{L}_T^2$ be the space of progressively measurable H with $\mathbb{E}\int_0^T H_s^2 ds < \infty$, with the norm on the right of (1.22). The map (1.21) extends uniquely to a linear isometry from $\mathcal{L}_T^2$ into $L^2(\mathbb{P})$, and the extension has the following properties.*

(i) $t \mapsto \int_0^t H\,dW$ has a continuous version and is a square integrable martingale with mean zero.

(ii) $\langle \int_0 H dW\rangle_t = \int_0^t H_s^2\,ds$; equivalently $\big(\int_0^t H dW\big)^2 - \int_0^t H^2 ds$ is a martingale.

(iii) If $H^{(n)} \to H$ in $\mathcal{L}_T^2$ then $\int_0 H^{(n)} dW \to \int_0 H dW$ uniformly on $[0,T]$ in probability.

The construction extends to progressively measurable H with $\int_0^T H_s^2 ds < \infty$ almost surely by localisation, and the resulting process is a continuous local martingale.

Proof sketch. Elementary processes are dense in $\mathcal{L}_T^2$: approximate first by bounded processes, then by processes continuous in L^2 in time, then by their left-endpoint step functions.

Proposition 1.29 makes (1.21) an isometry on a dense subspace, so it extends uniquely to the closure, which is $\mathcal{L}^2_T$. The martingale property passes to the limit because L^2 limits of martingales are martingales, and continuity of the version follows from Doob's inequality Theorem 1.21(iii) applied to the differences of the approximating sequence, which gives uniform convergence in probability and hence a continuous limit; that argument is also (iii). For (ii), (1.22) applied conditionally on $\mathcal{F}_s$ gives the martingale property of the compensated square, and the identification of the quadratic variation then follows from Lemma 1.27 with $H \equiv 1$, or directly from the uniqueness of the Doob–Meyer decomposition. Localisation replaces H by $H\mathbf{1}_{\{\int_0^t H^2 \le n\}}$ and lets $n \to \infty$.

This is a sketch: the density argument has three steps of which only the shape is given, and the Doob–Meyer uniqueness is quoted. Both are standard, and what the volume uses is the statement, together with Lemma 1.31, which is proved in full because Section 1.5.4 cites it inside a proof. ∎

Lemma 1.31 (Left-endpoint Riemann sums converge) [illegible]. *Let H be adapted, continuous and bounded by K on $[0,T]$, and for a partition Π let $H^{\Pi}_s = \sum_k H_{t_k} \mathbf{1}_{(t_k, t_{k+1}]}(s)$. Then $H^{\Pi} \to H$ in $\mathcal{L}^2_T$ and consequently*

$$\sum_k H_{t_k}(W_{t_{k+1}} - W_{t_k}) \longrightarrow \int_0^T H_s \, dW_s \quad \text{in } L^2(\mathbb{P}) \text{ as } |\Pi| \to 0. \tag{1.23}$$

Proof. Each H^{Π} is elementary, since H_{t_k} is $\mathcal{F}_{t_k}$-measurable and bounded. For fixed ω, $s \mapsto H_s(\omega)$ is continuous on the compact $[0,T]$, hence uniformly continuous, so $\sup_{s \le T} |H^{\Pi}_s(\omega) - H_s(\omega)| \to 0$ as $|\Pi| \to 0$. The integrand $|H^{\Pi} - H|^2$ is bounded by $4K^2$, so dominated convergence gives $\mathbb{E} \int_0^T |H^{\Pi}_s - H_s|^2 \, ds \to 0$, which is convergence in $\mathcal{L}^2_T$. By the isometry of Theorem 1.30 the integrals converge in $L^2(\mathbb{P})$, and by (1.21) the integral of H^{Π} is the left-hand side of (1.23). ∎

Where it fails. Continuity is used only for the uniform convergence $H^{\Pi} \to H$ and can be weakened to left-continuity. Adaptedness cannot be weakened at all: for $H_s = W_T$, a bounded and continuous but not adapted integrand, the left-endpoint sums converge to something that is not a martingale and the isometry fails, since $\mathbb{E}[W_T \Delta_k] \neq 0$. This is the same point as Proposition 1.23 seen from the other side — using information from the future is equivalent to moving the evaluation point — and it is the reason the word "adapted" appears in every hypothesis of Section 1.5.

What it does not say. It gives convergence in L^2 for a fixed terminal time, not uniform convergence in t, and it says nothing about the rate. Rates for Riemann-sum approximation of stochastic integrals are the subject of the strong-order analysis of Section 1.6.4 and are not obtained here.

1.5.4 Itô's Formula

Everything in this volume that converts a statement about trajectories into a statement about densities passes through one identity. The Fokker–Planck equation of Chapter 2 is derived from it; the time-reversal formula that makes diffusion models possible is derived from it; the Girsanov density whose logarithm measures the cost of steering a diffusion is an instance of it. It is worth stating with its hypotheses in full, once, so that later chapters can cite rather than restate it.

The identity is needed because the classical chain rule fails. If x is a differentiable path and f a differentiable function, then $\mathrm{d}f(x_t) = f'(x_t)\,\mathrm{d}x_t$, and the reason this works is that $|x_{t+h} - x_t| = O(h)$, so the second-order term in a Taylor expansion is $O(h^2)$ and vanishes in the limit. A Brownian path has $|W_{t+h} - W_t| \asymp \sqrt{h}$; its square is of order h, the same order as the increment of time itself. The second-order term therefore survives, and it survives not as a fluctuation but as a drift: by Theorem 1.25, the sum of squared increments converges to a deterministic limit. The correction term in Itô's formula is that limit.

Itô processes

Definition 1.32 (Itô process). Let $(\Omega, \mathcal{F}, (\mathcal{F}_t)_{t\geq 0}, \mathbb{P})$ be a filtered probability space satisfying the usual conditions, and let W be an m-dimensional $(\mathcal{F}_t)$-Brownian motion. An $\mathbb{R}^d$-valued process X is an *Itô process* if it admits the representation

$$X_t = X_0 + \int_0^t b_s\,\mathrm{d}s + \int_0^t \sigma_s\,\mathrm{d}W_s, \qquad t \in [0, T], \tag{1.24}$$

where X_0 is $\mathcal{F}_0$-measurable, b is an $\mathbb{R}^d$-valued $(\mathcal{F}_t)$-progressively measurable process with $\int_0^T |b_s|\,\mathrm{d}s < \infty$ almost surely, and σ is an $\mathbb{R}^{d\times m}$-valued progressively measurable process with $\int_0^T \|\sigma_s\|_F^2\,\mathrm{d}s < \infty$ almost surely.

The two integrability conditions are the minimal ones under which the two integrals in (1.24) are defined — pathwise in the first case, as the local L^2 limit constructed in Theorem 1.30 in the second. They are weaker than the conditions under which $\int_0^t \sigma_s\,\mathrm{d}W_s$ is a martingale; it is in general only a local martingale, and the distinction is not cosmetic. It is the first instance of a distinction this volume returns to repeatedly: a statement about paths and a statement about expectations are different statements with different hypotheses, and Proposition 1.35 below turns on exactly this point.

Write $a_s := \sigma_s\sigma_s^\top \in \mathbb{R}^{d\times d}$ for the *diffusion matrix*. It is symmetric and positive semidefinite, and it, rather than σ itself, is what appears in every formula below — a fact worth noticing early, since it is why a diffusion coefficient is identifiable only up to right multiplication by an orthogonal matrix.

The formula

Theorem 1.33 (Itô's formula) CLASSICAL. *Let X be the Itô process (1.24) and let $f \in C^{1,2}([0, T] \times \mathbb{R}^d)$. Then, almost surely, for every $t \in [0, T]$,*

$$\begin{aligned} f(t, X_t) = f(0, X_0) &+ \int_0^t \Big[\partial_s f(s, X_s) + \langle b_s, \nabla f(s, X_s)\rangle + \tfrac{1}{2}\operatorname{tr}\big(a_s \nabla^2 f(s, X_s)\big)\Big]\mathrm{d}s \\ &+ \int_0^t \Big\langle \sigma_s^\top \nabla f(s, X_s), \mathrm{d}W_s\Big\rangle, \end{aligned} \tag{1.25}$$

where ∇f and $\nabla^2 f$ denote the gradient and Hessian in the space variable. Equivalently, in differential form,

$$\mathrm{d}f(t, X_t) = \Big[\partial_t f + \langle b_t, \nabla f\rangle + \tfrac{1}{2}\operatorname{tr}(a_t \nabla^2 f)\Big]\mathrm{d}t + \Big\langle \sigma_t^{\mathsf{T}} \nabla f, \mathrm{d}W_t \Big\rangle. \tag{1.26}$$

Proof. We give the argument for $d = m = 1$ and $f = f(x)$; the general case differs only in bookkeeping, and Exercise 1.32 asks for it.

Step 1 (localisation). Let

$$\tau_n := \inf\Big\{ t \geq 0 : |X_t| \geq n \text{ or } \int_0^t (|b_s| + \sigma_s^2)\mathrm{d}s \geq n \Big\} \wedge T.$$

Each τ_n is a stopping time, $\tau_n \uparrow T$ almost surely by the integrability assumptions of Definition 1.32, and both sides of (1.25) are continuous in t. It therefore suffices to prove the identity for the stopped process X^{τ_n}, on which X is bounded and b, σ are square-integrable in the strong sense. We drop the index n and assume from now on that $|X_t| \leq K$ and $\mathbb{E}\int_0^T (|b_s| + \sigma_s^2)\,\mathrm{d}s < \infty$.

Step 2 (second-order Taylor expansion on a partition). Fix t and a partition $\Pi = \{0 = t_0 < t_1 < \cdots < t_N = t\}$ with mesh $\|\Pi\|$. Write $\Delta_k X := X_{t_{k+1}} - X_{t_k}$. Since $f \in C^2$ and X takes values in the compact set $[-K, K]$, Taylor's theorem with Lagrange remainder gives, for each k, a random point ξ_k on the segment between X_{t_k} and $X_{t_{k+1}}$ with

$$f(X_{t_{k+1}}) - f(X_{t_k}) = f'(X_{t_k})\,\Delta_k X + \tfrac{1}{2} f''(X_{t_k})\,(\Delta_k X)^2 + \tfrac{1}{2}\big[f''(\xi_k) - f''(X_{t_k})\big](\Delta_k X)^2. \tag{1.27}$$

Summing over k telescopes the left-hand side to $f(X_t) - f(X_0)$. We treat the three sums on the right in turn.

Step 3 (the first-order sum). By Definition 1.32, $\Delta_k X = \int_{t_k}^{t_{k+1}} b_s\,\mathrm{d}s + \int_{t_k}^{t_{k+1}} \sigma_s\,\mathrm{d}W_s$, so

$$\sum_k f'(X_{t_k})\,\Delta_k X = \sum_k f'(X_{t_k}) \int_{t_k}^{t_{k+1}} b_s\,\mathrm{d}s + \sum_k f'(X_{t_k}) \int_{t_k}^{t_{k+1}} \sigma_s\,\mathrm{d}W_s.$$

The first sum is a Riemann sum for $\int_0^t f'(X_s) b_s\,\mathrm{d}s$ and converges to it almost surely as $\|\Pi\| \to 0$, by continuity of $s \mapsto f'(X_s)$ and dominated convergence. The second sum is the stochastic integral of the elementary process $\sum_k f'(X_{t_k}) 1_{(t_k, t_{k+1}]}$, which is adapted, left-continuous and bounded; by Lemma 1.31 it converges in L^2, hence in probability, to $\int_0^t f'(X_s)\sigma_s\,\mathrm{d}W_s$.

Step 4 (the second-order sum). We claim

$$\sum_k f''(X_{t_k})\,(\Delta_k X)^2 \xrightarrow[\|\Pi\| \to 0]{\mathbb{P}} \int_0^t f''(X_s)\,\sigma_s^2\,\mathrm{d}s. \tag{1.28}$$

Expanding $(\Delta_k X)^2$ produces three terms. The drift–drift term is bounded by $\sup_k \Big|\int_{t_k}^{t_{k+1}} b_s\mathrm{d}s\Big| \cdot \int_0^t |b_s|\,\mathrm{d}s$, which tends to 0 almost surely because the first factor does. The cross term is bounded in the same way by $\sup_k \Big|\int_{t_k}^{t_{k+1}} b_s\,\mathrm{d}s\Big|$ times a factor with bounded second moment, and also vanishes. For the martingale–martingale term, write

$M_t := \int_0^t \sigma_s \mathrm{d}W_s$; the statement $\sum_k f''(X_{t_k})(\Delta_k M)^2 \to \int_0^t f''(X_s)\sigma_s^2\,\mathrm{d}s$ is Lemma 1.27 with the bounded adapted weight $f''(X_{t_k})$, whose proof rests on the orthogonality of the terms $(\Delta_k M)^2 - \int_{t_k}^{t_{k+1}} \sigma_s^2 \mathrm{d}s$ and is unaffected by a weight measurable with respect to $\mathcal{F}_{t_k}$.

Step 5 (the remainder). X has continuous paths on the compact interval $[0,t]$, hence uniformly continuous paths, and f'' is uniformly continuous on $[-K,K]$. Therefore $\varepsilon_\Pi := \max_k |f''(\xi_k) - f''(X_{t_k})| \to 0$ almost surely as $\|\Pi\| \to 0$. The remainder sum in (1.27) is bounded in absolute value by $\frac{1}{2}\varepsilon_\Pi \sum_k (\Delta_k X)^2$, and the sum $\sum_k (\Delta_k X)^2$ converges in probability to $\int_0^t \sigma_s^2\,\mathrm{d}s$ by (1.28) applied with $f'' \equiv 1$. A product of a sequence tending to 0 almost surely and a sequence converging in probability tends to 0 in probability. Passing to a subsequence along which all four limits hold almost surely and combining Steps 3–5 gives (1.25) for the fixed t; since both sides are continuous in t, the identity holds simultaneously for all t outside one null set. Undoing the localisation of Step 1 by letting $n \to \infty$ completes the proof. ■

Convention

Throughout this volume, an equation written $\mathrm{d}X_t = b_t\,\mathrm{d}t + \sigma_t\,\mathrm{d}W_t$ is an Itô equation. The Stratonovich integral, when it is meant, is written with an explicit circle, $\sigma_t \circ \mathrm{d}W_t$, and the two are related by Proposition 1.37. Three consequences of this choice are used everywhere and are fixed here once:

(i) the generator of X is $\mathcal{L} = \langle b, \nabla\rangle + \frac{1}{2}\operatorname{tr}(a\nabla^2)$, with the factor $\frac{1}{2}$, and $a = \sigma\sigma^\top$;

(ii) consequently the Fokker–Planck operator carries the same factor, $\mathcal{L}^*\rho = -\nabla\cdot(b\rho) + \frac{1}{2}\sum_{ij}\partial_{ij}(a_{ij}\rho)$;

(iii) the generative-modelling literature usually writes the forward process as $\mathrm{d}X_t = f(X_t,t)\mathrm{d}t + g(t)\mathrm{d}W_t$ with scalar g, so that $a = g^2 I$ and the reverse drift acquires the coefficient g^2, not $g^2/2$. That is the same convention: the $\frac{1}{2}$ of (i) is cancelled by the 2 produced when the second-order term is rewritten as a score. Section 3.1 carries this through explicitly, and no other normalisation appears anywhere in this book.

Both conventions are legitimate and neither is more fundamental. What is not legitimate is to move between them inside a single argument, which is the most common way an otherwise correct calculation in this subject acquires a factor of two.

Two immediate consequences

The first is the identity that turns a stochastic integral into a positive martingale, and it is the object on which the whole of Chapter 10 rests.

Corollary 1.34 (Doléans–Dade exponential) CLASSICAL. *Let θ be progressively measurable with $\int_0^T |\theta_s|^2\,\mathrm{d}s < \infty$ almost surely, and set*

$$Z_t := \exp\Big(\int_0^t \langle \theta_s, \mathrm{d}W_s\rangle - \frac{1}{2}\int_0^t |\theta_s|^2\,\mathrm{d}s\Big).$$

Then $\mathrm{d}Z_t = Z_t\,\langle\theta_t, \mathrm{d}W_t\rangle$, so Z is a positive local martingale with $Z_0 = 1$, and it is a supermartingale.

Proof. Apply Theorem 1.33 to $f(x) = e^x$ and the scalar Itô process $Y_t = \int_0^t \langle \theta_s, dW_s \rangle - \frac{1}{2}\int_0^t |\theta_s|^2\, ds$, which has $b_t = -\frac{1}{2}|\theta_t|^2$ and diffusion matrix $a_t = |\theta_t|^2$. The drift terms cancel: $e^{Y_t}(-\frac{1}{2}|\theta_t|^2 + \frac{1}{2}|\theta_t|^2) = 0$, leaving $dZ_t = Z_t \langle \theta_t, dW_t \rangle$. A positive local martingale is a supermartingale by Fatou's lemma applied along a localising sequence. ■

Whether Z is a *true* martingale — equivalently, whether $\mathbb{E} Z_T = 1$ — is a separate question with its own sufficient conditions (Novikov, Kazamaki), taken up in Theorem 10.1 where it decides whether a change of measure is legitimate. The gap between local and true martingale is not a technicality in this volume: it is the reason a path-space Kullback–Leibler divergence can be infinite, and therefore the reason some of the path-space variational statements in Chapter 10 carry hypotheses that look unmotivated until this point is recalled.

The second consequence is chosen for a different reason. It is elementary, and it contradicts an intuition that almost every reader arrives with.

Proposition 1.35 (The potential does not decrease along overdamped Langevin dynamics) DERIVED. *Let $V \in C^2(\mathbb{R}^d)$ and let X solve the overdamped Langevin equation*

$$dX_t = -\nabla V(X_t)\,dt + \sqrt{2}\,dW_t, \tag{1.29}$$

up to its explosion time. Then, pathwise and almost surely,

$$dV(X_t) = \left[-|\nabla V(X_t)|^2 + \Delta V(X_t)\right]dt + \sqrt{2}\,\langle \nabla V(X_t), dW_t \rangle. \tag{1.30}$$

If in addition $\mathbb{E}\int_0^t |\nabla V(X_s)|^2\, ds < \infty$, then

$$\frac{d}{dt}\mathbb{E}\, V(X_t) = \mathbb{E}\left[\Delta V(X_t)\right] - \mathbb{E}\left[|\nabla V(X_t)|^2\right], \tag{1.31}$$

and the right-hand side is not of one sign. For $V(x) = \frac{1}{2}|x|^2$ and $X_0 = 0$ one has $X_t \sim \mathcal{N}(0, (1 - e^{-2t})I)$ and

$$\mathbb{E}\, V(X_t) = \frac{d}{2}(1 - e^{-2t}), \tag{1.32}$$

which is strictly increasing in t, from 0 to $d/2$.

Proof. (1.30) is Theorem 1.33 applied to $f = V$, with $b = -\nabla V$ and $\sigma = \sqrt{2}I$, so that $a = 2I$ and $\frac{1}{2}\operatorname{tr}(a\nabla^2 V) = \Delta V$. Under the stated integrability the stochastic integral in (1.30) is a true martingale with zero mean, and (1.31) follows by taking expectations and differentiating. For the example, $\nabla V(x) = x$ and (1.29) is the Ornstein–Uhlenbeck equation of Proposition 1.43, whose solution from $X_0 = 0$ is the centred Gaussian with covariance $(1 - e^{-2t})I$; hence $\mathbb{E}V(X_t) = \frac{1}{2}\mathbb{E}|X_t|^2 = \frac{d}{2}(1 - e^{-2t})$. Alternatively, apply (1.31) directly: $\Delta V = d$ and $\mathbb{E}|\nabla V(X_t)|^2 = \mathbb{E}|X_t|^2 = d(1 - e^{-2t})$, giving $\frac{d}{dt}\mathbb{E}V(X_t) = d\,e^{-2t} > 0$. ■

Where the two displays part company. (1.30) is a pathwise identity and needs only $V \in C^2$. (1.31) is a statement about expectations and needs the stochastic integral to be a true martingale, not merely a local one; the stated square-integrability is one sufficient condition. For V growing fast enough that $\mathbb{E}\int_0^t |\nabla V(X_s)|^2\, ds = \infty$, the pathwise identity survives and the expectation identity may fail. This is the pathwise-versus-in-law distinction of

Definition 1.32, and it is the single most common way a correct calculation in this subject becomes a false statement about averages.

What this does not say. It does not say that Langevin dynamics fails to converge, nor that nothing decreases. What decreases along (1.29) is the free energy $\mathcal{F}[\rho] = \int V \mathrm{d}\rho + \int \rho \log \rho$, not $\int V \mathrm{d}\rho$; Section 2.2 proves this and identifies the dissipation rate. The two quantities differ by an entropy, and (1.32) is exactly the amount by which ignoring that entropy misleads.

> Operational reading
>
> Read (1.31) as a budget. The term $-\mathbb{E}\,|\nabla V|^2$ is the descent the drift buys; the term $\mathbb{E}[\Delta V]$ is what the noise spends. Where the landscape is steep relative to its curvature, descent wins and the average potential falls; near a minimum, where ∇V is small and $\Delta V > 0$, the noise wins and the average potential rises. Stationarity is the point at which the two exactly balance, which is the content of $\mathbb{E}_\pi[\Delta V] = \mathbb{E}_\pi|\nabla V|^2$ for the Gibbs measure $\pi \propto e^{-V}$ — an identity the reader can now verify by setting the left side of (1.31) to zero.
>
> That is one step beyond the theorem, and it is as far as this box goes. The step after it — reading (1.31) as a statement about a training loss under stochastic gradient descent — requires a diffusion approximation whose validity is a separate question with its own regime of validity, treated in Section 6.2. It is not licensed here.

Itô's formula is not a chain rule with an extra term. It is the chain rule for a different calculus. The map $f \mapsto f(X)$ it describes is not a derivation: applied to a product it gives $\mathrm{d}(XY) = X\mathrm{d}Y + Y\mathrm{d}X + \mathrm{d}\langle X, Y\rangle$, and the third term is not a correction to be neglected when the two processes are driven by the same noise. Stratonovich calculus restores the derivation property, at the cost of the martingale property; Section 1.5.5 makes the trade explicit.

The correction term is not a curvature penalty. It is $+\frac{1}{2}\operatorname{tr}(a\nabla^2 f)$, whose sign is the sign of the curvature of f in the directions the noise actually explores. For convex f it is positive, for concave f negative, and for f neither it varies along the path. Statements of the form "noise biases the dynamics towards flatter regions" do not follow from Itô's formula; they are claims about a stationary distribution and require the hypotheses under which one exists.

$C^{1,2}$ is a real hypothesis. It is not satisfied by the loss surface of a network with piecewise-linear activations, on which $\nabla^2 f$ fails to exist on a set that the process visits. The extension that covers this case — the Itô–Tanaka formula, with a local-time term — is not developed in this volume, and no result here should be applied to a non-smooth potential without it.

Where this is used

Theorem 1.33 is applied in four places later, and it is worth naming them so that the reader knows what the hypotheses are doing. Section 2.1 takes expectations of (1.26) against a test function to derive the Fokker–Planck equation, which is where the factor $\frac{1}{2}$ of the convention box first becomes visible in a PDE. Section 3.1 applies it to $\log \rho_t(X_t)$, with ρ_t the law of X_t, to obtain the evolution of the score — the calculation from which Anderson's time-reversal formula follows. Corollary 1.34 is the density in Girsanov's theorem, hence the object whose logarithm is the path-space relative entropy of Proposition 10.2. And Section 7.1 uses it on the value function of a controlled diffusion to produce the Hamilton–Jacobi–Bellman equation.

1.5.5 Stratonovich integration and the conversion

Proposition 1.23 showed that the limit of a Riemann sum against Brownian motion depends on the evaluation point, and that the analysis alone does not select one. Section 1.5.3 selected the left endpoint, and bought the martingale property with it. This subsection takes the midpoint instead, records exactly what changes, and states the conversion that lets a statement proved in one convention be used in the other.

Definition 1.36 (Stratonovich integral). Let X be a continuous semimartingale and $g \in C^1(\mathbb{R})$. The *Stratonovich integral* of $g(X)$ against a continuous semimartingale Y is

$$\int_0^t g(X_s) \circ dY_s := \lim_{|\Pi| \to 0} \sum_k \frac{g(X_{t_k}) + g(X_{t_{k+1}})}{2} (Y_{t_{k+1}} - Y_{t_k}), \tag{1.33}$$

the limit being in probability.

Proposition 1.37 (Itô–Stratonovich conversion) SERVES. *Let X solve the Itô equation $dX_t = b_t\,dt + \sigma_t\,dW_t$ in the sense of Definition 1.32, let $g \in C^1(\mathbb{R})$ with g' locally bounded, and suppose the localisation hypotheses of Theorem 1.33 hold. Then the limit in (1.33) with $Y = W$ exists and*

$$\int_0^t g(X_s) \circ dW_s = \int_0^t g(X_s)\,dW_s + \frac{1}{2} \int_0^t g'(X_s)\sigma_s\,ds. \tag{1.34}$$

Consequently the Stratonovich equation $dX_t = b_t\,dt + \sigma(X_t) \circ dW_t$, with $\sigma \in C^1$, is the Itô equation

$$dX_t = \left(b_t + \tfrac{1}{2}\sigma'(X_t)\sigma(X_t)\right)dt + \sigma(X_t)\,dW_t. \tag{1.35}$$

Proof. Write $g_k = g(X_{t_k})$ and $\Delta_k = W_{t_{k+1}} - W_{t_k}$. Then

$$\sum_k \frac{g_k + g_{k+1}}{2} \Delta_k = \sum_k g_k \Delta_k + \frac{1}{2} \sum_k (g_{k+1} - g_k) \Delta_k.$$

The first sum converges to $\int_0^t g(X)\,dW$ by Lemma 1.31, $s \mapsto g(X_s)$ being adapted, continuous and, after localisation, bounded. For the second, apply Theorem 1.33 to g and X: $g_{k+1} - g_k = \int_{t_k}^{t_{k+1}} g'(X_s)\,dX_s + \frac{1}{2}\int_{t_k}^{t_{k+1}} g''(X_s)\sigma_s^2 ds$ when $g \in C^2$, and in general the increment satisfies $g_{k+1} - g_k = \int_{t_k}^{t_{k+1}} g'(X_s)\sigma_s dW_s + O(\delta_k)$ in L^2. Multiplying by Δ_k and summing, the $O(\delta_k)$ terms contribute $O(\|\Pi\|^{1/2})$ in L^1 because $\mathbb{E}\,|\Delta_k| \le \delta_k^{1/2}$, while the stochastic terms give $\sum_k \int_{t_k}^{t_{k+1}} g'(X_s)\sigma_s dW_s \cdot \Delta_k$, which is the cross-variation sum for $(g(X), W)$ and converges to $\int_0^t g'(X_s)\sigma_s ds$ by Lemma 1.27 in polarised form. Halving gives (1.34). For (1.35), apply (1.34) with $g = \sigma$. ■

Proposition 1.38 (The Stratonovich chain rule) DERIVED. *Let X solve $\mathrm{d}X_t = b_t\,\mathrm{d}t + \sigma(X_t) \circ \mathrm{d}W_t$ with $\sigma \in C^1$, and let $f \in C^2(\mathbb{R})$. Then*

$$\mathrm{d}f(X_t) = f'(X_t) \circ \mathrm{d}X_t := f'(X_t)b_t\,\mathrm{d}t + f'(X_t)\sigma(X_t) \circ \mathrm{d}W_t. \tag{1.36}$$

There is no second-order term: the classical chain rule holds verbatim.

Proof. By (1.35) the Itô form of the equation has drift $b + \frac{1}{2}\sigma'\sigma$ and diffusion σ, so Theorem 1.33 gives

$$\mathrm{d}f(X_t) = f'(X_t)(b_t + \tfrac{1}{2}\sigma'\sigma)\mathrm{d}t + \tfrac{1}{2}f''(X_t)\sigma^2\,\mathrm{d}t + f'(X_t)\sigma\,\mathrm{d}W_t.$$

On the other side, (1.34) applied to $g = f'\sigma$ gives $f'\sigma \circ \mathrm{d}W = f'\sigma\,\mathrm{d}W + \frac{1}{2}(f''\sigma + f'\sigma')\sigma\,\mathrm{d}t$, so the right-hand side of (1.36) equals $f'b\,\mathrm{d}t + f'\sigma\,\mathrm{d}W + \frac{1}{2}f''\sigma^2\mathrm{d}t + \frac{1}{2}f'\sigma'\sigma\,\mathrm{d}t$. The two expressions agree term by term. ∎

Notes on Proposition 1.37 and Proposition 1.38

Assumed. A C^1 diffusion coefficient for the conversion of the equation, C^1 integrand for the conversion of the integral, and the localisation hypotheses under which Theorem 1.33 and Lemma 1.31 apply. In particular σ must be differentiable: the correction term contains σ', so a merely Lipschitz diffusion coefficient admits an Itô equation and no Stratonovich counterpart.

What is gained and what is lost. Gained: the chain rule (1.36), and with it coordinate invariance — a Stratonovich equation transforms under a change of variables exactly as an ordinary differential equation does, which is why it is the convention of choice on manifolds. Lost: the martingale property. The Stratonovich integral is not a martingale, its expectation is not zero, and none of the machinery of Theorem 1.21 applies to it. Every convergence and every bound in this volume is proved through the martingale property, which is why the Itô convention is the one fixed in Section 1.5.4.

Where the difference is invisible, and where it is not. For additive noise, σ constant, the correction vanishes and the two conventions agree; this covers the overdamped Langevin dynamics of Chapter 2 and most of Chapter 3, which is why the distinction can be ignored there and must not be ignored anywhere else. It does not vanish for state-dependent noise, for anisotropic noise with position-dependent covariance, or on a Riemannian manifold, where the Itô form is not coordinate-free and the correction term carries the Christoffel symbols.

Theorem 1.39 (Wong–Zakai) CLASSICAL. *Let $\sigma \in C_b^2(\mathbb{R})$ and $b \in C_b^1(\mathbb{R})$, let $W^{(\varepsilon)}$ be the piecewise linear interpolation of W on the mesh ε, and let $X^{(\varepsilon)}$ solve the ordinary differential equation $\dot{X}^{(\varepsilon)} = b(X^{(\varepsilon)}) + \sigma(X^{(\varepsilon)})\dot{W}^{(\varepsilon)}$ from X_0. Then, as $\varepsilon \downarrow 0$, $X^{(\varepsilon)}$ converges in probability, uniformly on compact time intervals, to the solution of the Stratonovich equation $\mathrm{d}X = b\,\mathrm{d}t + \sigma(X) \circ \mathrm{d}W$ — equivalently to the Itô equation* (1.35).

Proof roadmap. The interpolated driver is of bounded variation, so $X^{(\varepsilon)}$ is an ordinary differential equation solved pathwise and obeys the classical chain rule. Writing its solution through that chain rule and passing to the limit produces the Stratonovich integral, because the increment of the interpolation over a mesh interval is used symmetrically across the interval. The technical content is a uniform estimate on the difference of the two flows in terms of ε, obtained from a Grönwall argument on the difference of the integral equations. The full argument is quoted and not reproduced; nothing in this volume depends on it. □

What does not follow

Neither convention is more physical than the other. Theorem 1.39 is sometimes read as showing that Stratonovich is "the right one" because it is the limit of smooth noise. That reading omits the hypothesis. The limit is Stratonovich when the noise is smoothed *symmetrically in time*; smoothing it causally — so that the driver at time t depends only on the past — gives the Itô solution instead. Which limit is relevant is a modelling question about the physical or algorithmic process being idealised, and it is answered differently for a mechanical system driven by coloured noise and for a stochastic algorithm whose update uses only information already computed. This volume's objects are of the second kind.

The correction term is not a small error. In (1.35) the drifts differ by $\frac{1}{2}\sigma'\sigma$, which is of the same order as b itself, not asymptotically negligible in any parameter. Two papers writing the same equation in different conventions are studying different dynamics with different stationary laws, and the discrepancy is invisible in the notation. This is the normalisation substitution of Table 11.2, and it is the reason Section 1.5.4 fixes the convention once for the volume rather than per chapter.

A statement about the law does not fix the convention. Both equations in (1.35) generate a diffusion; they generate *different* diffusions with different generators. A reader who has only the law of the solution cannot recover which convention produced it, and a reader who has only the coefficients cannot compute the law without being told.

1.6 Stochastic Differential Equations

1.6.1 Strong solutions

Definition 1.40 (Strong solution). Given $b : [0,T] \times \mathbb{R}^d \to \mathbb{R}^d$, $\sigma : [0,T] \times \mathbb{R}^d \to \mathbb{R}^{d\times m}$, an m-dimensional Brownian motion W on a filtered space satisfying the usual conditions, and an $\mathcal{F}_0$-measurable X_0, a *strong solution* of

$$dX_t = b(t, X_t)\,dt + \sigma(t, X_t)\,dW_t, \qquad X\big|_{t=0} = X_0, \tag{1.37}$$

is a continuous $(\mathcal{F}_t)$-adapted process X satisfying Definition 1.32 with $b_t = b(t, X_t)$ and $\sigma_t = \sigma(t, X_t)$. "Strong" records that the Brownian motion and the filtration are given in advance and X is constructed on them.

Theorem 1.41 (Existence and uniqueness of strong solutions) CLASSICAL. *Suppose b and σ are measurable and satisfy, for a constant $K < \infty$ and all $t \in [0,T]$, $x, y \in \mathbb{R}^d$,*

$$|b(t,x) - b(t,y)| + \|\sigma(t,x) - \sigma(t,y)\|_F \le K\,|x - y|, \qquad |b(t,x)| + \|\sigma(t,x)\|_F \le K(1 + |x|), \tag{1.38}$$

and let $\mathbb{E}\,|X_0|^2 < \infty$. Then (1.37) has a strong solution, unique up to indistinguishability, with $\mathbb{E}\sup_{t\le T}|X_t|^2 < \infty$.

Proof. Define $X_t^0 = X_0$ and, inductively, $X_t^{n+1} = X_0 + \int_0^t b(s, X_s^n)ds + \int_0^t \sigma(s, X_s^n)dW_s$; each integral is defined because the linear growth bound and the previous step's square integrability put the integrands in $\mathcal{L}^2_T$. Put $D_n(t) = \mathbb{E}\sup_{s\le t}\left|X_s^{n+1} - X_s^n\right|^2$. Splitting the difference into its drift and martingale parts and using $(a+b)^2 \le 2a^2 + 2b^2$,

$$D_n(t) \le 2\,\mathbb{E}\sup_{s\le t}\Big(\int_0^s \big|b(r,X^n) - b(r,X^{n-1})\big|\,dr\Big)^2 + 2\,\mathbb{E}\sup_{s\le t}\Big|\int_0^s (\sigma(r,X^n) - \sigma(r,X^{n-1}))\,dW_r\Big|^2.$$

For the first term, Cauchy–Schwarz and (1.38) give the bound $2TK^2\int_0^t D_{n-1}(r)\,dr$. For the second, Doob's inequality Theorem 1.21(iii) with $p = 2$ bounds the supremum by four times the terminal second moment, and the Itô isometry (1.22) turns that into $8K^2\int_0^t D_{n-1}(r)\,dr$. Hence $D_n(t) \le C\int_0^t D_{n-1}$ with $C = 2K^2(T+4)$, and since $D_0(T) < \infty$ by linear growth, iterating gives $D_n(T) \le D_0(T)\,(CT)^n/n!$. The series $\sum_n D_n(T)^{1/2}$ therefore converges, so (X^n) is Cauchy in the complete space of continuous adapted processes with norm $(\mathbb{E}\sup_{t\le T}|\cdot|^2)^{1/2}$; its limit X is continuous and adapted, and passing to the limit in the recursion — legitimate by the same two estimates — shows X solves (1.37).

For uniqueness, let X, Y be two solutions and $u(t) = \mathbb{E}\sup_{s\le t}|X_s - Y_s|^2$, finite by the moment bound. The same two estimates give $u(t) \le C\int_0^t u(r)\,dr$, and Proposition 1.4 with $a = 0$ gives $u \equiv 0$; continuity of both paths upgrades this to indistinguishability. ■

Remark 1.42 (When the hypotheses fail). The Lipschitz condition may be localised: if (1.38)'s first inequality holds only on balls, with K depending on the radius, and the linear growth bound is replaced by a Lyapunov condition such as $\langle x, b(t,x)\rangle + \frac{1}{2}\|\sigma(t,x)\|_F^2 \le K(1+|x|^2)$, then a unique strong solution exists on $[0,T]$ without explosion; the proof is Theorem 1.41 on balls plus a stopping argument. Both weakenings are needed in this volume: the Langevin dynamics of Chapter 2 has a drift that is not globally Lipschitz whenever the potential grows faster than quadratically, and the reverse-time equations of Chapter 3 have drifts that are singular at one endpoint of the time interval and are treated there by a separate argument.

1.6.2 Two linear equations

Two explicitly solvable equations are used repeatedly in this volume and are recorded here once.

Proposition 1.43 (Ornstein–Uhlenbeck process) DERIVED. *Let $\theta > 0$, $\sigma > 0$ and let X solve*

$$dX_t = -\theta X_t\,dt + \sigma\,dW_t, \qquad X_0 = x_0 \in \mathbb{R}^d, \tag{1.39}$$

with W a d-dimensional Brownian motion. Then

$$X_t = e^{-\theta t}x_0 + \sigma\int_0^t e^{-\theta(t-s)}\,dW_s, \tag{1.40}$$

so $X_t \sim \mathcal{N}\big(e^{-\theta t}x_0, \frac{\sigma^2}{2\theta}(1-e^{-2\theta t})I\big)$, and $X_t \Rightarrow \mathcal{N}(0, \frac{\sigma^2}{2\theta}I)$ as $t \to \infty$. In the variance-preserving normalisation $\sigma^2 = 2\theta$ the limit is $\mathcal{N}(0, I)$ for every θ.

Proof. Apply Theorem 1.33 to $f(t,x) = e^{\theta t}x$ and the Itô process (1.39): since $\partial_t f = \theta e^{\theta t}x$, $\nabla f = e^{\theta t}I$ and $\nabla^2 f = 0$,

$$d(e^{\theta t}X_t) = \theta e^{\theta t}X_t\,dt + e^{\theta t}\big(-\theta X_t\,dt + \sigma\,dW_t\big) = \sigma e^{\theta t}\,dW_t.$$

and integrating from 0 to t and multiplying by $e^{-\theta t}$ gives (1.40). The stochastic integral in (1.40) is a Wiener integral of a deterministic function, hence centred Gaussian, with covariance $\sigma^2 \int_0^t e^{-2\theta(t-s)}\mathrm{d}s\, \mathrm{I} = \frac{\sigma^2}{2\theta}(1 - e^{-2\theta t})\mathrm{I}$ by the Itô isometry (1.22). Letting $t \to \infty$ gives the limit, and $\sigma^2 = 2\theta$ makes the limiting covariance the identity. ∎

Proposition 1.44 (Geometric Brownian motion) DERIVED. *Let $\mu, \varsigma \in \mathbb{R}$ and let S solve $\mathrm{d}S_t = \mu S_t \mathrm{d}t + \varsigma S_t \mathrm{d}W_t$ with $S_0 > 0$. Then*

$$S_t = S_0 \exp\Big(\big(\mu - \tfrac{1}{2}\varsigma^2\big)t + \varsigma W_t \Big) > 0, \tag{1.41}$$

so $\mathbb{E}S_t = S_0 e^{\mu t}$ while the median of S_t is $S_0 e^{(\mu - \varsigma^2/2)t}$. For $\mu < \frac{1}{2}\varsigma^2$ the process tends to 0 almost surely and its mean diverges if $\mu > 0$.

Proof. Apply Theorem 1.33 to $f(x) = \log x$, legitimate on $(0, \infty)$: with $b_t = \mu S_t$, $\sigma_t = \varsigma S_t$, $f' = 1/x$, $f'' = -1/x^2$, $\mathrm{d}\log S_t = (\mu - \frac{1}{2}\varsigma^2)\mathrm{d}t + \varsigma\, \mathrm{d}W_t$, and exponentiating gives (1.41); positivity is then automatic, which also justifies the use of log retrospectively by a localisation at the first time S reaches ε, whose limit as $\varepsilon \downarrow 0$ is the explosion time of $\log S$, almost surely infinite. The mean follows from $\mathbb{E}e^{\varsigma W_t} = e^{\varsigma^2 t/2}$, the median from the median of W_t being 0, and the almost-sure limit from the strong law applied to $W_t/t \to 0$. ∎

What the closed forms are used for. Proposition 1.43 is the forward process of every variance-preserving construction in Chapter 3 and the Langevin dynamics of a quadratic potential in Chapter 2; its explicit law is why those chapters can compute rather than estimate. That convenience is also a trap: a Gaussian target is the one case in which every geometry in Table 5.1 gives the same answer, so a construction validated only on (1.39) has been validated on the case that cannot distinguish the alternatives.

What they do not say. Proposition 1.44 separates the mean from the typical path by the factor $e^{\varsigma^2 t/2}$, and for $0 < \mu < \frac{1}{2}\varsigma^2$ the mean grows while almost every path decays to zero. An expectation is not a typical value, and no amount of averaging over paths recovers the pathwise behaviour. This is the pathwise-versus-in-law distinction of Table 11.2 in its most elementary form, and it is worth remembering that it appears already here, in a one-line closed-form solution.

1.6.3 Weak solutions and the martingale problem

Definition 1.40 requires the Brownian motion to be given first. Many of the equations in this volume arrive the other way round: a law is specified, or a drift is obtained from a time reversal or from an optimality condition, and no driving Brownian motion is designated in advance. The right notion is then the following.

Definition 1.45 (Weak solution and the martingale problem). A *weak solution* of (1.37) with initial law ν is a triple consisting of a filtered probability space, a Brownian motion W on it, and a continuous adapted X with $X_0 \sim \nu$ satisfying (1.37). *Uniqueness in law* means that any two weak solutions induce the same law on $C([0,T];\mathbb{R}^d)$.

Equivalently, let $\mathcal{L}_t = \langle b(t,\cdot), \nabla\rangle + \frac{1}{2}\operatorname{tr}\big(a(t,\cdot)\nabla^2\big)$ with $a = \sigma\sigma^\top$. A probability measure $\mathbb{Q}$ on $C([0,T];\mathbb{R}^d)$ solves the *martingale problem* for $(\mathcal{L}_t, \nu)$ if $\mathbb{Q} \circ X_0^{-1} = \nu$ and, for every

$f \in C_c^\infty(\mathbb{R}^d)$,

$$f(X_t) - \int_0^t \mathcal{L}_s f(X_s)\,\mathrm{d}s \tag{1.42}$$

is a Q-martingale.

Theorem 1.46 (Equivalence, and the gap between weak and strong) CLASSICAL. *A probability measure on $C([0,T];\mathbb{R}^d)$ is the law of a weak solution of (1.37) if and only if it solves the martingale problem for $(\mathcal{L}_t, \nu)$. Moreover strong existence implies weak existence; pathwise uniqueness implies uniqueness in law; and, by the Yamada–Watanabe theorem, weak existence together with pathwise uniqueness implies strong existence and uniqueness in law. The converses fail. The* Tanaka equation

$$\mathrm{d}X_t = \operatorname{sgn}(X_t)\,\mathrm{d}W_t, \qquad X_0 = 0, \qquad d = 1, \tag{1.43}$$

with $\operatorname{sgn}(0) := -1$ say, admits weak solutions and is unique in law — the law is that of a standard Brownian motion — but pathwise uniqueness fails and there is no strong solution adapted to the filtration of the driving Brownian motion.

Proof roadmap. That a weak solution solves the martingale problem is Theorem 1.33 applied to f, the stochastic integral being a martingale for bounded ∇f. The converse is a representation argument: the martingale (1.42) has quadratic variation $\int_0^t \langle \nabla f, a\nabla f\rangle(X_s)\mathrm{d}s$, and the martingale representation theorem produces a Brownian motion on a possibly enlarged space with respect to which X satisfies (1.37). The implications between the four notions are the Yamada–Watanabe theory. For (1.43): if X is any weak solution then $\langle X\rangle_t = \int_0^t \operatorname{sgn}(X_s)^2\mathrm{d}s = t$, so by Lévy's characterisation X is a Brownian motion, which gives uniqueness in law. Conversely, given a Brownian motion X, setting $W_t = \int_0^t \operatorname{sgn}(X_s)\mathrm{d}X_s$ produces a Brownian motion for which X solves (1.43); but $-X$ solves it with the *same* W, which is the failure of pathwise uniqueness. The driving noise turns out to be measurable with respect to $|X|$ alone — by Tanaka's formula $|X_t| = \int_0^t \operatorname{sgn}(X_s)\mathrm{d}X_s + L_t^0 = W_t + L_t^0$ with L^0 the local time at zero — so the sign of X carries information the Brownian filtration does not have, and no solution can be adapted to it. Note where the non-degeneracy lives: the coefficient carrying the discontinuity is σ, not the drift. The equation $\mathrm{d}X = \operatorname{sgn}(X)\mathrm{d}t + \mathrm{d}W$, with additive non-degenerate noise and a bounded measurable drift, is by contrast strongly well posed, and is not a counterexample to anything here. All of this is standard and is quoted.

What the volume uses is the equivalence in the direction from the martingale problem to the law, because that is the direction in which the Fokker–Planck equation of Section 2.1 is derived, and the fact that uniqueness in law is the notion available for the reverse-time equations of Chapter 3. □

What does not follow

A statement about the law is not a statement about the paths. Weak uniqueness says that any two solutions have the same law on path space; it does not say that two solutions driven by the same Brownian motion coincide, and (1.43) shows that they need not — there X and $-X$ solve the same equation with the same W. Every claim in this volume about a density, a marginal or a divergence is a claim about the law and needs only weak uniqueness. A claim about a pathwise coupling — two processes built on one Brownian motion and compared trajectory by trajectory, as in Proposition 8.2 — needs pathwise uniqueness. Couplings constructed at the level of laws, such as an optimal transport plan between two marginals, need neither. Which one a result requires is recorded in its scope box, and the two are not interchangeable.

The martingale problem does not know the diffusion coefficient. The generator sees $a = \sigma\sigma^\top$ and not σ, so σ and σQ for orthogonal Q give the same generator, the same martingale problem and the same law. Any statement identifying a diffusion coefficient from data or from a law is therefore identifying it at best up to right multiplication by an orthogonal matrix, a point already flagged after Definition 1.32 and used in Chapter 3.

1.6.4 Discretisation: two different orders

Definition 1.47 (Strong and weak order). Let $\bar{X}^\eta$ be a scheme with step η on $[0, T]$. It has *strong order* p if $\left(\mathbb{E}\,|X_T - \bar{X}^\eta_T|^2\right)^{1/2} \le C\eta^p$, and *weak order* q if $|\mathbb{E}f(X_T) - \mathbb{E}f(\bar{X}^\eta_T)| \le C_f\eta^q$ for all f in a specified class of smooth test functions, with C and C_f independent of η.

The two are different statements about different objects. Strong order compares trajectories driven by the same Brownian motion; weak order compares expectations, and a scheme with no pathwise accuracy at all can have weak order one. That distinction is the one on which Section 6.2 rests, and it is the reason the modified-equation analysis there is a statement about test functions and not about paths.

Proposition 1.48 (Euler–Maruyama and Milstein) CLASSICAL. *Under (1.38), with additionally b, σ Hölder-$\frac{1}{2}$ in t, the Euler–Maruyama scheme*

$$\bar{X}_{k+1} = \bar{X}_k + \eta\, b(t_k, \bar{X}_k) + \sigma(t_k, \bar{X}_k)(W_{t_{k+1}} - W_{t_k}) \tag{1.44}$$

has strong order $\frac{1}{2}$ and, for $b, \sigma \in C^4_b$, weak order 1 with respect to the test class C^4_{pol} of four-times continuously differentiable f whose derivatives up to order four have polynomial growth — *the class must be named, because Definition 1.47 quantifies over one and the constant C_f depends on it. Adding the Milstein correction $\frac{1}{2}\sigma'\sigma\big((\Delta W)^2 - \eta\big)$ in $d = m = 1$ raises the strong order to* 1 *and leaves the weak order at* 1*; in $d, m > 1$ the same correction requires either simulated Lévy areas or a commutativity condition on the diffusion columns, and is not asserted here. See Särkkä and Solin [24] for the schemes and their orders.*

The exponents look arbitrary and are not. The following isolates where $\frac{1}{2}$ comes from, in the one-dimensional autonomous case, and it is worth seeing because the same computation explains why the Milstein term is the term it is.

Proposition 1.49 (Where the exponent comes from) DERIVED. *Let $d = m = 1$, $b \equiv 0$ and $\sigma \in C^2_b$ with σ, σ' bounded by K. Over one step from t_k to $t_{k+1} = t_k + \eta$, with $X_{t_k} = \bar{X}_k$,*

$$X_{t_{k+1}} - \bar{X}_{k+1} = \sigma'(\bar{X}_k)\sigma(\bar{X}_k)\int_{t_k}^{t_{k+1}}(W_s - W_{t_k})\,dW_s + R_k, \qquad \left(\mathbb{E}R_k^2\right)^{1/2} \le C\eta^{3/2}, \tag{1.45}$$

and the displayed term has L^2 norm $\sigma'\sigma\,\eta/\sqrt{2}$ and mean zero. Consequently the local error is of order η in L^2 and of order η^2 in expectation; since the local errors are martingale increments they accumulate in quadrature over $N = T/\eta$ steps, giving a global strong error of order $\sqrt{N\eta^2} = \sqrt{T\eta}$ and a global weak error of order $N\eta^2 = T\eta$.

Proof. Exactly, $X_{t_{k+1}} - \bar{X}_{k+1} = \int_{t_k}^{t_{k+1}} (\sigma(X_s) - \sigma(\bar{X}_k))\,dW_s$. Apply Theorem 1.33 to σ along X: $\sigma(X_s) - \sigma(\bar{X}_k) = \int_{t_k}^{s} \sigma'(X_r)\sigma(X_r)\,dW_r + \frac{1}{2}\int_{t_k}^{s} \sigma''\sigma^2\,dr$. Substituting, the second integral contributes a term bounded in L^2 by $C\eta^{3/2}$ by the Itô isometry, and replacing $\sigma'\sigma(X_r)$ by its value at $\bar{X}_k$ costs another $C\eta^{3/2}$ because $\sigma'\sigma$ is Lipschitz and $\mathbb{E}|X_r - \bar{X}_k|^2 \le C(r - t_k)$. What remains is $\sigma'\sigma(\bar{X}_k)\int_{t_k}^{t_{k+1}} (W_s - W_{t_k})\,dW_s$. Its L^2 norm is computed from the isometry: $\mathbb{E}\big(\int_{t_k}^{t_{k+1}} (W_s - W_{t_k})\,dW_s\big)^2 = \int_{t_k}^{t_{k+1}} \mathbb{E}(W_s - W_{t_k})^2\,ds = \eta^2/2$; its mean is zero because it is a stochastic integral. The accumulation statement is the observation that a sum of N orthogonal mean-zero terms of L^2 norm $c\eta$ has L^2 norm $c\eta\sqrt{N}$, together with $N = T/\eta$. ■

Notes on Proposition 1.48 and Proposition 1.49

Assumed. Proposition 1.48: the global Lipschitz and linear growth conditions (1.38), Hölder-$\frac{1}{2}$ regularity in t for the strong order, and $b, \sigma \in C_b^4$ for the weak order — a genuinely stronger hypothesis, and the reason the two orders are quoted under different assumptions rather than as two properties of one theorem. Proposition 1.49: one dimension, no drift, bounded σ with two bounded derivatives, and exact Brownian increments. The purpose is to exhibit the mechanism, not to prove Proposition 1.48, whose hypotheses are weaker and whose accumulation argument is a Grönwall estimate rather than the exact orthogonality used here.

Why the quadrature step is not a heuristic and where it is. The local errors of a stochastic integral against non-overlapping intervals are orthogonal, which is exactly Proposition 1.29, so their sum has L^2 norm $\sqrt{N}$ times the individual norm rather than N times it — this is rigorous. What is heuristic in the display is the treatment of the propagated error: the true global error also contains the accumulated effect of earlier errors transported by the flow, which contributes the Grönwall factor and is suppressed above. That factor is again exponential in T, as in Corollary 1.5 and Proposition 1.11.

What it does not say. It says nothing about a fixed step size at large T, nor about the stationary law of the scheme. The Euler–Maruyama scheme applied to (1.39) has a stationary law that is Gaussian with the wrong variance, by a relative error of order η that does not decrease as T grows; that bias is invisible to Definition 1.47 because both orders are defined at fixed T. Chapter 5 treats it as a separate error source and Section 4.1 quantifies it.

Table 1.2: The two orders. They answer different questions and are not comparable; a scheme is chosen by deciding which question is being asked. The last column is the one that decides whether a reported order is relevant to a given use.

Order	Compares	Euler–Maruyama	What it is needed for
Strong, p	Trajectories driven by the same Brownian motion, in L^2	$p = \frac{1}{2}$; $p = 1$ with the Milstein term	Couplings, variance reduction across schemes, pathwise error bounds, anything comparing two simulations run on the same noise

continued overleaf

Table 1.2: (continued)

Order	Compares	Euler–Maruyama	What it is needed for
Weak, q	Expectations of smooth test functions at a fixed terminal time	$q = 1$; unchanged by the Milstein term	Estimates of $\mathbb{E} f(X_T)$, densities, divergences — and every statement in Section 6.2, which is a weak statement and not a pathwise one
Neither	The stationary law at fixed η as $T \to \infty$	Bias of order η, not vanishing in T	Sampling. Both orders are defined at fixed T and say nothing here; Section 4.1 treats it separately

1.7 Chapter Summary

Chapter 1 supplies the state-space and path-space calculus the rest of the volume depends on. Existence and uniqueness for ordinary differential equations, the matrix exponential and its two abscissas, convergence of the Euclidean gradient flow under strong convexity and under the Polyak–Łojasiewicz inequality, the construction of Brownian motion, the Itô integral and formula, and existence theory for stochastic differential equations are all classical and are quoted or proved in the forms later chapters actually use.

Two distinctions set here are used everywhere afterwards. The first separates *asymptotic* from *finite-time* behaviour: the spectral abscissa governs the former, the numerical abscissa detects the onset of the latter, and neither predicts the size of a transient. The second separates statements about *paths* from statements about *laws*: Theorem 1.46 gives uniqueness in law for the Tanaka equation while pathwise uniqueness fails, so a claim about a density needs only the first and a claim about a trajectory-by-trajectory coupling needs the second. The chapter does not attempt a complete treatment of stochastic analysis; it fixes conventions, proves what is load-bearing, and names its sources for the rest.

Exercises

§1.1

Tier I Direct application.

1.1 *Explicit blow-up*

Solve $\dot{x} = x^p$ on $\mathbb{R}$ from $x_0 > 0$ for $p > 1$ and find the blow-up time. Show that $p = 1$ is the boundary case and explain, in terms of Theorem 1.2, why nothing goes wrong with local existence at any p.

1.2 *Contraction under one-sided Lipschitz*

Prove the contraction estimate quoted in the notes on Corollary 1.5: if $\langle b(x) - b(y), x - y \rangle \leq$

$-\mu\,|x-y|^2$ then $|x(t)-y(t)| \le |x_0-y_0|\,e^{-\mu t}$.

Hint. Differentiate $|x(t)-y(t)|^2$ and use Grönwall on the result, not on $|x-y|$.

Tier II Completing the derivation.

1.3 *Two different equations*

Let x solve $\dot{x} = b(x)$ and $\tilde{x}$ solve $\dot{\tilde{x}} = \tilde{b}(\tilde{x})$ from the same initial condition, with b globally L-Lipschitz and $\sup_{|y|\le R} |b(y)-\tilde{b}(y)| \le \delta$ on a ball containing both trajectories. Prove $|x(t)-\tilde{x}(t)| \le \delta L^{-1}(e^{Lt}-1)$. Then state what this says about a learned vector field with training error δ, and identify the two quantities a paper reporting such an error must also report for the bound to mean anything.

1.4 *Grönwall with a time-dependent rate*

Extend Proposition 1.4 to $u(t) \le a + \int_0^t \ell(s)u(s)ds$ with $\ell \ge 0$ integrable, obtaining $u(t) \le a \exp \int_0^t \ell$. Identify the step of the original proof that needs changing and the step that does not.

Tier III Probing the boundary.

1.5 *Where uniqueness fails, and how badly*

For $\dot{x} = |x|^\alpha$ with $\alpha \in (0,1)$ and $x_0 = 0$, describe the full set of solutions and show that it is uncountable. Then compute, for α close to 1, the local Lipschitz constant of $|x|^\alpha$ on $[\varepsilon, 1]$ and observe that it diverges as $\varepsilon \to 0$. Say what this shows about relying on a Lipschitz constant estimated away from the region a trajectory actually visits.

1.6 *Spectrum against transient*

For $A = \begin{pmatrix} -1 & M \\ 0 & -2 \end{pmatrix}$ compute $\sup_{t\ge0} \|e^{tA}\|$ as a function of M and the time at which it is attained, in the Euclidean operator norm throughout. Then find the numerical abscissa and check that its sign predicts initial growth while the spectral abscissa does not. Verify that the numerical abscissa does *not* predict the *size* of $\sup_t \|e^{tA}\|$ — compare the two as M grows — and state what a stability claim about a finite-length computation can and cannot quote from either.

Tier IV Computational.

1.7 *Measuring the Grönwall constant*

Take $b(x) = \sin(kx)$ on $\mathbb{R}$ for $k \in \{1,5,25\}$, so that $L = k$.

(a) Integrate from a grid of initial conditions and measure the growth of $|x(t)-y(t)|$ for nearby pairs. Compare the measured exponent with L and report the ratio.

(b) Repeat with $b(x) = -\sin(kx)$ and explain the difference in one sentence.

(c) State which of the two experiments Corollary 1.5 describes and which it merely bounds, and say what would have to be measured for the bound to be tight.

Acceptance criteria: the ratio in (a) is reported as a number for each k; (c) distinguishes the sharp case from the loose one.

§1.2

Tier I Direct application.

1.8 *The quadratic case in closed form*

For $f(\theta) = \frac{1}{2}\langle\theta, H\theta\rangle$ with $H \succ 0$, solve (1.5) explicitly and verify (1.9) with $\mu = \lambda_{\min}(H)$. Show that the bound on $f - f_*$ is attained exactly when θ_0 lies in the eigenspace of $\lambda_{\min}$, and that it is loose by a factor $e^{-2(\lambda-\mu)t}$ in the eigendirection of λ.

1.9 *Natural gradient is a change of metric, not of objective*

For $G \succ 0$ constant, show that $\dot{\theta} = -G^{-1}\nabla f(\theta)$ is the Euclidean gradient flow of f in the coordinates $\phi = G^{1/2}\theta$, and that it dissipates f at rate $\langle\nabla f, G^{-1}\nabla f\rangle$ rather than $|\nabla f|^2$. State the one-sentence version of what changed and what did not.

Tier II Completing the derivation.

1.10 *Łojasiewicz gives convergence of the trajectory*

Assume $|\nabla f(x)| \geq c\,(f(x) - f_*)^{1-\theta}$ near a limit point, with $\theta \in (0, \frac{1}{2}]$. Show that the arc length $\int_0^\infty |\dot{\theta}|\,dt$ is finite, and deduce that $\theta(t)$ converges to a single point θ_∞. Then sharpen the estimate to the tail bound $|\theta(t) - \theta_\infty| \leq \int_t^\infty |\dot{\theta}| \leq (\theta c)^{-1}\big(f(\theta(t)) - f_*\big)^\theta$, and combine it with the value rate of Table 1.1 to obtain the distance rate in that table's third column. Say why the constant depends on the entry time into the region where the inequality holds, and therefore on the initialisation. Identify where $\theta \leq \frac{1}{2}$ is used and what happens at $\theta = \frac{1}{2}$.

Hint. Estimate $-\frac{d}{dt}(f - f_*)^\theta$ from below by a multiple of $|\dot{\theta}|$.

1.11 *Discrete Grönwall*

State and prove the discrete inequality used at the end of the proof of Proposition 1.11: if $a_{k+1} \leq (1 + c)a_k + b$ with $a_0 = 0$ then $a_k \leq b\big((1 + c)^k - 1\big)/c$. Then explain why replacing $(1 + c)^k$ by e^{ck} is legitimate and why it is the step that makes the horizon appear in the exponent.

Tier III Probing the boundary.

1.12 *Dissipation without convergence, and what bounded orbits force*

Two parts, and the second is the point.

(a) Write down, in polar coordinates, a C^∞ function on $\mathbb{R}^2$ that is bounded below and whose gradient flow from some θ_0 stays in a bounded region and does not converge — a circular trough with a slowly rotating angular perturbation. Verify that Proposition 1.8 holds, and identify which of its conclusions your example is compatible with.

(b) Prove that in your example $|\nabla f(\theta(t))| \to 0$ *necessarily*, and more generally that this holds along every precompact orbit: use $\int_0^\infty |\nabla f|^2 < \infty$, boundedness of θ, and uniform continuity of ∇f on the closure of the orbit. Conclude that a trajectory along which $|\nabla f|$ returns infinitely often to a fixed positive value must be unbounded, and say which hypothesis of your argument fails for such a trajectory.

1.13 *Where the horizon eats the bound*

Fix $\beta = 1$ and $M = 1$ in (1.11). For accuracy 10^{-3}, compute the largest admissible η as a function of T for $T \in \{1, 10, 100\}$, and the corresponding number of steps. Compare with the step count that gradient descent actually needs on a 1-smooth 1-strongly convex quadratic. Explain the discrepancy in one sentence, and say which of the two numbers is the honest one for each purpose.

1.14 *A rate that does not transfer*

Take f satisfying (1.10) with $\mu = 1$ and β-smooth with $\beta = 100$. Write down the continuous-time time to accuracy from Theorem 1.10 and the discrete iteration count from the corresponding descent lemma. Show that they differ by a factor of β, identify where that factor enters, and state which of Table 1.1's columns records it.

Tier IV Computational.

1.15 *Flow, small step, and fixed step*

Take $f(\theta) = \frac{1}{2}\theta^\top H\theta$ in $\mathbb{R}^2$ with $H = \mathrm{diag}(1, \kappa)$ for $\kappa \in \{2, 20, 200\}$.

(a) Integrate (1.5) accurately and confirm (1.9) with $\mu = 1$.

(b) Run explicit Euler at $\eta \in \{10^{-3}, 10^{-2}, 10^{-1}\}$ and measure $\max_k |\theta_k - \theta(k\eta)|$ over $T = 1$. Confirm the linear order in η and report the measured constant against the bound (1.11).

(c) Increase η past $2/\kappa$ and describe what happens. Say which of Proposition 1.11 and Table 1.1 still applies, and which does not.

(d) State, for each of (a), (b) and (c), which regime of Table 6.1 it belongs to.

Acceptance criteria: the measured constant in (b) is reported alongside the bound, and the ratio is stated; (d) is answered with one row name per part.

§1.3

Tier I Direct application.

1.16 *Discrete Grönwall, restated*

Prove: if $a_{t+1} \le (1+c)a_t + b$ with $a_0 = 0, c > 0$, then $a_t \le b((1+c)^t - 1)/c \le b(e^{ct} - 1)/c$. Identify where this is used in the proofs of Proposition 1.11 and Proposition 1.12, and what plays the role of b in each.

1.17 *The adjoint of a linear field*

For $F(x, \theta) = Ax$ with constant A, solve (1.15) explicitly and verify Proposition 1.13 against a direct computation of $\nabla_{x_0}\mathcal{L}(e^{A}x_0)$.

Tier II Completing the derivation.

1.18 *Where the parameter regularity enters*

Redo the proof of Proposition 1.12 with (H2) weakened to $|\theta_{\ell+1} - \theta_\ell| \leq \Delta$ for all ℓ, with Δ not tied to L. Show that the resulting bound is $O(1/L) + O(\Delta)$ and identify the term that does not vanish. State the condition on Δ under which the continuum limit exists.

1.19 *Gradients of the parameters, discretely*

Derive the discrete analogue of (1.16) for (1.12) directly by the chain rule, and compare with the discretisation of (1.16) by the same rule that produced (1.12). Show that the two differ at order $1/L$, and say which of them a backpropagation implementation computes.

Tier III Probing the boundary.

1.20 *A family with no continuum limit*

Construct a sequence of residual networks satisfying (H1), with θ_ℓ alternating between two fixed values, whose forward maps converge as $L \to \infty$ to the flow of a field that is *not* $F(\cdot, \theta)$ for any single θ. Identify the limiting field and explain why Proposition 1.12 is not contradicted.

Hint. Average the two fields, and note that the averaging is exact only in the limit and only because the two layer maps are applied alternately.

1.21 *Backward error in the adjoint*

Take a stiff linear field $\dot{x} = Ax$ with $\alpha(A) < 0$ and large $\|A\|$. Integrate forwards to $s = 1$, then reconstruct $x(0)$ by integrating backwards from $x(1)$ with the same solver and tolerance, and report the error. Repeat for several tolerances and describe the scaling. State which constant in Corollary 1.5 your measurement is estimating.

Tier IV Computational.

1.22 *Does depth behave like time?*

Train a residual network of depth $L = 32$ on a small two-dimensional classification problem, with the $1/L$ scaling of (1.12).

(a) Measure $\max_\ell |\theta_{\ell+1} - \theta_\ell|$ after training and compare with $1/L$. Report whether (H2) holds empirically, with the constant.

(b) Retrain at $L \in \{8, 16, 32, 64, 128\}$ from the same initialisation scheme and measure whether the forward maps converge, using a fixed set of test inputs.

(c) Add a penalty on $\sum_\ell |\theta_{\ell+1} - \theta_\ell|^2$ and repeat (b). Report the change.

(d) State what (a)–(c) together do and do not establish about Proposition 1.12, and which part of it is an empirical question rather than a mathematical one.

Acceptance criteria: (a) reports a constant, not an impression; (d) distinguishes the hypothesis from the conclusion.

§1.4

Tier I Direct application.

1.23 *Using the invariances*

Use Proposition 1.16(iii) to prove that $\lim_{t\to\infty} W_t/t = 0$ almost surely, and (i) to show that $\sup_{s\le t} W_s$ has the same law as $\sqrt{t}\,\sup_{s\le 1} W_s$. State which invariance each conclusion used and why the other will not do.

1.24 *Second moment of the exit time*

In the setting of Proposition 1.22 with $a = b$, apply Theorem 1.21 to a suitable polynomial in W_t and t to compute $\mathbb{E}\tau^2$, and deduce $\operatorname{Var}\tau$.

Hint. $W_t^4 - 6tW_t^2 + 3t^2$ is a martingale; verify this first.

Tier II Completing the derivation.

1.25 *Exponential martingale and the hitting-time transform*

For $\lambda > 0$ let $M_t = \exp(\lambda W_t - \frac{1}{2}\lambda^2 t)$ and let τ_a be the first hitting time of $a > 0$. Show that M^{τ_a} is bounded, deduce $\mathbb{E}[e^{-\lambda^2\tau_a/2}] = e^{-\lambda a}$, and invert to obtain the density of τ_a. Identify where boundedness was needed and what it replaced.

1.26 *Why the usual conditions*

Give an example of a filtration that is not right-continuous and an adapted process for which the hitting time of an open set fails to be a stopping time. Then say which step of the proof of Proposition 1.22 would break.

Tier III Probing the boundary.

1.27 *Almost sure is not integrable*

Show directly, without Proposition 1.22, that $\mathbb{P}(\tau_a < \infty) = 1$ and $\mathbb{E}\tau_a = \infty$ for the one-sided hitting time. Then write down the false conclusion that Theorem 1.21(ii) yields if applied without checking uniform integrability, and identify the exact quantity that fails to converge in L^1.

1.28 *Hölder at the boundary exponent*

Show that almost surely W is not $\frac{1}{2}$-Hölder on any interval, so that the exponent $\frac{1}{2}$ belongs to neither part of Theorem 1.17. Say what this implies for a numerical scheme whose error analysis assumes $\frac{1}{2}$-Hölder paths.

Hint. Use Theorem 1.25 and the fact that a $\frac{1}{2}$-Hölder path would have quadratic variation bounded by a constant times the interval length with a constant depending on the path, then consider a sequence of nested intervals.

Tier IV Computational.

1.29 *Measuring the exit time and its tail*

Simulate W by a random walk with step $\sqrt{\eta}$ for $\eta \in \{10^{-2}, 10^{-3}, 10^{-4}\}$.

(a) With $a = b = 1$, estimate $\mathbb{E}\tau$ from 10^5 runs and compare with Proposition 1.22. Report the bias against η and its order.

(b) With one barrier at $a = 1$, estimate $\mathbb{E}\tau_a$ from the same number of runs at each η, and plot the estimate against the number of runs. Describe what you see and explain it in one sentence.

(c) State which of (a) and (b) is estimating a finite quantity, and what the plot in (b) would look like if you did not already know the answer.

Acceptance criteria: the order of the bias in (a) is stated with its evidence; (c) is answered in two sentences.

§1.5.4

Tier I Direct application.

1.30 *Geometric Brownian motion*

Let $dS_t = \mu S_t\,dt + \sigma S_t\,dW_t$ with $S_0 > 0$ and μ, σ constant. Apply Theorem 1.33 to $f(x) = \log x$ to show that $S_t = S_0 \exp\big((\mu - \frac{1}{2}\sigma^2)t + \sigma W_t\big)$. Explain in one sentence why $\mathbb{E}S_t = S_0 e^{\mu t}$ while the median of S_t grows at rate $\mu - \frac{1}{2}\sigma^2$, and identify which of the two an investor experiences.

Hint. The correction $-\frac{1}{2}\sigma^2$ is the second-order term of (1.25) for $f = \log$, where $f'' < 0$.

1.31 *Ornstein–Uhlenbeck by integrating factor*

For $dX_t = -\theta X_t\,dt + \sqrt{2\theta}\,dW_t$ with $\theta > 0$, apply Theorem 1.33 to $f(t, x) = e^{\theta t}x$ and deduce the closed form $X_t = e^{-\theta t}X_0 + \sqrt{2\theta}\int_0^t e^{-\theta(t-s)}dW_s$. Compute $\mathrm{Var}(X_t)$ and verify that it converges to 1 for every θ. State what this says about the role of θ in the variance-preserving forward process of Section 3.2.

Tier II Completing the derivation.

1.32 *The general case*

Carry out the proof of Theorem 1.33 for general d and m and for $f = f(t, x)$. Identify precisely where the cross terms $\Delta_k X^i \Delta_k X^j$ with $i \neq j$ enter, and show that they contribute $\int_0^t a_s^{ij}\partial_{ij}f\,ds$ and not twice that — i.e. account for the $\frac{1}{2}$ in front of the trace.

1.33 *Itô product rule*

Let X, Y be scalar Itô processes with diffusion coefficients σ^X, σ^Y driven by the same Brownian motion. Prove $d(X_t Y_t) = X_t\,dY_t + Y_t\,dX_t + \sigma_t^X\sigma_t^Y\,dt$ by applying Theorem 1.33 to $f(x, y) = xy$, and again by applying it to $\frac{1}{2}[(x+y)^2 - x^2 - y^2]$. Explain why the second route makes it obvious that the extra term is a quadratic covariation.

1.34 *Why localisation is needed and what it costs*

In Step 1 of the proof of Theorem 1.33, the stopping times τ_n were used to reduce to a bounded process. Show that the conclusion for X^{τ_n} implies the conclusion for X on $\{t < \tau_n\}$, and that $\mathbb{P}(\tau_n < T) \to 0$. Then explain why this argument delivers the pathwise identity (1.25) but does *not* deliver $\mathbb{E}f(t, X_t) = \mathbb{E}f(0, X_0) + \mathbb{E}\int_0^t \mathcal{L}f \, ds$ without a further hypothesis.

Tier III Probing the boundary. Each of these asks where a statement stops being true.

1.35 *When the expectation identity fails*

Construct $V \in C^2(\mathbb{R})$, growing faster than *any* polynomial, and an initial condition, such that the solution of (1.29) exists for all time but $\mathbb{E}\int_0^t |V'(X_s)|^2 \, ds = \infty$ for every $t > 0$. Verify that (1.30) still holds pathwise. Then say, in one sentence, which of the two displays in Proposition 1.35 a numerical experiment would appear to confirm, and why that appearance is misleading.

Hint. Explosion is not the only obstruction; a heavy-tailed initial law with a light-tailed stationary law will do it at small t.

1.36 *The same equation in the other calculus*

Write (1.29) as a Stratonovich equation and recompute $dV(X_t)$ using the Stratonovich chain rule. The drift you obtain differs from (1.30). Both are correct statements about different objects: identify what changed, and state which of the two drifts appears in the Fokker–Planck equation of Section 2.1. Then explain why the distinction is invisible here but not on a Riemannian manifold, where the diffusion coefficient depends on position.

1.37 *A non-smooth potential*

Let $V(x) = |x|$ on $\mathbb{R}$. Show that (1.30) is false as stated, that the process $|X_t|$ nevertheless satisfies a stochastic differential equation, and identify the additional term. Say what this implies about applying any result of this volume to a network with piecewise-linear activations.

Tier IV Computational.

1.38 *Verifying (1.32), and measuring the discretisation bias*

Simulate (1.29) with $V(x) = \frac{1}{2}|x|^2$ by Euler–Maruyama from $X_0 = 0$, for $d \in \{1, 10, 100\}$ and step sizes $\eta \in \{10^{-1}, 10^{-2}, 10^{-3}\}$, and estimate $\mathbb{E}V(X_t)$ from 10^4 independent runs.

(a) Plot the estimate against $\frac{d}{2}(1 - e^{-2t})$ and confirm the agreement improves linearly in η.

(b) The stationary value of $\mathbb{E}V$ under exact dynamics is $d/2$. Measure the stationary value under Euler–Maruyama at each η and show that the bias is positive and of order η. Give the sign a one-sentence explanation in terms of (1.31).

(c) Repeat (b) for the Metropolis-adjusted scheme of Section 4.1 and report the bias. State what you have and have not shown about which algorithm is preferable.

Acceptance criteria: the linear-in-η rate in (a) is visible over at least two decades; the sign of the bias in (b) is stated with its reason; the claim in (c) is stated with its class restriction.

§1.5.5

Tier I Direct application.

1.39 *The simplest conversion*

Compute $\int_0^t W_s \circ \mathrm{d}W_s$ directly from (1.33) and from (1.34), and check that both give $\frac{1}{2}W_t^2$. Then verify Proposition 1.23 against your answer.

1.40 *Geometric Brownian motion in two conventions*

Write $\mathrm{d}X = \mu X\,\mathrm{d}t + \sigma X \circ \mathrm{d}W$ in Itô form and solve it. Compare with the solution of the Itô equation with the same coefficients and report the ratio of the two at time t. State which of the two has $\mathbb{E}X_t = X_0 e^{\mu t}$.

Tier II Completing the derivation.

1.41 *The cross-variation step*

Fill in the step of the proof of Proposition 1.37 that identifies $\sum_k \left(\int_{t_k}^{t_{k+1}} g'(X)\sigma\,\mathrm{d}W\right)\Delta_k$ as a cross-variation sum, and state the polarised form of Lemma 1.27 that it needs. Identify where adaptedness is used.

1.42 *Multidimensional correction*

For $\mathrm{d}X = b\,\mathrm{d}t + \sum_j \sigma_j(X) \circ \mathrm{d}W^j$ in $\mathbb{R}^d$ with independent W^j, derive the Itô form and show that the correction is $\frac{1}{2}\sum_j (\partial\sigma_j)\sigma_j$. Verify that it is not in general the gradient of anything, and say what that implies about the existence of a free energy for such a dynamics — a question taken up in Chapter 2.

Tier III Probing the boundary.

1.43 *Causal smoothing gives Itô*

Replace the symmetric interpolation of Theorem 1.39 by the causal smoothing $\dot{W}_t^{(\varepsilon)} = \varepsilon^{-1}(W_t - W_{t-\varepsilon})$ and redo the heuristic computation of the limit. Show that the correction term is now absent, and identify the exact place where the symmetry of the interpolation entered the original argument.

1.44 *A Lipschitz diffusion with no Stratonovich form*

Take $\sigma(x) = |x|$, which is Lipschitz and not C^1. Show that the Itô equation $\mathrm{d}X = \sigma(X)\mathrm{d}W$ has a unique strong solution by Theorem 1.41, and that (1.35) is not defined. Then say what a numerical scheme using the midpoint rule on this equation actually converges to, if anything.

Tier IV Computational.

1.45 *Measuring the correction*

Take $\mathrm{d}X = \sigma(X)\mathrm{d}W$ with $\sigma(x) = 1 + \frac{1}{2}\sin x$ and $X_0 = 0$.

(a) Simulate with Euler–Maruyama (left endpoint) and with the midpoint rule at $\eta \in \{10^{-2}, 10^{-3}, 10^{-4}\}$, and estimate $\mathbb{E}X_T$ at $T = 1$ from 10^5 paths.

(b) The Itô solution has $\mathbb{E}X_T = 0$. Report the midpoint estimate and compare it with $\frac{1}{2}\mathbb{E}\int_0^T \sigma'(X)\sigma(X)ds$ estimated from the same runs.

(c) Repeat with $\sigma \equiv 1$ and report both estimates.

(d) State what (c) shows about when the distinction can be ignored, and name one construction later in this volume where it can and one where it cannot.

Acceptance criteria: (b) reports the two numbers side by side with their Monte-Carlo errors; (d) names both constructions.

§1.6

Tier I Direct application.

1.46 *Moments of the Ornstein–Uhlenbeck process*

From Proposition 1.43 compute $\mathbb{E}|X_t|^2$ and verify directly that it solves the ordinary differential equation obtained by taking expectations in (1.39) after applying Theorem 1.33 to $|x|^2$. Confirm that the noise contributes $+d\sigma^2$ and identify which term of Itô's formula it comes from.

1.47 *Mean against median*

For Proposition 1.44 with $\mu = 0.05$ and $\varsigma = 0.4$, compute $\mathbb{E}S_T$ and the median of S_t at $t = 10$, and the probability that $S_{10} < S_0$. State in one sentence which of the three numbers an investor experiences.

Tier II Completing the derivation.

1.48 *Localisation in Theorem 1.41*

Carry out the argument sketched in Remark 1.42: assuming the local Lipschitz condition and the Lyapunov bound, construct the solution on $[0, T]$ and prove non-explosion.

Hint. Apply Theorem 1.33 to $\log(1+|x|^2)$ and use the Lyapunov bound to get a supermartingale.

1.49 *The Milstein term is not a guess*

Add the term $\frac{1}{2}\sigma'\sigma((\Delta W)^2 - \eta)$ to (1.44) and show, using (1.45), that the leading local error is cancelled. Then verify that the added term has mean zero, and deduce why the weak order is unchanged.

Tier III Probing the boundary.

1.50 *Weak accuracy without pathwise accuracy*

Construct a scheme for $dX = dW$ that has weak order 1 and no strong order at all — for instance by replacing each Brownian increment with an independent two-point random variable of matching first two moments. Verify both claims, and state which entry of Table 1.2 each verification belongs to.

1.51 *A stationary bias that no order sees*

Apply (1.44) to (1.39) with $\sigma^2 = 2\theta$, whose exact stationary law is $\mathcal{N}(0,1)$. Compute the stationary variance of the scheme exactly as a function of η and θ, and show that the relative error is $\theta\eta/(2-\theta\eta)$. Explain why neither order in Definition 1.47 detects it, and identify the quantifier in Definition 1.47 that excludes it.

1.52 *Tanaka's example, numerically*

Simulate (1.43) from $X_0 = 0$ with (1.44), twice on the same Brownian path but with different tie-breaking at $X = 0$. Report the two trajectories — they should separate — and then report the empirical law of X_T in both runs, which should agree and should be $\mathcal{N}(0,T)$. Say which of the two comparisons Theorem 1.46 predicts to agree and which not, and connect the answer to the distinction in the box following it. Then verify numerically that $\int_0^t \operatorname{sgn}(X_s)\,\mathrm{d}X_s$ is the same process in both runs, which is why no solution can be recovered from the driving noise.

Tier IV Computational

1.53 *Measuring both orders, and the constant*

Take $d = m = 1$, $b \equiv 0$, $\sigma(x) = 1 + \frac{1}{2}\tanh x$, $X_0 = 0$, $T = 1$.

(a) Generate reference paths at $\eta_{ref} = 2^{-16}$ and measure the strong error at $\eta = 2^{-k}$, $k = 6, \ldots, 12$, using the *same* Brownian path for scheme and reference. Fit the exponent and report it with a confidence interval.

(b) Measure the one-step L^2 error at $X_0 = 0$ and compare its leading coefficient with $|\sigma'\sigma|/\sqrt{2}$ from Proposition 1.49. Report the ratio.

(c) Measure the weak error for $f(x) = x^2$ and $f(x) = \cos x$, using independent noise for scheme and reference, and fit the exponent.

(d) Repeat (a) with independent noise for scheme and reference. Report the exponent you obtain and explain, in one sentence, why it is not the strong order.

Acceptance criteria: the fitted exponents in (a) and (c) are reported with uncertainty; the ratio in (b) is a number; (d) is answered in one sentence.

CHAPTER 2

Fokker–Planck Equations, Reversible Diffusions, and Wasserstein Gradient Flows

Chapter 1 followed a single trajectory. This chapter follows its law.

The change of object is the one that makes the rest of the volume possible, and it is worth being precise about what is gained and what is given up. A trajectory is a random element of a path space; its law at a fixed time is a deterministic element of a space of measures. Statements about the second are statements about a deterministic partial differential equation, and can be proved by the methods of analysis rather than of probability. What is given up is everything the individual path knew: two diffusions can have the same marginal laws at every time and different path measures (Proposition 1.43 and the notes there already exhibit the elementary version of this), and no result about ρ_t recovers a statement about a coupling or a simulation.

Section 2.1 derives the equation for ρ_t and sorts out three conditions that are routinely merged: that π is stationary, that the dynamics is reversible with respect to π, and that ρ_t equals π. Section 2.2 shows that for reversible diffusions with gradient drift the curve $t \mapsto \rho_t$ is the steepest descent of an explicit functional in an explicit geometry, and — equally importantly — exhibits a family of diffusions for which it is not. Section 2.3 recasts all of this in operator language, which is what Section 2.4 needs in order to say where convergence rates come from. They come from hypotheses on the target, and the last section is a catalogue of those hypotheses ordered by strength.

2.1 The Fokker–Planck Equation

2.1.1 The generator and its adjoint

Definition 2.1 (Infinitesimal generator). Let $b : [0,T] \times \mathbb{R}^d \to \mathbb{R}^d$ and $\sigma : [0,T] \times \mathbb{R}^d \to \mathbb{R}^{d\times m}$ be measurable, and write $a = \sigma\sigma^\top$. The *generator* of the diffusion (1.37) is the second-order operator

$$\mathcal{L}_t f = \langle b(t,\cdot), \nabla f\rangle + \tfrac{1}{2}\operatorname{tr}\left(a(t,\cdot)\,\nabla^2 f\right), \qquad f \in C_c^\infty(\mathbb{R}^d), \tag{2.1}$$

with the factor $\frac{1}{2}$ fixed by the Itô convention of Section 1.5.4. Its *formal adjoint* is

$$\mathcal{L}_t^* \rho = -\nabla\cdot\left(b(t,\cdot)\rho\right) + \tfrac{1}{2}\sum_{ij}\partial_{ij}\left(a_{ij}(t,\cdot)\rho\right), \tag{2.2}$$

defined so that $\int (\mathcal{L}_t f)\rho = \int f(\mathcal{L}_t^*\rho)$ for $f \in C_c^\infty$ and ρ twice differentiable.

The generator is the object the martingale problem of Definition 1.45 was written in terms of, and by Theorem 1.46 it determines the law of the diffusion. It does not determine σ: only a appears in (2.1).

Theorem 2.2 (Kolmogorov's backward equation) CLASSICAL. *Let b, σ satisfy (1.38) and be time-independent, and suppose $f \in C_b^2(\mathbb{R}^d)$. Then $u(t,x) := \mathbb{E}\big[f(X_t) \mid X_0 = x\big]$ satisfies $\partial_t u = \mathcal{L}u$ with $u(0,\cdot) = f$, in the sense of distributions; if in addition a is uniformly elliptic and the coefficients are Hölder continuous, u is a classical solution.*

Proof roadmap. Apply Theorem 1.33 to f along the diffusion started at x, take expectations, and differentiate in t to get $\partial_t u(t,x) = \mathbb{E}[\mathcal{L}f(X_t)]$; the Markov property turns the right-hand side into $\mathcal{L}u(t,x)$ once $u(t,\cdot)$ is known to be twice differentiable, which is exactly what the ellipticity and Hölder hypotheses buy through Schauder theory. Without them the identity holds in the distributional sense and no more. The regularity theory is quoted; nothing in this volume uses the classical form of the statement. □

The equation this chapter is about is the adjoint one, and it needs no regularity theory at all if it is stated in the right form.

Proposition 2.3 (Fokker–Planck equation) DERIVED. *Let X solve (1.37) with coefficients satisfying (1.38), and let $\rho_t = \mathrm{Law}(X_t)$. Then for every $f \in C_c^\infty(\mathbb{R}^d)$ and every t,*

$$\int_{\mathbb{R}^d} f\, \mathrm{d}\rho_t = \int_{\mathbb{R}^d} f\, \mathrm{d}\rho_0 + \int_0^t \int_{\mathbb{R}^d} \mathcal{L}_s f\, \mathrm{d}\rho_s\, \mathrm{d}s. \tag{2.3}$$

If moreover ρ_s has a density that is $C^{1,2}$ in (s,x), then that density satisfies

$$\partial_t \rho_t = \mathcal{L}_t^\dagger \rho_t \tag{2.4}$$

pointwise.

No converse is asserted. Equation (2.3) is a statement about the one-time marginals, and recovering from it a process — a measure on paths solving the martingale problem — is a separate theorem, not a restatement. Under hypotheses ensuring a superposition principle for the forward equation, every weakly continuous solution of (2.3) with the required integrability is the family of one-time marginals of some solution of the martingale problem, and well-posedness of that problem then makes the marginal family unique. Without a superposition principle or an equivalent duality theorem, the passage from (2.3) back to a path measure is not available, and nothing in this volume uses it.

Proof. By Theorem 1.33 applied to f, $f(X_t) = f(X_0) + \int_0^t \mathcal{L}_s f(X_s)\,\mathrm{d}s + \int_0^t \langle \sigma_s^\top \nabla f(X_s), \mathrm{d}W_s \rangle$. The integrand of the stochastic integral is bounded, because ∇f has compact support and σ has linear growth, so the stochastic integral is a true martingale with zero mean by Theorem 1.30. Taking expectations and using Fubini, justified by $|\mathcal{L}_s f| \le C$ uniformly, gives (2.3).

For (2.4), differentiate (2.3) in t to get $\int f\, \partial_t \rho_t = \int (\mathcal{L}_t f) \rho_t$, and integrate the right-hand side by parts twice, the boundary terms vanishing because f has compact support: the result is $\int f\, \mathcal{L}_t^\dagger \rho_t$. Since this holds for every $f \in C_c^\infty$ and both sides are continuous, the two densities agree pointwise.

■

> **Interpretation**
>
> Equation (2.3) is a conservation law. Probability is neither created nor destroyed, so its density can only change by flowing, and the equation says exactly how much flows: a drift part that carries mass along the vector field, and a diffusion part that spreads it. Writing it in the flux form (2.6) makes this explicit and is the form used whenever the question is *where the probability is going* rather than how fast it is changing — which is why stationarity, detailed balance and equilibrium are distinguished in Proposition 2.6 by conditions on the flux rather than on the density.

Which form is primary. The weak form (2.3) is the theorem; the strong form is a corollary available under regularity. Starting from a Dirac initial condition, no density exists at $t = 0$ and (2.4) is meaningless there, while (2.3) holds. For degenerate a — as in the kinetic dynamics of Chapter 4, where the noise acts only on the velocity variable — a density may exist for $t > 0$ by Hörmander's theorem or may fail to exist at all, and the weak form is what survives. Every argument in this volume that can be run weakly is run weakly.

What it does not say. It is a statement about one-dimensional marginals. Two diffusions with the same ρ_t for every t can have different path measures, different couplings and different exit-time laws; Proposition 2.16 exhibits a family with the same ρ_t at equilibrium and demonstrably different behaviour away from it. Anything requiring the path law needs Definition 1.45 and not this proposition.

2.1.2 Stationarity, reversibility, and the probability flux

Three conditions on a measure π are used constantly and are not the same condition. This subsection separates them, and the separation is the reason several arguments elsewhere in the literature do not go through.

Definition 2.4 (Stationary, reversible, and the flux). Fix time-independent coefficients. A probability measure π is *stationary* for $\mathcal{L}$ if $\int \mathcal{L} f \, d\pi = 0$ for every $f \in C_c^\infty$; equivalently $\mathcal{L}^* \pi = 0$ in the distributional sense. It is *reversible*, and the dynamics is said to satisfy *detailed balance* with respect to it, if $\mathcal{L}$ is symmetric in $L^2(\pi)$:

$$\int f \, \mathcal{L} g \, d\pi = \int g \, \mathcal{L} f \, d\pi \qquad \text{for all } f, g \in C_c^\infty. \tag{2.5}$$

For a measure with density ρ, the *probability flux* is

$$j[\rho] := b\rho - \tfrac{1}{2}\nabla\cdot(a\rho), \qquad \text{so that} \qquad \partial_t \rho_t = -\nabla\cdot j[\rho_t]. \tag{2.6}$$

For the rest of this chapter we take $a = 2I$, so that the equation is $dX_t = b(X_t)dt + \sqrt{2}\, dW_t$, the generator is $\mathcal{L} = \langle b, \nabla \rangle + \Delta$, and the flux is $j[\rho] = b\rho - \nabla\rho$. The general case changes no conclusion below and obscures every computation.

Proposition 2.5 (Reversible means gradient drift) [illegible]. *Let $\pi = Z^{-1} e^{-V}$ with $V \in C^2$ and $e^{-V} \in L^1$, and set*

$$c := b + \nabla V, \tag{2.7}$$

the non-gradient part of the drift. Then, with $\mathcal{L}$ as above and subject to the integrations by parts below being justified:

(i) *for all* $f, g \in C_c^\infty$,

$$\int f\,\mathcal{L}g\,\mathrm{d}\pi = -\int \langle \nabla f, \nabla g\rangle\,\mathrm{d}\pi + \int f\,\langle c, \nabla g\rangle\,\mathrm{d}\pi, \tag{2.8}$$

of which the first term is symmetric in (f, g);

(ii) *the flux at* π *is* $j[\pi] = \pi c$;

(iii) π *is stationary if and only if* $\nabla\cdot(\pi c) = 0$;

(iv) π *is reversible if and only if* $c = 0$ π*-almost everywhere, that is, if and only if* $b = -\nabla V$.

Proof. (i) Since $\Delta g = e^{V}\nabla\cdot(e^{-V}\nabla g) + \langle \nabla V, \nabla g\rangle$,

$$f\,\mathcal{L}g\,e^{-V} = f\,\langle b + \nabla V, \nabla g\rangle e^{-V} + f\,\nabla\cdot(e^{-V}\nabla g),$$

and integrating the second term by parts gives $-\int \langle \nabla f, \nabla g\rangle\, e^{-V}$. Dividing by Z gives (2.8).

(ii) $j[\pi] = b\pi - \nabla\pi = b\pi + \pi\nabla V = \pi c$, using $\nabla\pi = -\pi\nabla V$.

(iii) Taking $f \equiv 1$ in (2.8) — legitimate by approximation, since the integrands are integrable — gives $\int \mathcal{L}g\,\mathrm{d}\pi = \int \langle c, \nabla g\rangle\,\mathrm{d}\pi = -\int g\,\nabla\cdot(\pi c)$. This vanishes for all $g \in C_c^\infty$ if and only if $\nabla\cdot(\pi c) = 0$, which by (ii) is $\nabla\cdot j[\pi] = 0$.

(iv) By (i) the first term of (2.8) is symmetric, so reversibility holds if and only if $B(f, g) := \int f\,\langle c, \nabla g\rangle\,\mathrm{d}\pi$ is symmetric. But

$$B(f, g) + B(g, f) = \int \langle c, \nabla(fg)\rangle\,\mathrm{d}\pi = -\int fg\,\nabla\cdot(\pi c),$$

which vanishes for all f, g when π is stationary, by (iii). Hence, given stationarity, B is antisymmetric; a bilinear form that is both symmetric and antisymmetric is zero, so reversibility forces $B \equiv 0$.

It remains to conclude $c = 0$ from $B \equiv 0$, and this is done one coordinate at a time rather than by producing a potential for c. Fix i and $f \in C_c^\infty(\mathbb{R}^d)$. Choose $\chi \in C_c^\infty(\mathbb{R}^d)$ with $\chi \equiv 1$ on a neighbourhood of $\operatorname{supp} f$, and set $g(x) = x_i\chi(x)$, which is again in C_c^∞. On $\operatorname{supp} f$ one has $\nabla g = e_i$, so

$$0 = B(f, g) = \int f\,\langle c, \nabla g\rangle\,\mathrm{d}\pi = \int f\,c_i\,\mathrm{d}\pi.$$

As $f \in C_c^\infty$ was arbitrary, $c_i = 0$ π-almost everywhere; and as i was arbitrary, $c = 0$ π-almost everywhere. The converse is immediate from (2.8) with $c = 0$.

The cutoff is what makes this work with no hypothesis on c beyond local integrability against π. One might be tempted instead to choose g with $\nabla g = c$ and read off $\int f\,|c|^2\,\mathrm{d}\pi = 0$; but a continuous vector field need not be a gradient even locally — that requires c to be curl free, which is not assumed here and is exactly what fails for the rotational fields of Definition 2.15. ∎

The identity $j[\pi] = \pi c$ makes the difference between (iii) and (iv) visible: stationarity says the flux at π is divergence free, reversibility says it is zero. A rotational flux that circulates without accumulating anywhere is stationary and not reversible, and Definition 2.15 is exactly that construction.

Proposition 2.6 (Three statements that are not one) REVISED. *With the notation above, the following are distinct, and no implication holds among them beyond those stated.*

(a) π is stationary*: a property of the pair $(\mathcal{L}, \pi)$, equivalent to $\nabla \cdot j[\pi] = 0$.*

(b) The dynamics is reversible with respect to π*: a property of the pair $(\mathcal{L}, \pi)$, equivalent to $j[\pi] = 0$, and strictly stronger than (a).*

(c) $\rho_t = \pi$*: a property of the initial condition, not of the pair $(\mathcal{L}, \pi)$. If $\rho_0 = \pi$ and (a) holds then $\rho_t = \pi$ for all t. Conversely, if the adjoint semigroup P_t^* is injective on the relevant class of laws — a backward-uniqueness hypothesis, not a consequence of irreducibility — then $\rho_t = \pi$ at some $t > 0$ forces $\rho_0 = \pi$. In the reversible case $b = -\nabla V$ the flux is*

$$j[\rho_t] = -\rho_t \nabla \log \frac{\rho_t}{\pi}, \tag{2.9}$$

and it vanishes identically exactly when $\rho_t = \pi$.

Proof. (a) and (b) are Proposition 2.5(iii)–(iv), and (b) implies (a) because $j[\pi] = 0$ implies $\nabla \cdot j[\pi] = 0$; the converse fails by Definition 2.15. For (c), if $\rho_0 = \pi$ then $\rho_t = \pi$ solves (2.3) and uniqueness gives $\rho_t = \pi$. For the flux formula, $j[\rho] = -\nabla V \rho - \nabla \rho = -\rho \nabla(V + \log \rho) = -\rho \nabla \log(\rho/\pi)$, which vanishes identically only if ρ/π is constant, that is only if $\rho = \pi$. The converse in (c) is exactly injectivity of P_t^*: if $\rho_t = P_t^* \rho_0 = \pi = P_t^* \pi$ and P_t^* is injective then $\rho_0 = \pi$. ■

Remark 2.7 (Why irreducibility is not the right hypothesis here). An earlier formulation of (c) excluded $\rho_t = \pi$ at finite t by appealing to irreducibility together with "reversibility of the equation in time at the deterministic level". Neither step survives inspection. The Fokker–Planck equation is parabolic and *smoothing*: its backward initial-value problem is ill posed in general, so there is no deterministic time-reversal to invoke. And irreducibility is a statement about the support of the transition kernel — every open set is reached with positive probability — which says nothing about whether two distinct initial laws can be carried to the same law by P_t^*. What is needed is injectivity of P_t^*, which for many familiar diffusions does hold (for the Ornstein–Uhlenbeck semigroup it follows from the strict positivity of the Mehler kernel's Fourier multiplier), but which is a hypothesis with its own proof obligation. This volume uses only the safe direction: $\rho_0 = \pi \Rightarrow \rho_t = \pi$, and $j[\rho_t] = 0 \Leftrightarrow \rho_t = \pi$ in the reversible case.

Scope of Proposition 2.5 and Proposition 2.6

Assumed. Constant isotropic diffusion $a = 2I$; $V \in C^2$ with e^{-V} integrable; enough decay for two integrations by parts, which for ε of at most polynomial growth and π with Gaussian-type tails is automatic. For (iv), only local integrability of ε against π: the cutoff-coordinate argument in the proof assumes no potential for ε, and none exists in general.

Where the characterisation changes. With state-dependent a the condition $b = -\nabla V$ is replaced by $b = \frac{1}{2}\nabla \cdot a - \frac{1}{2} a \nabla V$, and the "gradient drift" formulation is no longer coordinate free; the correct invariant statement is the vanishing of $j[\pi]$, which is why (2.6) rather than the drift is the object this subsection is built on. On a manifold the same is true with the Riemannian divergence.

What this does not say. It does not say that non-reversible dynamics is worse, or slower, or lacks a Lyapunov functional. Proposition 2.16 shows that an admissible non-reversible perturbation dissipates the same free energy at exactly the same instantaneous rate; what it loses is the gradient-flow structure, not the decrease.

What would falsify it. A drift with $c \neq 0$ on a set of positive π-measure for which (2.5) nevertheless holds.

What does not follow

Reversibility is a property of the generator, not of the current law. It is a statement about $\mathcal{L}$ and π, quantified over test functions, and it holds at every time or at none. It says nothing about ρ_t. In particular "the process is at equilibrium" and "the process is reversible" are different assertions, and only the first is a statement about the state of the system.

The flux does not vanish before equilibrium. By (2.9) the flux of a reversible diffusion at time t is $-\rho_t \nabla \log(\rho_t/\pi)$, which is zero precisely when $\rho_t = \pi$ — a statement at a fixed time, and one that says nothing about whether $\rho_t = \pi$ can occur at finite t from $\rho_0 \neq \pi$; that is Remark 2.7. It is this non-zero flux that carries the free energy downhill in Proposition 2.12, so an argument that sets it to zero has assumed the conclusion. Detailed balance is therefore not available as a step in any argument about ρ_t at finite t.

A time-inhomogeneous forward process is at no time stationary. The noising processes of Chapter 3 have time-dependent coefficients and a law that moves at every instant. Whatever justifies their reverse-time formula, it is not detailed balance: that formula is proved by time-reversal theory for diffusions, under its own hypotheses, and applies to processes that are never in equilibrium. Reading it as a consequence of a vanishing flux is the most common way this chapter's material is misapplied, and Proposition 2.6 exists to make the three claims separable on the page.

Stationarity does not identify the dynamics. Infinitely many generators share a given stationary π — one for every admissible c — and they have different mixing rates, different path measures and different finite-time behaviour. A construction that specifies only the target has specified almost nothing about the process that reaches it, which is the whole design space of Chapter 4.

2.2 Free Energy and the Wasserstein Gradient Flow

Proposition 2.3 derived the Fokker–Planck equation as an identity about expectations: it says how $\mathbb{E} f(X_t)$ moves, for each test function f. This section says something different in kind. It says that the curve $t \mapsto \rho_t$ traced by the law of a reversible diffusion is not merely a solution of a partial differential equation but the *steepest descent* of an explicit functional, in an explicit geometry on the space of probability measures. Once that is proved, three things follow that are not visible from the PDE: a Lyapunov functional that is canonical rather than guessed, a dissipation identity that converts functional inequalities directly into convergence rates, and an implicit time-discretisation — the JKO scheme — that is unconditionally stable where the explicit one is not.

The result is genuinely beautiful, and for that reason it is also the single most over-extended statement in this subject. It holds for reversible diffusions with gradient drift and for the specific functional identified below. It does not hold for non-reversible diffusions, for kinetic dynamics, for controlled dynamics, or for mean-field games, all of which appear later in this volume. Section 2.2.5 makes that boundary sharp, by exhibiting a family of diffusions that dissipates this very free energy at exactly the same instantaneous rate and is still not a W_2 gradient flow of it.

2.2.1 The free energy functional

Throughout this section $V \in C^2(\mathbb{R}^d)$ satisfies $e^{-V} \in L^1(\mathbb{R}^d)$; we write $Z := \int_{\mathbb{R}^d} e^{-V}$ and $\pi := Z^{-1}e^{-V}$ for the associated Gibbs measure. The dynamics is the overdamped Langevin equation

$$dX_t = -\nabla V(X_t)\,dt + \sqrt{2}\,dW_t, \tag{2.10}$$

whose Fokker–Planck equation, from Proposition 2.3, is

$$\partial_t \rho_t = \nabla\cdot(\rho_t \nabla V) + \Delta\rho_t. \tag{2.11}$$

Definition 2.8 (Free energy). For $\rho \in \mathcal{P}_2(\mathbb{R}^d)$ absolutely continuous with density again written ρ,

$$\mathcal{F}[\rho] := \underbrace{\int_{\mathbb{R}^d} V\rho\,dx}_{\text{potential energy}} + \underbrace{\int_{\mathbb{R}^d} \rho\log\rho\,dx}_{\text{negative entropy}}, \tag{2.12}$$

with $\mathcal{F}[\rho] := +\infty$ if ρ is not absolutely continuous or if the two integrals are not separately defined. The first term is written $\mathcal{V}[\rho]$ and the second $\mathcal{H}[\rho]$.

Remark 2.9 (Free energy and relative entropy differ by a constant). Since $\log\pi = -V - \log Z$,

$$\mathcal{F}[\rho] = \mathrm{KL}(\rho\,\|\,\pi) - \log Z. \tag{2.13}$$

The two functionals therefore have the same gradient, the same sublevel sets and the same minimiser π, and every statement below may be read in either form. They are not interchangeable in one respect: $\mathrm{KL}(\cdot\,\|\,\pi)$ is non-negative and vanishes only at π, whereas $\mathcal{F}$ has no distinguished value, and $\log Z$ is exactly the quantity that a sampling algorithm does not have access to. Where a claim is about the value of the functional rather than its gradient — as in any bound that is used numerically — this volume writes $\mathrm{KL}(\cdot\,\|\,\pi)$.

Convention

Three normalisations are fixed here and used without further comment.

(i) A gradient flow always names its metric. "ρ_t is a gradient flow" is not a well-formed statement. The heat equation $\partial_t\rho = \Delta\rho$ is the W_2 gradient flow of the negative entropy $\mathcal{H}$, and it is equally the L^2 gradient flow of the Dirichlet energy $\frac{1}{2}\int|\nabla\rho|^2$: same equation, two metrics, two functionals, and the two structures give different discretisations and different convergence arguments. Every occurrence of "gradient flow" in this volume is followed by the metric and the functional, or it is a defect.

(ii) The log-Sobolev constant. π satisfies a logarithmic Sobolev inequality with constant

$\alpha > 0$, written $\mathrm{LSI}(\alpha)$, if

$$\mathrm{KL}(\nu \,\|\, \pi) \le \frac{1}{2\alpha} \mathcal{I}(\nu \,\|\, \pi) \qquad \text{for all } \nu \in \mathcal{P}_2(\mathbb{R}^d),\ \nu \ll \pi, \tag{2.14}$$

where $\mathcal{I}(\nu \,\|\, \pi) := \int |\nabla \log(d\nu/d\pi)|^2 \, d\nu$ is the relative Fisher information. With this normalisation the Bakry–Émery criterion reads: $\nabla^2 V \succeq \alpha I$ implies $\mathrm{LSI}(\alpha)$, and the free energy decays at rate 2α. The competing convention $\mathrm{KL}(\cdot \,\|\, \pi) \le \frac{C}{2}\mathcal{I}(\cdot \,\|\, \pi)$ with $C = 1/\alpha$ is equally common in the literature; every rate in this book is stated in the convention (2.14), and a rate quoted from a source using the other one is converted at the point of citation, not carried across.

(iii) Fisher information. $\mathcal{I}(\cdot \,\|\, \pi)$ always denotes the *relative* Fisher information above, never the Fisher information matrix of a parametric family. The latter appears in Chapter 5 under the symbol $G(\theta)$ and is an object on a parameter space, not on $\mathcal{P}_2(\mathbb{R}^d)$. The two are related only through a specific parametrisation and never by default.

2.2.2 Curves in Wasserstein space

To speak of steepest descent one needs a notion of velocity. The following is the standard one; it is stated here in full rather than cited, because every hypothesis in Theorem 2.13 refers back to it.

Theorem 2.10 (Benamou–Brenier form of the metric, and the tangent space) CLASSICAL. *Write $AC^p([0,1];\mathcal{P}_2(\mathbb{R}^d))$ for the curves whose metric derivative lies in $L^p(0,1)$; the exponent is not decorative, and the two clauses below use different ones.*

(i) (Characterisation, $p = 1$.) *A curve $(\rho_t)_{t\in[0,1]} \subset \mathcal{P}_2(\mathbb{R}^d)$ lies in $AC^1([0,1];\mathcal{P}_2(\mathbb{R}^d))$ if and only if there is a Borel vector field (v_t) with $\int_0^1 \|v_t\|_{L^2(\rho_t)}\,dt < \infty$ solving the continuity equation*

$$\partial_t \rho_t + \nabla\cdot(\rho_t v_t) = 0 \tag{2.15}$$

in the sense of distributions. The curve lies in $AC^2([0,1];\mathcal{P}_2(\mathbb{R}^d))$ if and only if the field can be chosen with $\int_0^1 \|v_t\|^2_{L^2(\rho_t)}\,dt < \infty$.

(ii) (Minimal representative.) *Among all such v_t there is one of minimal norm for almost every t; it satisfies $\|v_t\|_{L^2(\rho_t)} = |\rho'|(t)$, the metric derivative of the curve, and it belongs to the closed subspace*

$$\mathrm{Tan}_\rho \mathcal{P}_2(\mathbb{R}^d) := \overline{\{\nabla\varphi : \varphi \in C_c^\infty(\mathbb{R}^d)\}}^{L^2(\rho)}. \tag{2.16}$$

(iii) (Benamou–Brenier, $p = 2$.)

$$W_2^2(\mu,\nu) = \inf \int_0^1 \|v_t\|^2_{L^2(\rho_t)}\,dt, \tag{2.17}$$

the infimum being over AC^2 curves joining μ to ν and their admissible velocity fields.

Everything downstream — the metric derivative in (2.21), the JKO scheme of Theorem 2.13, the curve-of-maximal-slope formulation — works in the quadratic regime, that is with AC^2 and (2.17).

Part (i) is stated with both exponents because the L^1-in-time characterisation is the one that is true for a general absolutely continuous curve, and quoting it while using (2.17) is a regime substitution of exactly the kind Table 11.2 catalogues.

Proof roadmap. The correspondence between absolutely continuous curves and solutions of (2.15) is proved by mollification in space and time, which produces a smooth approximating family for which the velocity can be defined classically, followed by a compactness argument in $L^2(\rho_t)$; the minimal velocity is obtained by projecting any admissible v_t onto the subspace (2.16), the projection being admissible because (2.15) tests v_t only against gradients. The identification of the infimum with W_2^2 is the Benamou–Brenier theorem. The full argument occupies a chapter of the standard references and is not reproduced; the statement is used below only through the two facts that the velocity of an absolutely continuous curve may be taken tangential, and that its $L^2(\rho)$ norm is the metric speed. □

The projection appearing in the roadmap is used explicitly in Proposition 2.16; write Π_ρ for the orthogonal projection of $L^2(\rho;\mathbb{R}^d)$ onto $\mathrm{Tan}_\rho\mathcal{P}_2(\mathbb{R}^d)$.

2.2.3 The gradient flow structure

The first observation is elementary and is the whole mechanism.

Proposition 2.11 (Flux form of the Fokker–Planck equation) DERIVED. *Let ρ_t be a positive classical solution of (2.11) with $\rho_t \in \mathcal{P}_2(\mathbb{R}^d)$ and $\mathcal{F}[\rho_t] < \infty$. Then*

$$\partial_t\rho_t = \nabla\cdot\Big(\rho_t\nabla\frac{\delta\mathcal{F}}{\delta\rho}[\rho_t]\Big) = -\nabla\cdot(\rho_t v_t), \qquad v_t = -\nabla\log\frac{\rho_t}{\pi} = -\nabla(V+\log\rho_t), \tag{2.18}$$

where $\delta\mathcal{F}/\delta\rho\,[\rho] = V + \log\rho + 1$ is the first variation of (2.12). In particular $v_t \in \mathrm{Tan}_{\rho_t}\mathcal{P}_2(\mathbb{R}^d)$.

Proof. For the first variation, let $\rho^\varepsilon = \rho + \varepsilon\chi$ with $\chi \in C_c^\infty$, $\int\chi = 0$, and $\rho > 0$ on the support of χ. Then $\frac{d}{d\varepsilon}\big|_{\varepsilon=0}\mathcal{F}[\rho^\varepsilon] = \int(V+\log\rho+1)\chi$, which identifies $\delta\mathcal{F}/\delta\rho = V + \log\rho + 1$. Since $\rho\nabla\log\rho = \nabla\rho$, we have $\nabla\cdot(\rho\nabla(V+\log\rho+1)) = \nabla\cdot(\rho\nabla V) + \Delta\rho$, which is the right-hand side of (2.11). The additive constant 1 is annihilated by the gradient, which is why the flow sees $\mathcal{F}$ and $\mathrm{KL}(\cdot\,\|\,\pi)$ identically, as Remark 2.9 asserted. Finally v_t is a gradient of a function, hence lies in $\mathrm{Tan}_{\rho_t}\mathcal{P}_2(\mathbb{R}^d)$ by (2.16). ■

The velocity field in (2.18) is worth pausing on. It is minus the gradient of $\log(\rho_t/\pi)$: the drift $-\nabla V$ that the particle feels, plus the term $-\nabla\log\rho_t$ that no particle feels. The second term is the score of the current law, and its appearance here — as the part of the velocity of the distribution that is invisible at the level of a single trajectory — is the same object that Chapter 3 learns from data in order to run time backwards. That is a correspondence between two occurrences of one field, not an equivalence between two problems; Chapter 11 records it with its exact scope.

Proposition 2.12 (Energy dissipation identity) DERIVED. *Under the hypotheses of Proposition 2.11, and assuming in addition that $\rho_t\,|\nabla\log(\rho_t/\pi)|\,(1+|V|+|\log\rho_t|)\to 0$ as $|x|\to\infty$*

for each t, so that the boundary terms below vanish,

$$\frac{\mathrm{d}}{\mathrm{d}t}\mathcal{F}[\rho_t] = \frac{\mathrm{d}}{\mathrm{d}t}\mathrm{KL}(\rho_t \,\|\, \pi) = -\int_{\mathbb{R}^d} \rho_t \left|\nabla \log \tfrac{\rho_t}{\pi}\right|^2 \mathrm{d}x = -\mathcal{I}(\rho_t \,\|\, \pi) = -\|v_t\|^2_{L^2(\rho_t)}. \tag{2.19}$$

Proof. Differentiate under the integral sign and use (2.18):

$$\frac{\mathrm{d}}{\mathrm{d}t}\mathcal{F}[\rho_t] = \int \frac{\delta \mathcal{F}}{\delta \rho}[\rho_t]\, \partial_t \rho_t = \int (V + \log \rho_t + 1)\, \nabla \cdot (\rho_t \nabla (V + \log \rho_t)).$$

Integrating by parts, the boundary term vanishes by the stated decay and

$$\frac{\mathrm{d}}{\mathrm{d}t}\mathcal{F}[\rho_t] = -\int \rho_t \left|\nabla (V + \log \rho_t)\right|^2 = -\int \rho_t \left|\nabla \log \tfrac{\rho_t}{\pi}\right|^2,$$

using $\nabla(V + \log \rho) = \nabla \log(\rho/\pi)$ once more. The equality with $\frac{\mathrm{d}}{\mathrm{d}t}\mathrm{KL}(\rho_t \,\|\, \pi)$ is (2.13), and the last equality is the definition of v_t. ■

Identity (2.19) is the reason this section exists. Its right-hand side is simultaneously the dissipation of the functional and the squared speed of the curve, and a curve whose energy falls at exactly the rate of its squared speed is, in a metric space, precisely a curve of steepest descent. That heuristic is made a theorem by the following, which is the rigorous content of the gradient-flow claim.

Theorem 2.13 (Fokker–Planck as the W_2 gradient flow of $\mathcal{F}$) CLASSICAL. *Assume $V \in C^2(\mathbb{R}^d)$ with $\nabla^2 V \succeq -\kappa I$ for some $\kappa \geq 0$, $e^{-V} \in L^1$, and let $\rho_0 \in \mathcal{P}_2(\mathbb{R}^d)$ with $\mathcal{F}[\rho_0] < \infty$. For $\tau > 0$ define the minimising-movement sequence*

$$\rho_\tau^{n+1} \in \operatorname*{arg\,min}_{\rho \in \mathcal{P}_2(\mathbb{R}^d)} \left\{ \frac{1}{2\tau} W_2^2(\rho, \rho_\tau^n) + \mathcal{F}[\rho] \right\}, \qquad \rho_\tau^0 = \rho_0. \tag{2.20}$$

Then for every $\tau > 0$ and every n the minimising set in (2.20) is non-empty, and the minimiser is unique when $\kappa = 0$, or when $\kappa > 0$ and $\tau < 1/\kappa$; the piecewise-constant interpolants $\rho_\tau^t := \rho_\tau^{\lceil t/\tau \rceil}$ converge, as $\tau \downarrow 0$ and locally uniformly in t, in W_2 to a curve $(\rho_t)_{t \geq 0}$; and (ρ_t) is the unique distributional solution of (2.11) with initial datum ρ_0. Equivalently, (ρ_t) is the unique curve of maximal slope of $\mathcal{F}$ in $(\mathcal{P}_2(\mathbb{R}^d), W_2)$: it is absolutely continuous and satisfies the energy dissipation equality

$$\mathcal{F}[\rho_s] - \mathcal{F}[\rho_t] = \frac{1}{2}\int_s^t |\rho'|^2(r)\,\mathrm{d}r + \frac{1}{2}\int_s^t |\partial \mathcal{F}|^2(\rho_r)\,\mathrm{d}r, \qquad 0 \leq s \leq t, \tag{2.21}$$

where $|\partial \mathcal{F}|(\rho) = \mathcal{I}(\rho \,\|\, \pi)^{1/2}$ is the metric slope.

Proof roadmap. Existence of a minimiser in (2.20) follows from lower semicontinuity of $\mathcal{F}$ and of $W_2^2(\cdot, \rho_\tau^n)$ together with compactness of the sublevel sets, and needs no restriction on τ. Uniqueness is a separate matter, and the curves along which convexity is tested are not the ones a Hilbert-space intuition would suggest. The map $\mu \mapsto W_2^2(\mu, \rho_\tau^n)$ is *not* 1-convex along arbitrary W_2 geodesics: $\mathcal{P}_2(\mathbb{R}^d)$ is not a Hilbert space, and the parallelogram-law argument that gives strong convexity of a squared distance there has no counterpart here. The correct device is the *generalized geodesic* with base measure ρ_τ^n — the curve $\mu_s = ((1-s)T_0 + sT_1)_\# \rho_\tau^n$ built from the optimal maps T_i from the base to the two endpoints. Along such curves the squared-distance term to the base does have the required $\frac{1}{\tau}$-convexity, because it becomes a genuine quadratic in s under a single push-forward, and $\mathcal{F}$ inherits the $(-\kappa)$-convexity supplied by $\nabla^2 V \succeq -\kappa I$. The JKO objective is therefore strictly convex *in the generalized-geodesic sense* when $\frac{1}{\tau} - \kappa > 0$, which gives uniqueness of the resolvent for $\tau < 1/\kappa$. At larger steps the sum of a convex and a concave term need not be convex and the resolvent need not be single valued, which is why the statement asserts uniqueness only in the stated range. Writing the optimality condition of (2.20) against a smooth perturbation transported by a vector field gives, after one integration by parts, the discrete equation $(\mathrm{id} - T^{n+1})/\tau = \nabla(V + \log \rho_\tau^{n+1})$ with T^{n+1} the optimal map from ρ_τ^{n+1} to ρ_τ^n, which is the implicit Euler scheme for (2.18). The a priori estimates obtained by testing (2.20) with $\rho = \rho_\tau^n$ — a bound on $\sum_n W_2^2(\rho_\tau^{n+1}, \rho_\tau^n)/\tau$ and monotonicity of $\mathcal{F}$ along the sequence — give equicontinuity in W_2; a refined version of the Arzelà–Ascoli theorem for curves in a metric space extracts a limit, and the discrete optimality condition passes to the limit against smooth test functions. That the limit is the *unique* curve of maximal slope, and that (2.21) holds with equality rather than inequality, is *not* a consequence of the energy identity by itself: it is the uniqueness theory for gradient flows of semiconvex functionals in a metric space, which supplies the chain rule for $\mathcal{F}$ along absolutely continuous curves and, with it, the equivalence of the curve-of-maximal-slope, EDE and EVI formulations. That is where semiconvexity is used a second time — there in the limit $\tau \downarrow 0$, where the restriction $\tau < 1/\kappa$ is eventually satisfied, so the limit statement needs no small-step proviso of its own.

This is a roadmap and not a proof: each of the three steps just named is a technical result of independent difficulty, and the statement is quoted from the literature rather than established here. What this volume uses downstream is (2.18) and (2.19), both proved above in full; no argument in later chapters depends on (2.21) beyond its use as an organising picture. □

2.2.4 From dissipation to a rate

The gradient-flow structure by itself gives monotonicity and no rate. A rate requires an inequality relating the functional to its dissipation, and such an inequality is an assumption on π, not a consequence of the dynamics. That sentence is the reason Section 2.4 exists, and the following corollary is the template it instantiates.

Corollary 2.14 (Exponential decay under a log-Sobolev inequality) [illegible]. *Assume the hypotheses of Proposition 2.12 and that π satisfies* LSI(α) *in the normalisation (2.14). Then for all $t \geq 0$*

$$\mathrm{KL}(\rho_t \,\|\, \pi) \;\leq\; e^{-2\alpha t}\, \mathrm{KL}(\rho_0 \,\|\, \pi), \tag{2.22}$$

and consequently, since LSI(α) *implies the Talagrand inequality* $W_2^2(\nu, \pi) \leq \frac{2}{\alpha}\mathrm{KL}(\nu \,\|\, \pi)$,

$$W_2(\rho_t, \pi) \;\leq\; \sqrt{\tfrac{2}{\alpha}\mathrm{KL}(\rho_0 \,\|\, \pi)}\; e^{-\alpha t}. \tag{2.23}$$

Proof. Write $h(t) = \mathrm{KL}(\rho_t \,\|\, \pi)$. By Proposition 2.12, $h'(t) = -\mathcal{I}(\rho_t \,\|\, \pi)$, and by (2.14),

$\mathcal{I}(\rho_t \,\|\, \pi) \geq 2\alpha h(t)$. Hence $h' \leq -2\alpha h$, and Grönwall's lemma gives (2.22). Substituting into the Talagrand inequality and taking square roots gives (2.23). Note that the two decay rates differ by a factor of two because W_2 is comparable to the square root of the relative entropy, not to the relative entropy itself; this is the most frequent source of mis-stated rates in this subject and is examined in Table 2.2. ∎

Notes on Proposition 2.11, Proposition 2.12 and Corollary 2.14

Assumed. For all three: ρ_t a positive classical solution of (2.11) in $\mathcal{P}_2(\mathbb{R}^d)$ with $\mathcal{F}[\rho_t] < \infty$, and enough decay at infinity for the two integrations by parts. For the rate in addition: LSI(α) for π with $\alpha > 0$, in the normalisation (2.14); $\mathrm{KL}(\rho_0 \,\|\, \pi) < \infty$, which already excludes some natural initialisations, for instance a Dirac.

Where it fails. LSI(α) with α bounded below is a strong hypothesis. By Bakry–Émery it holds with α the strong convexity modulus of V; under bounded perturbation $V = V_0 + u$ with u bounded it survives with $\alpha_0 e^{-\operatorname{osc}(u)}$ (Proposition 2.28), a bound that degrades exponentially in the perturbation and is essentially sharp. For a symmetric double well with barrier height Δ at temperature 1, the constant is exponentially small, $\alpha \lesssim e^{-\Delta}$ by Eyring–Kramers together with $\alpha \leq \lambda$ (Example 2.29), and (2.22) then says nothing useful on any accessible time scale — which is a statement about the guarantee, not a proof that the dynamics is that slow. Multimodality is not a failure of the theorem; it is the theorem reporting the truth through a small constant.

Dimension. Nothing in (2.22) depends explicitly on d. That is not a dimension-free convergence result: α may degrade with d, and $\mathrm{KL}(\rho_0 \,\|\, \pi)$ typically grows linearly in d for product-like initialisations, so the time to reach a fixed accuracy grows like $\alpha^{-1} \log d$. Any statement of the form "Langevin mixes in time proportional to d" is a statement about a discretisation, not about (2.22); Section 4.1 separates the two and this volume never merges them.

2.2.5 The boundary: which dynamics are Wasserstein gradient flows

Theorem 2.13 is a statement about (2.10), a diffusion that is reversible with respect to π and whose drift is a gradient. It is tempting to promote it into a general principle — that the evolution of a probability law in time is always the steepest descent of some energy in W_2 — and this section is the reason not to.

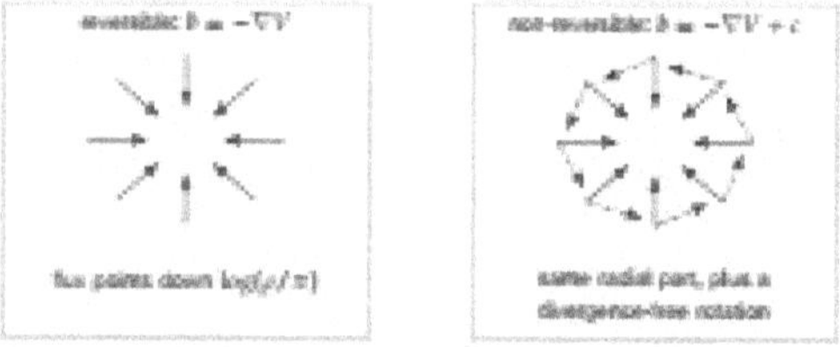

Figure 2.1: Two diffusions with the same stationary law π and the same instantaneous free-energy dissipation. The added rotational component is divergence free, so it changes neither π nor $\frac{d}{dt}\mathcal{F}$; it is orthogonal to the radial part in $L^2(\rho)$, so it strictly increases the metric speed (Proposition 2.16). What this figure does not imply: that the right-hand flow converges faster — that is a statement about trajectories, not about instantaneous rates — or that it is not the gradient flow of some other functional, which remains open (Open Problem 11.2).

The obstruction is not exotic. It appears in the smallest possible perturbation of (2.10) that keeps the same stationary measure.

Definition 2.15 (Admissible non-reversible perturbation). A vector field $b \in C^1(\mathbb{R}^d; \mathbb{R}^d)$ is an *admissible perturbation* of (2.10) if

$$\nabla \cdot (b e^{-V}) = 0 \quad \text{on } \mathbb{R}^d. \tag{2.24}$$

The perturbed dynamics is $dX_t = (-\nabla V(X_t) + b(X_t))dt + \sqrt{2}\,dW_t$, with Fokker–Planck equation

$$\partial_t \rho_t = \nabla \cdot (\rho_t \nabla V) + \Delta \rho_t - \nabla \cdot (\rho_t b). \tag{2.25}$$

Condition (2.24) says exactly that π remains stationary: the extra flux πb is divergence free. The perturbed process is no longer reversible with respect to π, and an admissible perturbation can improve asymptotic convergence in important classes while preserving the invariant law; Section 4.3 exhibits a class where the gain is computed exactly, and Corollary 2.26 shows what is *not* improved. Proposition 2.16 below proves neither an improvement nor its absence: it isolates a different fact, that the instantaneous free-energy dissipation can be identical while the Wasserstein geometry is not. What is relevant here is the structure.

Proposition 2.16 (Same dissipation, different geometry) DERIVED. *Let b be admissible in the sense of Definition 2.15 and let ρ_t be a positive classical solution of (2.25) with enough decay for the integrations by parts below. Then:*

(i) the velocity field of the curve is $u_t = -\nabla \log(\rho_t/\pi) + b$, and its two summands are orthogonal in $L^2(\rho_t; \mathbb{R}^d)$;

(ii) the free energy dissipates at exactly the rate of the reversible case,

$$\frac{d}{dt}\mathrm{KL}(\rho_t \,\|\, \pi) = -\mathcal{I}(\rho_t \,\|\, \pi); \tag{2.26}$$

(iii) the metric speed of the curve is

$$|\rho'|^2(t) = \mathcal{I}(\rho_t \,\|\, \pi) + \|\Pi_{\rho_t} b\|^2_{L^2(\rho_t)}, \tag{2.27}$$

so that whenever $\Pi_{\rho_t} b \neq 0$ the energy dissipation equality (2.21) fails, and (ρ_t) is not the W_2 gradient flow of $\mathcal{F}$.

Proof. (i) Rewriting (2.25) as $\partial_t \rho_t = -\nabla \cdot (\rho_t u_t)$ with $u_t = -\nabla(V + \log \rho_t) + b = -\nabla \log(\rho_t/\pi) + b$ is Proposition 2.11 with the extra flux carried along. For orthogonality, set $h_t := \rho_t/\pi$ and compute

$$(-\nabla \log h_t, b)_{L^2(\rho_t)} = -\int \pi h_t \nabla \log h_t \cdot b = -\int \pi \nabla h_t \cdot b = \int h_t \nabla \cdot (\pi b) = 0,$$

the last step by (2.24) and the boundary term vanishing by assumption.

(ii) Differentiating $\mathrm{KL}(\rho_t \,\|\, \pi) = \int \rho_t \log h_t$ and using $\partial_t \rho_t = -\nabla\cdot(\rho_t u_t)$,

$$\frac{\mathrm{d}}{\mathrm{d}t}\mathrm{KL}(\rho_t \,\|\, \pi) = \int \log h_t \, \partial_t \rho_t = \int \rho_t \nabla \log h_t \cdot u_t = -\mathcal{I}(\rho_t \,\|\, \pi) + \langle \nabla \log h_t, b \rangle_{L^2(\rho_t)},$$

and the second term vanishes by (i). This is (2.26).

(iii) By Theorem 2.10 the metric speed is the $L^2(\rho_t)$ norm of the tangential part of any admissible velocity, so $|\rho'|^2(t) = \|\Pi_{\rho_t} u_t\|^2$. Now $-\nabla \log h_t$ is already tangential, hence fixed by Π_{ρ_t}, and it is orthogonal to b by (i); since Π_{ρ_t} is an orthogonal projection and $\nabla \log h_t \in \mathrm{Tan}_{\rho_t}\mathcal{P}_2(\mathbb{R}^d)$, orthogonality to b implies orthogonality to $\Pi_{\rho_t} b$. Pythagoras gives (2.27). Finally, (2.21) for a curve of maximal slope forces $|\rho'|^2 = |\partial \mathcal{F}|^2 = \mathcal{I}(\rho_t \,\|\, \pi)$, which contradicts (2.27) whenever $\Pi_{\rho_t} b \neq 0$. ■

Interpretation

Free-energy dissipation does not determine Wasserstein geometry. Adding the divergence-free field b leaves the stationary law untouched and leaves the rate at which $\mathcal{F}$ decreases untouched as well; what it changes is the path the law takes to get there. In the language of Theorem 2.10, the curve now moves faster than it descends: its metric speed strictly exceeds its metric slope, and the excess is exactly the part of the velocity that is orthogonal to $\nabla \log(\rho_t/\pi)$ in $L^2(\rho_t)$.

This is the cleanest place in the volume to see that "dissipates $\mathcal{F}$" and "is the W_2 gradient flow of $\mathcal{F}$" are different statements — the first is a scalar identity, the second a statement about a geometry. The freedom this exposes is not a pathology: Section 4.3 exploits exactly this extra motion to converge faster, and Figure 2.1 draws the two velocity fields side by side.

Scope of Proposition 2.16

Assumed. $b \in C^1$ with $\nabla\cdot(b e^{-V}) = 0$; a positive classical solution of (2.25) with enough decay for two integrations by parts; $\rho_t \in \mathcal{P}_2(\mathbb{R}^d)$ with finite relative entropy.

Where it fails. The orthogonality in (i), and with it the whole result, uses $\nabla\cdot(\pi b) = 0$ exactly. A perturbation that merely leaves π approximately stationary gives an extra term in (2.26) of the size of that error, and the clean statement — same dissipation, faster travel — degrades continuously rather than surviving.

What this does not say. It does not say the perturbed flow converges faster; that requires an argument about the trajectory, not about the instantaneous rate, and is taken up in Section 4.3. Nor does it say the flow is not the gradient flow of some other functional in some geometry: only the pair (free energy $\mathcal{F}$, standard W_2 metric) is ruled out, and the negation is of that pair and not of the phrase "gradient flow". The general question is Open Problem 11.2.

What would falsify it. An admissible b and a law ρ with $\Pi_\rho b \neq 0$ for which $|\rho'|^2(t) = \mathcal{I}(\rho_t \,\|\, \pi)$.

Example 2.17 (A two-dimensional instance, computed). Take $d = 2$, $V(x) = \frac{1}{2}|x|^2$, so $\pi = \mathcal{N}(0, I)$, and let $b(x) = Jx$ with $J = \begin{pmatrix} 0 & -1 \\ 1 & 0 \end{pmatrix}$. Then $\nabla\cdot(Jx e^{-|x|^2/2}) = e^{-|x|^2/2}(\mathrm{tr}\, J - x^{\top} J x) = 0$, since an antisymmetric matrix has zero trace and zero associated quadratic form; b is admissible. Gaussian laws remain Gaussian under (2.25), so take $\rho = \mathcal{N}(0, \Sigma)$ with $\Sigma = \mathrm{diag}(\sigma_1^2, \sigma_2^2)$. Testing against the gradient of $\varphi(x) = x_1 x_2$, which lies in $\mathrm{Tan}_\rho \mathcal{P}_2(\mathbb{R}^d)$,

$$\langle b, \nabla \varphi \rangle_{L^2(\rho)} = \mathbb{E}_\rho\left[-X_2 \cdot X_2 + X_1 \cdot X_1\right] = \sigma_1^2 - \sigma_2^2.$$

So $\Pi_\rho b \neq 0$ as soon as $\sigma_1 \neq \sigma_2$, and by Proposition 2.16 the curve strictly outruns its own dissipation. At $\rho = \pi$ the test integral vanishes, as it must: at equilibrium the rotational flux moves no mass.

Proposition 2.16 is worth restating in words, because it is easy to misremember in either direction. The perturbed dynamics is *not* a W_2 gradient flow of $\mathcal{F}$, and it *does* dissipate $\mathcal{F}$ at exactly the same instantaneous rate as the unperturbed one. Both halves are surprising. The first says that "the free energy decreases" is far weaker than "the flow is steepest descent". The second says that the faster convergence of non-reversible samplers cannot be explained by a larger instantaneous dissipation — at a given law they dissipate identically. The speed-up comes from the trajectory: the extra rotation changes which laws ρ_t the curve visits, and hence the Fisher information it encounters later. That is a genuinely different mechanism from steepest descent, and no amount of reweighting turns it into one.

The same question can be asked of the other continuous-time dynamics in this volume, and the answers differ. The table below is the section's summary and is the row set that Chapter 11 inherits.

Table 2.1: Which continuous-time dynamics are W_2 gradient flows of a free energy. The second column answers the question as posed; the third names what is true instead, and where in this volume it is proved. Only the first row is the subject of Theorem 2.13.

Dynamics	W_2 gradient flow?	What holds instead
Reversible diffusion, gradient drift (2.10)	Yes	Steepest descent of $\mathcal{F}$; Theorem 2.13. Rates require a functional inequality, which is a hypothesis on π.
Non-reversible perturbation (2.25)	Not of $\mathcal{F}$	Not the W_2 gradient flow of the displayed free energy $\mathcal{F}$ in general: same stationary law and the same dissipation identity (2.26), but the curve travels strictly faster than it dissipates, (2.27). Proposition 2.16 does not rule out a W_2 gradient-flow representation with a different functional; that question is open (Open Problem 11.2).
Underdamped Langevin and Hamiltonian samplers	No	A conservative part plus a dissipative part acting only in the velocity variable. The free energy still decreases, but the dissipation is degenerate — it controls only $\nabla_v \log(f/M)$ — so Grönwall does not close and convergence is proved by hypocoercivity instead Section 4.2.
Controlled diffusions; mean-field games	No	A forward Fokker–Planck equation coupled to a backward Hamilton–Jacobi–Bellman equation. No single functional decreases along the coupled system in general; a variational structure exists for potential and entropy-regularised mean-field games only Chapter 9.

What does not follow

Reversibility of the dynamics is not stationarity of the law. (2.10) is reversible with respect to π; this is a property of its generator and of π, and it holds at every time. It does not say that $\rho_T = \pi$, and it does not say that the probability flux vanishes. The flux is $j_t = -\rho_t \nabla \log(\rho_t/\pi)$, and it is zero precisely when $\rho_t = \pi$ — that is, never before equilibrium. Detailed balance is therefore not available as a step in any argument about ρ_t at finite t, and in particular it is not the source of the reverse-time formula of Chapter 3; that formula is proved by time-reversal theory for diffusions, under its own hypotheses, and holds for time-inhomogeneous processes that are at no time in equilibrium.

Convexity of $\mathcal{F}$ in ρ is not geodesic convexity. $\mathcal{F}$ is convex along linear interpolations $(1-s)\rho_0 + s\rho_1$ for every V, because both $\mathcal{V}$ and $\mathcal{H}$ are. Convexity along W_2 geodesics — displacement convexity — holds if and only if V is convex, and it is a different property with different consequences. The two are routinely conflated, and the conflation is the mechanism by which global convergence gets claimed for non-convex problems; Chapter 8 returns to this in the mean-field setting, where the cost of the confusion is highest.

Two rate mechanisms, and displacement convexity belongs to one of them. Displacement convexity is the hypothesis behind the metric route: the evolution variational inequality, contraction of W_2 between two solutions, and the geodesic-convexity form of (2.21). The rates of Section 2.4 are a *different* mechanism: they follow from a functional inequality for π applied to the dissipation identity, and they do not require V to be convex — by Proposition 2.28 a bounded non-convex perturbation of a convex V retains LSI, with a worse constant, and (2.22) holds verbatim. Every rate in this volume should be read with its mechanism named: EVI and contraction on one side, LSI or Poincaré on the other, and discretisation error on neither.

The gradient-flow structure supplies no rate. Theorem 2.13 gives monotonicity of $\mathcal{F}$ and nothing quantitative. Every rate in this volume comes from an inequality between the functional and its dissipation, and every such inequality is an assumption on the target. A sentence of the form "because it is a gradient flow, it converges exponentially" is missing its hypothesis.

None of this is a statement about an algorithm. (2.19) describes the exact continuous-time flow. An implementable scheme is a discretisation, and its bias is not controlled by anything proved here; (2.20) is itself only implementable when the W_2 term can be computed. Chapter 4 treats discretisation as a separate source of error with separate hypotheses.

Operational reading

Read (2.20) as a proximal-point step. The scheme is $\rho^{n+1} = \mathrm{prox}^{W_2}_{\tau\mathcal{F}}(\rho^n)$: it descends $\mathcal{F}$ while paying a transport penalty for moving, exactly as the proximal algorithm in a Hilbert space descends a convex function while paying a squared-norm penalty. This buys the same thing it buys there — energy descent at every step size, because the new iterate is defined by a minimisation rather than by an extrapolation, and $\mathcal{F}[\rho^{n+1}] \leq \mathcal{F}[\rho^n]$ follows from testing (2.20) with $\rho = \rho^n$ whatever τ is. It does not buy a single-valued proximal map: by Theorem 2.13 the minimiser is guaranteed unique only for $\tau < 1/\kappa$, and unconditional descent and unconditional well-posedness are different properties. Whereas the explicit counterpart, the unadjusted Langevin algorithm of Section 4.1, is stable only for step sizes small relative to the smoothness of V and carries a bias that does not vanish as the number of steps grows.

That is one step beyond the theorem, and this box stops there. Which of the two is preferable for a given sampling problem depends on the cost of the proximal step, the oracle available, and the target class, and is decided in Section 4.4 against an explicit table of assumptions. Nothing in this section licenses a comparison of algorithms.

The section therefore contributes two rows to the Equivalence Atlas of Chapter 11. They are written now rather than reconstructed at the end, and they are set as a list rather than a table because four columns of prose do not fit the measure inside a box without falling below the character floor.

Atlas entries generated by this section

R1. Fokker–Planck flow of a reversible diffusion = W_2 gradient flow of the free energy. EXACT

Exact conditions. Drift of the form $-\nabla V$ with $V \in C^2$ and $\nabla^2 V \succeq -\kappa I$; constant isotropic diffusion coefficient; the functional $\mathcal{F}$ of (2.12) and no other; $\rho_0 \in \mathcal{P}_2(\mathbb{R}^d)$ with $\mathcal{F}[\rho_0] < \infty$. Uniqueness of the JKO step holds for $\tau < 1/\kappa$ and is proved along generalized geodesics based at ρ_k^τ, not along arbitrary W_2 geodesics; $\mathcal{P}_2(\mathbb{R}^d)$ is not a Hilbert space and the squared distance is not 1-convex along the latter.

What goes wrong outside them. For an admissible non-reversible perturbation the dissipation identity survives unchanged while the gradient-flow structure does not (Proposition 2.16); for kinetic and Hamiltonian dynamics both fail, and the dissipation becomes degenerate; for state-dependent diffusion coefficients the functional acquires an extra term and the drift is no longer a W_2 gradient.

R2. Sampling algorithm = gradient descent on the space of measures. ANALOGY

Exact conditions. None. The relation holds between the continuous-time flow and steepest descent, not between an algorithm and steepest descent.

What goes wrong outside them. An implementable scheme is a discretisation with a bias of its own, controlled by nothing proved here; and for non-reversible or kinetic samplers — which is to say, for most samplers that are actually fast — the underlying flow is not steepest descent of any free energy in W_2.

2.3 Generators, Semigroups, and Dirichlet Forms

Section 2.1 and Section 2.2 described the same object twice: once as a partial differential equation for ρ_t and once as a curve of steepest descent. There is a third description, in terms of operators acting on functions, and it is the one in which convergence rates are proved. The three are equivalent for reversible diffusions. They are not equivalent in general, and the last part of this section says exactly what the operator description loses.

2.3.1 The semigroup and the Dirichlet form

Throughout this section the coefficients are time independent, $a = 2I$, and π is a stationary measure.

Definition 2.18 (Markov semigroup, Dirichlet form, carré du champ). For bounded measurable f set $P_t f(x) = \mathbb{E}[f(X_t) \mid X_0 = x]$. The *Dirichlet form* associated with $(\mathcal{L}, \pi)$ is

$$\mathcal{E}(f,g) := -\int f\,\mathcal{L}g\,\mathrm{d}\pi, \qquad f,g \in C_c^\infty, \tag{2.28}$$

and the *carré du champ* operator is

$$\Gamma(f,g) := \tfrac{1}{2}\Big(\mathcal{L}(fg) - f\mathcal{L}g - g\mathcal{L}f\Big), \qquad \Gamma(f) := \Gamma(f,f). \tag{2.29}$$

The iterated operator is $\Gamma_2(f) := \frac{1}{2}\mathcal{L}\Gamma(f) - \Gamma(f, \mathcal{L}f)$.

Proposition 2.19 (Elementary properties) CLASSICAL. *$(P_t)_{t\geq 0}$ is a semigroup of positivity-preserving contractions with $P_t 1 = 1$; if π is stationary it extends to a contraction semigroup on $L^2(\pi)$, strongly continuous, with generator $\mathcal{L}$ on a domain containing C_c^∞.*

Proof roadmap. The semigroup property is the Markov property of Theorem 1.46; positivity and $P_t 1 = 1$ are immediate from the definition; the $L^2(\pi)$ contraction follows from Jensen and stationarity, $\int (P_t f)^2 \mathrm{d}\pi \leq \int P_t(f^2)\mathrm{d}\pi = \int f^2 \mathrm{d}\pi$. Strong continuity and the identification of the generator are the Hille–Yosida theory, quoted. □

Proposition 2.20 (The three objects, computed) DERIVED. *Let $\mathcal{L} = \langle b, \nabla\rangle + \Delta$ with $b = -\nabla V$ and $\pi = Z^{-1}e^{-V}$. Then for $f, g \in C_c^\infty$,*

$$\Gamma(f,g) = \langle \nabla f, \nabla g\rangle, \qquad \mathcal{E}(f,g) = \int \langle \nabla f, \nabla g\rangle\,\mathrm{d}\pi = \int \Gamma(f,g)\,\mathrm{d}\pi, \tag{2.30}$$

and consequently, for every f in the domain,

$$\frac{\mathrm{d}}{\mathrm{d}t}\int (P_t f)^2\,\mathrm{d}\pi = -2\mathcal{E}(P_t f, P_t f) = -2\int \|\nabla P_t f\|^2\,\mathrm{d}\pi \leq 0. \tag{2.31}$$

Proof. For the first identity, $\Delta(fg) = f\Delta g + g\Delta f + 2\langle \nabla f, \nabla g\rangle$ and $\langle b, \nabla(fg)\rangle = f\langle b, \nabla g\rangle + g\langle b, \nabla f\rangle$; substituting into (2.29) the first-order terms cancel and the Laplacian terms leave $\langle \nabla f, \nabla g\rangle$. The second identity is (2.8) with $c = 0$. For (2.31), differentiate under the integral, $\frac{\mathrm{d}}{\mathrm{d}t}\int (P_t f)^2 \mathrm{d}\pi = 2\int P_t f\, \mathcal{L}P_t f\,\mathrm{d}\pi = -2\mathcal{E}(P_t f, P_t f)$, which is (2.28). ■

Equation (2.31) is the L^2 counterpart of the entropy dissipation identity (2.19): the same structure, with the variance in place of the relative entropy and the Dirichlet form in place of the relative Fisher information. Section 2.4 exploits the parallel, and the two inequalities that close each of the two identities into a rate are the first two rows of Table 2.2.

Where the formulas change. With state-dependent a the carré du champ is $\Gamma(f) = \frac{1}{2}\langle \nabla f, a\nabla f\rangle$, not $\|\nabla f\|^2$, and the Dirichlet form is the associated weighted energy; on a Riemannian manifold it is the intrinsic gradient. Writing $|\nabla f|^2$ in either setting is the geometry substitution of Table 11.2 in operator clothing, and it changes every constant that follows.

What is lost without reversibility. Equation (2.28) sees only the symmetric part of $\mathcal{L}$: by (2.8) the non-gradient part c contributes the antisymmetric form B, and $\mathcal{E}(f,f) = -\int f\mathcal{L}f\,d\pi$ is unchanged when $c \neq 0$, because $B(f,f) = 0$. So two generators with the same π and different c have the same Dirichlet form. Every conclusion drawn from the Dirichlet form alone is therefore a conclusion about the whole family, and cannot distinguish the reversible member from the others — which is precisely why Proposition 2.16 needed a different argument, and why the faster convergence of non-reversible samplers is invisible to this section's methods.

2.3.2 Curvature

Proposition 2.21 (Bochner's formula) CLASSICAL. *For $\mathcal{L} = \Delta - \langle\nabla V, \nabla\rangle$ with $V \in C^3$ and $f \in C^\infty(\mathbb{R}^d)$,*

$$\Gamma_2(f) = \|\nabla^2 f\|_F^2 + \langle\nabla f, \nabla^2 V\,\nabla f\rangle. \tag{2.32}$$

Proof. By Proposition 2.20, $\Gamma(f) = |\nabla f|^2$. The Euclidean Bochner identity is $\frac{1}{2}\Delta|\nabla f|^2 = \|\nabla^2 f\|_F^2 + \langle\nabla f, \nabla\Delta f\rangle$. Hence

$$\tfrac{1}{2}\mathcal{L}|\nabla f|^2 = \|\nabla^2 f\|_F^2 + \langle\nabla f, \nabla\Delta f\rangle - \langle\nabla V, \nabla^2 f\,\nabla f\rangle.$$

On the other side, $\nabla\mathcal{L}f = \nabla\Delta f - \nabla^2 V\nabla f - \nabla^2 f\nabla V$, so $\Gamma(f, \mathcal{L}f) = \langle\nabla f, \nabla\Delta f\rangle - \langle\nabla f, \nabla^2 V\nabla f\rangle - \langle\nabla f, \nabla^2 f\nabla V\rangle$. Subtracting, and using the symmetry of $\nabla^2 f$ to cancel the two terms involving ∇V, gives (2.32). ■

Definition 2.22 (Curvature condition). $(\mathcal{L}, \pi)$ satisfies the *curvature–dimension condition* $\mathrm{CD}(K,\infty)$ for $K \in \mathbb{R}$ if $\Gamma_2(f) \geq K\,\Gamma(f)$ for all f in a core.

Corollary 2.23 (The criterion in $\mathbb{R}^d$) DERIVED. *For $\mathcal{L} = \Delta - \langle\nabla V, \nabla\rangle$ on $\mathbb{R}^d$, $\mathrm{CD}(K,\infty)$ holds if and only if $\nabla^2 V \succeq KI$ everywhere.*

Proof. If $\nabla^2 V \succeq KI$ then (2.32) gives $\Gamma_2(f) \geq \langle\nabla f, \nabla^2 V\nabla f\rangle \geq K|\nabla f|^2$. Conversely, fix x and $v \in \mathbb{R}^d$ and take f affine near x with $\nabla f = v$; then $\nabla^2 f = 0$ at x and (2.32) reduces to $\Gamma_2(f)(x) = \langle v, \nabla^2 V(x)v\rangle$, so the condition forces $\langle v, \nabla^2 V(x)v\rangle \geq K|v|^2$. ■

Where it changes. On a Riemannian manifold (2.32) acquires the Ricci curvature, $\Gamma_2(f) = \|\nabla^2 f\|^2 + \langle\nabla f, (\mathrm{Ric} + \nabla^2 V)\nabla f\rangle$, and Corollary 2.23 becomes the Bakry–Émery condition $\mathrm{Ric} + \nabla^2 V \succeq K$. The Euclidean statement is the special case $\mathrm{Ric} = 0$, and reading it as the general one silently sets the curvature of the state space to zero.

What it does not say. $\mathrm{CD}(K,\infty)$ with $K > 0$ is sufficient for the inequalities of Section 2.4 and not necessary: a potential that is convex outside a compact set and arbitrary inside it satisfies LSI and fails $\mathrm{CD}(K,\infty)$ for every $K > 0$.

What does not follow

Curvature is a pointwise condition and is destroyed by bounded perturbations. $\mathrm{CD}(K,\infty)$ with $K > 0$ requires $\nabla^2 V \succeq KI$ *everywhere*. Adding to V a smooth bump of height 10^{-6} and width 10^{-3} makes $\nabla^2 V$ negative somewhere and destroys the condition outright, while changing the measure π by a factor within $e^{\pm 10^{-6}}$. The two inequalities of the next section are more robust exactly because they are integrated rather than pointwise, and Proposition 2.28 quantifies how much they lose. Reporting a curvature bound for a loss landscape, or for any potential that is not verified convex, is therefore reporting a condition that almost certainly fails.

The last column of Table 2.2 is not a footnote. It records which hypotheses survive a bounded perturbation. That column is the difference between a hypothesis one can hope to verify for a target of interest and one that can be verified only for the Gaussian case.

2.4 Functional Inequalities and Convergence Rates

Proposition 2.12 and (2.31) give two exact dissipation identities and no rate. A rate needs an inequality bounding a functional by its own dissipation, and every such inequality is a hypothesis about π. This section catalogues the hypotheses, orders them, and — in Table 2.2 — records for each what it does not give.

2.4.1 The inequalities

Definition 2.24 (Poincaré, log-Sobolev, Talagrand). Let $\pi \in \mathcal{P}_2(\mathbb{R}^d)$ have full support. Then π satisfies

(i) the *Poincaré inequality* $\mathrm{P}(\lambda)$ with $\lambda > 0$ if

$$\mathrm{Var}_\pi(f) \le \frac{1}{\lambda}\int |\nabla f|^2 \, d\pi \qquad \text{for all } f \in C_c^\infty; \tag{2.33}$$

(ii) the *logarithmic Sobolev inequality* $\mathrm{LSI}(\alpha)$ if (2.14) holds, that is $\mathrm{KL}(\nu \,\|\, \pi) \le \frac{1}{2\alpha}\mathcal{I}(\nu \,\|\, \pi)$ for all $\nu \ll \pi$ in $\mathcal{P}_2(\mathbb{R}^d)$;

(iii) the *Talagrand inequality* $\mathrm{T}_2(\alpha)$ if

$$W_2^2(\nu, \pi) \le \frac{2}{\alpha}\mathrm{KL}(\nu \,\|\, \pi) \qquad \text{for all } \nu \in \mathcal{P}_2(\mathbb{R}^d). \tag{2.34}$$

Proposition 2.25 (What each inequality closes) [illegible]. *Let $\mathcal{L} = \Delta - \langle \nabla V, \nabla \rangle$ be reversible with respect to π, and let ρ_t solve (2.11) with $\mathrm{KL}(\rho_0 \,\|\, \pi) < \infty$.*

(i) Under $\mathrm{P}(\lambda)$, for every $f \in L^2(\pi)$,

$$\mathrm{Var}_\pi(P_t f) \le e^{-2\lambda t}\,\mathrm{Var}_\pi(f), \tag{2.35}$$

and, provided $\chi^2(\rho_0 \| \pi) < \infty$ — equivalently $d\rho_0/d\pi \in L^2(\pi)$, which the standing hypothesis $\mathrm{KL}(\rho_0 \,\|\, \pi) < \infty$ does not imply — the χ^2 divergence of ρ_t from π decays at rate 2λ.

(ii) Under $\mathrm{LSI}(\alpha)$, $\mathrm{KL}(\rho_t \,\|\, \pi) \le e^{-2\alpha t}\mathrm{KL}(\rho_0 \,\|\, \pi)$ *— this is Corollary 2.14 — and under* $\mathrm{T}_2(\alpha)$ *in addition* $W_2(\rho_t, \pi) \le \sqrt{2\alpha^{-1}\mathrm{KL}(\rho_0 \,\|\, \pi)}\, e^{-\alpha t}$.

(iii) $\mathrm{LSI}(\alpha)$ *implies* $\mathrm{P}(\alpha)$.

Proof. (i) By (2.31) and (2.33) applied to $P_t f$, $\frac{d}{dt}\mathrm{Var}_\pi(P_t f) = -2\int |\nabla P_t f|^2\,d\pi \le -2\lambda\,\mathrm{Var}_\pi(P_t f)$ — note $\mathrm{Var}_\pi(P_t f) = \int (P_t f)^2 d\pi - (\int f\,d\pi)^2$ since P_t preserves the mean — and Grönwall gives (2.35). The χ^2 statement is (2.35) applied to $f = d\rho_0/d\pi$ together with the duality $\chi^2(\rho_t \| \pi) = \mathrm{Var}_\pi(d\rho_t/d\pi)$.

(ii) is Corollary 2.14 and its corollary, restated here for comparison.

(iii) Take $\nu = f^2\pi$ with $\int f^2 d\pi = 1$. Then $\mathrm{KL}(\nu \,\|\, \pi) = \int f^2 \log f^2 d\pi$ and $\mathcal{I}(\nu \,\|\, \pi) = \int |\nabla \log f^2|^2 f^2 d\pi = 4\int |\nabla f|^2\,d\pi$, so (2.14) reads $\int f^2 \log f^2 d\pi \le \frac{2}{\alpha}\int |\nabla f|^2\,d\pi$. Apply this to $f = 1 + \varepsilon g$ with $\int g\,d\pi = 0$, normalised, and expand in ε: the left side is $2\varepsilon^2\,\mathrm{Var}_\pi(g) + O(\varepsilon^3)$ and the right side is $\frac{2}{\alpha}\varepsilon^2 \int |\nabla g|^2\,d\pi$. Dividing by $2\varepsilon^2$ and letting $\varepsilon \to 0$ gives (2.33) with $\lambda = \alpha$. ■

The proof of (i) used the dissipation identity (2.31) and the Poincaré inequality, and nothing else. Both are properties of π and of the Dirichlet *energy* $g \mapsto \mathcal{E}(g, g)$, and by the notes of Proposition 2.20 that energy does not see the non-gradient part of the drift — a statement about the diagonal of the form, not about the form. The following is therefore immediate, and it is worth stating because it is easy to assume the opposite.

Corollary 2.26 (Non-reversible perturbations inherit the rates) DERIVED. *Let* $\mathcal{L} = \Delta - \langle \nabla V, \nabla \rangle$ *be reversible with respect to* π *and let* c *be admissible in the sense of Definition 2.15, so that* $\mathcal{L}^c = \mathcal{L} + \langle c, \nabla \rangle$ *has the same stationary measure. Then the symmetric Dirichlet energy is unchanged,*

$$\mathcal{E}^c(g, g) = \mathcal{E}(g, g) \qquad \textit{for every } g \textit{ in the common form domain,} \tag{2.36}$$

and consequently:

(i) under $\mathrm{P}(\lambda)$ *for* π, *the semigroup* P_t^c *generated by* $\mathcal{L}^c$ *satisfies* $\mathrm{Var}_\pi(P_t^c f) \le e^{-2\lambda t}\,\mathrm{Var}_\pi(f)$ *for every* $t \ge 0$ *— a finite-time bound, with the same constant as in the reversible case;*

(ii) under $\mathrm{LSI}(\alpha)$, *the law* ρ_t *of the perturbed diffusion satisfies* $\mathrm{KL}(\rho_t \,\|\, \pi) \le e^{-2\alpha t}\mathrm{KL}(\rho_0 \,\|\, \pi)$, *by Proposition 2.16(ii) and the same Grönwall argument.*

Proof. For (i), $\frac{d}{dt}\int (P_t^c f)^2 d\pi = 2\int P_t^c f\, \mathcal{L}^c P_t^c f\,d\pi = -2\mathcal{E}^c(P_t^c f, P_t^c f)$, and by (2.8) the non-gradient part contributes $B(g, g) = \int g\,\langle c, \nabla g\rangle\,d\pi = \frac{1}{2}\int \langle c, \nabla(g^2)\rangle\,d\pi = -\frac{1}{2}\int g^2 \nabla\cdot(\pi c) = 0$, which is (2.36), so (2.33) applies verbatim. Note what this argument gives and what it does not: it shows that the *quadratic* form vanishes on the diagonal, $B(g, g) = 0$, and therefore says nothing about $B(f, g)$ for $f \ne g$. The full bilinear forms $\mathcal{E}^c$ and $\mathcal{E}$ are *not* equal — $\mathcal{E}^c$ is not even symmetric, which is the analytic content of non-reversibility — and every estimate below uses only the diagonal. Since P_t^c preserves the mean, the variance obeys the same Grönwall estimate as in Proposition 2.25(i). Part (ii) is Proposition 2.16(ii), which gives $\frac{d}{dt}\mathrm{KL}(\rho_t \,\|\, \pi) = -\mathcal{I}(\rho_t \,\|\, \pi)$ unchanged, with (2.14). ■

Theorem 2.27 (Bakry–Émery, and Otto–Villani) CLASSICAL. *If* $\mathrm{CD}(K, \infty)$ *holds with* $K > 0$ *then* π *satisfies* $\mathrm{LSI}(K)$ *and* $\mathrm{P}(K)$. *If* π *satisfies* $\mathrm{LSI}(\alpha)$ *then it satisfies* $\mathrm{T}_2(\alpha)$. *Neither converse* holds.

Proof sketch. For Bakry–Émery, consider the semigroup interpolation

$$\Lambda(s) = \int P_s((P_{t-s}f)\log P_{t-s}f)\,\mathrm{d}\pi, \qquad \Lambda'(s) = \int P_s\Big(\frac{\Gamma(P_{t-s}f)}{P_{t-s}f}\Big)\mathrm{d}\pi.$$

Then $\mathrm{CD}(K,\infty)$ gives the differential inequality $\frac{\mathrm{d}}{\mathrm{d}s}(e^{2Ks}\Gamma(P_s f)) \le 0$ in the appropriate sense; integrating from 0 to ∞ produces (2.14) with $\alpha = K$. The Poincaré statement follows from Proposition 2.25(iii), or directly by the same interpolation with the square in place of the entropy.

Otto–Villani is proved by running the gradient flow from ν and integrating the metric speed: by Theorem 2.10 and Proposition 2.12, $|\rho'|(t) = \mathcal{I}(\rho_t \,\|\, \pi)^{1/2}$, so $W_2(\nu,\pi) \le \int_0^\infty \mathcal{I}(\rho_t \,\|\, \pi)^{1/2}\,\mathrm{d}t$. The step that must be done carefully is the bound on the integrand, because (2.14) bounds the Fisher information from *below* and what is needed is an upper bound on its square root. Write $I_t = \mathcal{I}(\rho_t \,\|\, \pi)$ and $H_t = \mathrm{KL}(\rho_t \,\|\, \pi)$. Then $\dot H_t = -I_t$ by Proposition 2.12, and (2.14) reads $I_t \ge 2\alpha H_t$. Dividing the identity $I_t^{1/2} = I_t / I_t^{1/2}$ by the *lower* bound $I_t^{1/2} \ge (2\alpha H_t)^{1/2}$ turns that lower bound into the upper bound one wants:

$$I_t^{1/2} \le \frac{I_t}{(2\alpha H_t)^{1/2}} = \frac{-\dot H_t}{(2\alpha H_t)^{1/2}} = -\sqrt{\frac{2}{\alpha}}\,\frac{\mathrm{d}}{\mathrm{d}t} H_t^{1/2}. \tag{2.37}$$

Integrating (2.37) from 0 to ∞ and using $H_t \to 0$ gives $W_2(\nu,\pi) \le \sqrt{2\mathrm{KL}(\nu \,\|\, \pi)/\alpha}$, which is (2.34). The exponential decay of H_t is *not* what does the work here — it is a consequence of the same inequality — and an argument that invokes it in place of (2.37) has not closed. Exercise 2.19 asks for the computation with the constants tracked.

Both are sketches. The interpolation argument needs a regularity discussion that is standard and omitted; the second requires justifying the differentiation of W_2 along the flow, the absolute continuity of $t \mapsto H_t^{1/2}$, and the positivity of H_t needed to divide in (2.37) — if H_t vanishes at a finite time the flow has already reached π and the bound is trivial there. The statements are used below only to order the rows of Table 2.2. ∎

Proposition 2.28 (Holley–Stroock perturbation) ESTABLISHED. *Let π_0 satisfy* $\mathrm{LSI}(\alpha_0)$ *and let* $\mathrm{d}\pi = Z^{-1}e^{-u}\mathrm{d}\pi_0$ *with u bounded. Then π satisfies* $\mathrm{LSI}(\alpha)$ *with*

$$\alpha \ge \alpha_0 e^{-\mathrm{osc}(u)}, \qquad \mathrm{osc}(u) = \sup u - \inf u \le 2\|u\|_\infty, \tag{2.38}$$

and the exponential dependence is essentially sharp. The same statement holds for the Poincaré constant. The perturbation result is due to Holley and Stroock; see Bakry, Gentil and Ledoux [3] for a textbook account together with the Eyring–Kramers asymptotics quoted in Example 2.29.

Example 2.29 (A double well, and what the constant reports). Let $V(x) = \frac{1}{4\varepsilon}(x^2-1)^2$ on $\mathbb{R}$, a symmetric double well with barrier height $\Delta = 1/(4\varepsilon)$ measured from the wells.

Note first what cannot be done. V is *not* a bounded perturbation of a quadratic: its quartic tails minus any quadratic are unbounded, so Proposition 2.28 does not apply to the pair (V, quadratic) at all, however large one is willing to let $\mathrm{osc}(u)$ be. Holley–Stroock is usable here only against a reference with the *same* quartic growth — for instance the single-well

$V_0(x) = \frac{1}{4\varepsilon}(x^2+1)^2$, which is strictly convex, differs from V by $u = V - V_0 = -x^2/\varepsilon$, and that difference is again unbounded. Producing a genuinely bounded u requires modifying V_0 only inside a compact barrier region, and then $\operatorname{osc}(u)$ is of order Δ by construction. This is worth spelling out because "bounded perturbation of a Gaussian" is the standard informal gloss on a double well, and it is false.

What is true, and is quoted rather than proved here, is the Eyring–Kramers asymptotics: the spectral gap λ of $\mathcal{L}$ is exponentially small in Δ. Since $\alpha \le \lambda$ by Proposition 2.25(iii), the log-Sobolev constant is exponentially small too, and the *guaranteed* entropy rate $e^{-2\alpha t}$ of Corollary 2.14 is therefore useless on any time scale shorter than e^{Δ}.

That is a statement about the available guarantee, and it should not be inflated into a statement about the dynamics. Concluding that the actual relative-entropy halving time *is* of order e^{Δ} requires a matching lower bound — in the same divergence, and for a specified class of initial data — which the upper bound does not supply and which this volume does not prove. Metastability results of that kind exist and have their own hypotheses; what Corollary 2.14 contributes is one side of the inequality only.

With that distinction in place, the reading is the theorem reporting the truth rather than failing. Any claim of polynomial-time sampling for a target with a barrier must either change the dynamics — Section 4.3 — or change the target — Chapter 3.

2.4.2 The hierarchy

Table 2.2: Hypotheses that produce a rate, in decreasing strength. Every row is an assumption on the target π, not a property of an algorithm. The third column is the one most often assumed rather than checked, and the fourth decides whether a hypothesis is verifiable for anything other than a Gaussian.

Hypothesis	What decays, in which metric, at what rate	What it does not give	Survives a bounded perturbation of V?
$\mathrm{CD}(K,\infty)$, $K > 0$; equivalently $\nabla^2 V \succeq KI$	Implies all rows below with $\alpha = \lambda = K$ (Theorem 2.27)	Nothing about non-reversible or kinetic dynamics; no statement at finite step size	No. A C^2 bump of arbitrarily small height destroys it pointwise

continued overleaf

Table 2.2: (continued)

Hypothesis	What decays, in which metric, at what rate	What it does not give	Survives a bounded perturbation of V?
$\mathrm{LSI}(\alpha)$	$\mathrm{KL}(\rho_t \,\Vert\, \pi) \le e^{-2\alpha t}\mathrm{KL}(\rho_0 \,\Vert\, \pi)$ (Corollary 2.14); and, since $\mathrm{LSI}(\alpha)$ implies $\mathrm{T}_2(\alpha)$, $W_2(\rho_t, \pi) \le \sqrt{2\mathrm{KL}(\rho_0 \,\Vert\, \pi)/\alpha}\, e^{-\alpha t}$	No sharper χ^2 mechanism than the one it already implies: by Proposition 2.25(iii) $\mathrm{LSI}(\alpha) \Rightarrow \mathrm{P}(\alpha)$, so variance and χ^2 do decay at rate 2α, inherited through Poincaré and requiring $\chi^2(\rho_0 \Vert \pi) < \infty$. Nothing beyond Pinsker in TV, and no metric conversion not supplied by a further inequality; the W_2 statement is the entropy statement converted, not a second mechanism	Yes, with $\alpha_{B} e^{-\mathrm{osc}(u)}$ (Proposition 2.28)
$\mathrm{T}_2(\beta)$ — *a converter, not a decay mechanism*	*Nothing on its own.* Combined with an entropy rate: if $\mathrm{KL}(\rho_t \,\Vert\, \pi) \le e^{-2\alpha t}\mathrm{KL}(\rho_0 \,\Vert\, \pi)$ then $W_2(\rho_t, \pi) \le \sqrt{2\mathrm{KL}(\rho_0 \,\Vert\, \pi)/\beta}\, e^{-\alpha t}$	No time decay of anything: T_2 compares two functionals at one instant and contains no dynamics. It does not imply LSI	Yes, but not by Holley–Stroock, which is stated for LSI; the perturbation result for T_2 needs its own source and its own constant
$\mathrm{P}(\lambda)$	$\mathrm{Var}_\pi(P_t f) \le e^{-2\lambda t}\,\mathrm{Var}_\pi(f)$, equivalently χ^2 decay (Proposition 2.25)	No entropy rate: P does not imply LSI, and the gap can be arbitrarily large	Yes, with $\lambda_{B} e^{-\mathrm{osc}(u)}$
Admissible non-reversible perturbation of a reversible $\mathcal{L}$	*At least* the reversible guarantee, for every t: $e^{-2\lambda t}$ in variance and $e^{-2\alpha t}$ in entropy (Corollary 2.26). The spectral abscissa of the linearised moment dynamics may be strictly better (Proposition 4.14), which is a statement about a matrix and not about the law in any metric	The sharp constant. The Dirichlet form does not see c (Proposition 2.20 and the notes there), so no inequality of this table can detect the improvement	Yes, with the same constants, since they are those of π

Scope of Proposition 2.25, Corollary 2.26, Theorem 2.27, Proposition 2.28 and Table 2.2

Two converses, and they fail separately. Where the text says a converse fails, it means two distinct statements: a target may satisfy LSI without any positive Bakry–Émery curvature, and a target may satisfy T_2 without satisfying LSI. Neither of these is the converse of LSI $\Rightarrow$ P, which is a third statement and also false (Proposition 2.25(iii) and the row below it). Reading one sentence as covering all three is the most common way this hierarchy is misquoted.

Assumed. Reversibility with $a = 2I$ throughout; π with full support and finite second moment; for each row, the inequality named in it, which is a hypothesis and never a conclusion of the dynamics. For Proposition 2.28, a perturbation that is bounded — not merely integrable, and not merely small in L^2: an unbounded u with small norm can destroy the inequality entirely, and the bound sees only $\operatorname{osc}(u)$.

Where the constants go. None of the rates displays the dimension, and none is dimension free. The constant κ typically degrades with d — for a product measure it does not, which is exactly why product measures mislead — and $\mathrm{KL}(\rho_0 \,\|\, \pi)$ typically grows linearly in d for a product-like initialisation, so the time to fixed accuracy grows like $\kappa^{-1}(d + \log(1/\varepsilon))$. A claim that Langevin dynamics "mixes in time proportional to d" is a claim about a discretisation with a particular initialisation, and Section 4.1 keeps the two separate.

The factor of two. The entropy decays at rate 2κ and the W_2 distance at rate κ, because W_2 is comparable to the square root of the entropy and not to the entropy. Quoting one rate for the other is the most frequent arithmetic error in this subject and is invisible without the metric written down.

What the transfer corollary does and does not give. Corollary 2.26 transfers a rate that depends on π only through the Dirichlet form, and every rate in this section is of that kind. It does *not* transfer sharpness: the perturbed generator may decay strictly faster, and by the notes on Proposition 2.20 no inequality here can detect that. It also says nothing about kinetic dynamics, whose Dirichlet form is degenerate and therefore not the one appearing above.

What none of this gives. No row of Table 2.2 says anything about a discretisation, about a finite particle system, or about kinetic dynamics. Chapter 5 and Chapter 4 treat those as separate error sources with separate hypotheses.

What would falsify them. A target satisfying LSI(κ) along whose reversible dynamics the relative entropy decays more slowly than $e^{-2\kappa t}$; equality for a Gaussian shows the rate cannot be improved under the hypothesis alone.

Operational reading

Read Table 2.2 downwards as decreasing information and increasing verifiability. The top row is checkable by differentiating V twice and almost never holds for a target anyone wants to sample; the bottom rows hold much more widely and are almost never checkable in closed form. The practical consequence is that a convergence rate for a real target is established either by exhibiting a bounded perturbation of a case that is understood — which is (2.38), with its exponential price — or not at all.

That is one step beyond the table, and this box stops there. Which sampler to run given that the constant is unknown, and what can be said without it, is Section 4.4; nothing in this chapter compares algorithms.

2.5 Chapter Summary

Chapter 2 is the law-level spine of the volume. It moves from a single trajectory to the evolution of its law, establishes the Fokker–Planck equation and its flux form, separates three statements that are routinely merged — stationarity, detailed balance, and $\rho_t \to \pi$ —

and identifies the Fokker–Planck flow of a reversible diffusion as the W_2 gradient flow of the free energy.

The chapter's boundary is drawn by two constructions rather than by caveats. Proposition 2.16 exhibits a diffusion with the same stationary law and the same instantaneous free-energy dissipation as the gradient flow which is nevertheless not one: its curve outruns its own dissipation. And Corollary 2.26 shows that the Dirichlet *energy* is unchanged by an admissible perturbation, so the finite-time guarantees survive while the geometry does not. Rates come from two separate mechanisms — the metric route through displacement convexity, and the functional-inequality route through LSI or Poincaré — and this volume names which is in use at every occurrence. Whether the non-reversible curve is the gradient flow of some *other* functional is open (Open Problem 11.2).

Exercises

§2.1

Tier I Direct application.

2.1 *The Ornstein–Uhlenbeck generator*

For $b(x) = -\theta x$ and $a = \sigma^2 I$, write down $\mathcal{L}$ and $\mathcal{L}^*$, verify that $\pi = \mathcal{N}(0, \frac{\sigma^2}{2\theta} I)$ satisfies $\mathcal{L}^*\pi = 0$, and check by Proposition 2.5 that the dynamics is reversible. Confirm the answer against Proposition 1.43.

2.2 *The heat equation as a special case*

Take $b \equiv 0$, $a = 2I$. Show that (2.4) is $\partial_t\rho = \Delta\rho$, that no probability measure is stationary on $\mathbb{R}^d$, and that the flux is $-\nabla\rho$. Explain in one sentence why the absence of a stationary measure does not contradict Proposition 2.5.

Tier II Completing the derivation.

2.3 *Adjoint with state-dependent diffusion*

Derive (2.2) from (2.1) by integrating by parts twice, and write out the drift of the equivalent form $\partial_t\rho = -\nabla\cdot(\bar{b}\rho) + \frac{1}{2}\sum\partial_{ij}(a_{ij}\rho)$ explicitly in $d = 1$. Identify the extra term that appears when a is not constant and say which convention of Section 1.5.4 it depends on.

2.4 *Reversibility with general a*

Redo Proposition 2.5 for a smooth, symmetric and uniformly positive definite, and show that $\pi \propto e^{-V}$ is reversible if and only if $b = \frac{1}{2}\nabla\cdot a - \frac{1}{2}a\nabla V$. Verify that this reduces to $b = -\nabla V$ when $a = 2I$.

Tier III Probing the boundary.

2.5 *Stationary without detailed balance, explicitly*

In $d = 2$ let $V(x) = \frac{1}{2}|x|^2$ and $c(x) = Jx$ with J antisymmetric. Verify $\nabla\cdot(\pi c) = 0$ directly, compute $j[\pi]$, and confirm that it is a non-zero divergence-free field. Then verify by explicit integration that (2.5) fails for $f(x) = x_1, g(x) = x_2$, and compute the size of the asymmetry.

2.6 *Where the weak form survives and the strong form does not*

Let $\rho_0 = \delta_0$ and let X be a standard Brownian motion. Verify (2.3) directly for $f \in C_c^\infty$ using the Gaussian density, and explain precisely at which time and in what sense (2.4) fails. Then construct a degenerate a in $d = 2$ for which no density exists at any time, and check that (2.3) still holds.

Tier IV Computational.

2.7 *Watching the flux*

Take $V(x) = \frac{1}{4}(x^2 - 1)^2$ in $d = 1$ and $b = -V'$.

(a) Solve (2.4) on a grid from $\rho_0 = \mathcal{N}(-1, 0.01)$ and plot $j[\rho_t]$ at several times. Report the time at which $\sup_x |j[\rho_t](x)|$ has fallen by a factor of 100.

(b) Add the rotational perturbation of the previous exercise in $d = 2$ with $V(x) = \frac{1}{4}(x_1^2 - 1)^2 + \frac{1}{2}x_2^2$ and repeat. Plot $j[\pi]$ and confirm numerically that it is divergence free and non-zero.

(c) State which of (a) and (b) has $j[\pi] = 0$, which has $\nabla \cdot j[\pi] = 0$, and which has $j[\rho_t] = 0$ at any finite t.

Acceptance criteria: (b) reports the numerical divergence with its grid error; (c) is answered with one sentence per condition.

§2.2

Tier I Direct application.

2.8 *First variation with an interaction term*

Let $\mathcal{F}_W[\rho] = \int V\rho + \int \rho \log \rho + \frac{1}{2} \iint W(x-y)\rho(x)\rho(y)$ with W even and C^1. Compute $\delta\mathcal{F}_W / \delta\rho$ and write down the corresponding W_2 gradient flow. Verify that it is the McKean–Vlasov equation of Chapter 8, and identify which hypothesis of Theorem 2.13 is no longer available.

Hint. The factor $\frac{1}{2}$ disappears because W is even and both arguments are perturbed.

2.9 *The Gaussian case in closed form*

For $V(x) = \frac{1}{2}|x|^2$ and $\rho_t = \mathcal{N}(m_t, \Sigma_t)$, solve (2.11) for m_t, Σ_t, and compute $\mathrm{KL}(\rho_t \,\|\, \pi)$ and $\mathcal{I}(\rho_t \,\|\, \pi)$ explicitly. Verify (2.19) by direct differentiation, and check that (2.22) holds with $\lambda = 1$ and with equality when $\Sigma_0 = I$ and $m_0 \neq 0$.

Tier II Completing the derivation.

2.10 *The optimality condition of the JKO step*

Derive the Euler–Lagrange equation quoted in the roadmap of Theorem 2.13: perturb the minimiser of (2.20) by the flow of a smooth compactly supported vector field, differentiate the two terms at time zero, and conclude $(\mathrm{id} - T^{n+1})/\tau = \nabla(V + \log \rho^{n+1})$. Identify the step at which the first variation of the W_2 term requires the optimal map to be unique.

2.11 *Talagrand from log-Sobolev, with the constants tracked*

Assume LSI(α) in the normalisation (2.14). Prove $W_2^2(\nu, \pi) \leq \frac{2}{\alpha}\mathrm{KL}(\nu \,\|\, \pi)$ by running the flow from ν, integrating $|\rho'|(t) \leq \mathcal{I}(\rho_t \,\|\, \pi)^{1/2}$ in t, and using (2.22). Then redo the computation in the convention $\mathrm{KL}(\cdot \,\|\, \pi) \leq \frac{C}{2}\mathcal{I}(\cdot \,\|\, \pi)$ and state the constant you obtain. The point of the exercise is the second half.

Tier III Probing the boundary. Each of these asks where a statement stops being true.

2.12 *Making Proposition 2.16 quantitative*

In the setting of Example 2.17, compute $\|\Pi_\rho b\|^2_{L^2(\rho)}$ exactly for $\rho = \mathcal{N}(0, \Sigma)$ with Σ diagonal, and exhibit the gap between $|\rho'|^2$ and the dissipation as a function of σ_1^2/σ_2^2. Then explain why the gap vanishes at $\rho = \pi$ and why that is consistent with the perturbed process converging *faster*.

Hint. Restrict to linear vector fields; on Gaussians the tangent space intersected with linear fields is the symmetric matrices acting on $x - m$.

2.13 *One equation, two gradient-flow structures*

Show that $\partial_t \rho = \Delta \rho$ is the W_2 gradient flow of $\mathcal{H}[\rho] = \int \rho \log \rho$ and also the L^2 gradient flow of $\mathcal{D}[\rho] = \frac{1}{2}\int |\nabla \rho|^2$. Compute the dissipation identity in each structure and observe that they are different identities about the same solution. State, in one sentence, what this shows about the sentence "the heat equation is a gradient flow".

2.14 *Where Grönwall stops closing*

For the kinetic Fokker–Planck equation $\partial_t f + v \cdot \nabla_x f - \nabla V \cdot \nabla_v f = \nabla_v \cdot (vf + \nabla_v f)$ with the Gibbs law $\mathcal{M}(x, v) \propto e^{-V(x) - |v|^2/2}$, show that $\frac{d}{dt}\mathrm{KL}(f_t \,\|\, \mathcal{M}) = -\int f_t |\nabla_v \log(f_t/\mathcal{M})|^2$. Explain why no logarithmic Sobolev inequality can be applied to close this into an exponential rate, and state what additional structure a hypocoercivity argument supplies. Do *not* attempt the rate itself.

2.15 *A functional that decreases but is not the driver*

Exhibit a functional $\mathcal{G} \neq \mathcal{F}$ that is also monotone non-increasing along (2.11), and show that (2.11) is not the W_2 gradient flow of $\mathcal{G}$. Conclude that monotonicity of a functional along a flow identifies neither the functional nor the geometry.

Tier IV Computational.

2.16 *JKO against ULA on a double well*

Take $d = 1$ and $V(x) = \frac{1}{4}(x^2 - 1)^2$, so π is bimodal with barrier height $\frac{1}{4}$.

(a) Implement the JKO step (2.20) on a grid using entropic regularisation of the W_2 term with parameter ε, and the unadjusted Langevin algorithm with step η. Take $\tau = \eta$ throughout.

(b) Plot $\mathrm{KL}(\rho_T\,\|\,\pi)$ against t for $\eta \in \{10^{-1}, 10^{-2}, 10^{-3}\}$ for both schemes. Report the asymptotic bias of each as a function of the step size, and its order.

(c) Repeat (b) with the barrier height raised to 2. Report how the time to a fixed accuracy scales, and compare it with the bound (2.22) using a numerically estimated α.

(d) Two of the errors you have measured are of different kinds: discretisation bias and slow mixing. Say which figures show which, and state which of the two (2.22) speaks about.

Acceptance criteria: the order of the bias in (b) is stated for both schemes with the evidence that fixes it; the entropic parameter ϵ is reported and shown not to drive the conclusion; part (d) is answered in one sentence each.

§2.3 and §2.4

Tier I Direct application.

2.17 *The Gaussian case, all four rows*

For $V(x) = \frac{\mu}{2}|x|^2$ compute $\nabla^2 V$, and hence K, and verify directly that $\pi = \mathcal{N}(0, \mu^{-1}I)$ satisfies $\mathrm{P}(\mu)$ and $\mathrm{LSI}(\mu)$ with equality attained in each. Confirm that all four rows of Table 2.2 give the same constant here, and say in one sentence why this makes the Gaussian a poor test case.

2.18 *Carré du champ with state-dependent noise*

For $\mathcal{L} = \langle b, \nabla\rangle + \frac{1}{2}\mathrm{tr}(a\nabla^2)$ compute $\Gamma(f)$ and verify $\Gamma(f) = \frac{1}{2}\langle\nabla f, a\nabla f\rangle$. Then state what $\mathcal{E}(f, f)$ becomes and identify which constant in (2.33) changes.

Tier II Completing the derivation.

2.19 *Otto–Villani with the constants*

Complete the second half of the proof of Theorem 2.27: run the flow from ν, use $|\rho'|(t) = \mathcal{I}(\rho_t\,\|\,\pi)^{1/2}$ and $\frac{d}{dt}\mathrm{KL}(\rho_t\,\|\,\pi) = -\mathcal{I}(\rho_t\,\|\,\pi)$, and carry out (2.37) in full to obtain (2.34). Then answer the question the step is designed to expose: (2.14) is a *lower* bound on the Fisher information, so say precisely how a lower bound on I_t produces an upper bound on $I_t^{1/2}$, and why the exponential decay of $\mathrm{KL}(\rho_t\,\|\,\pi)$ — which is also available — does not by itself bound $\int_0^\infty I_t^{1/2}\,dt$. Track every constant and state where the factor of two between the two rates of Proposition 2.25(iii) is created.

2.20 *Linearisation, in the other convention*

Redo Proposition 2.25(iii) in the convention $\mathrm{KL}(\cdot\,\|\,\pi) \le \frac{C}{2}\mathcal{I}(\cdot\,\|\,\pi)$ and report the Poincaré constant you obtain. Then state the conversion rule between the two conventions in one line, and check it against Section 2.2.1.

Tier III Probing the boundary.

2.21 *Poincaré without log-Sobolev*

Let $\pi \propto e^{-|x|}$ on $\mathbb{R}$. Show that π satisfies a Poincaré inequality and no logarithmic Sobolev inequality.

Hint. Test the log-Sobolev inequality with $f = e^{\gamma x/2}$ on the right half-line and let $\gamma \uparrow 1$; for the Poincaré inequality use the Muckenhoupt criterion or a direct one-dimensional estimate.

Then state which row of Table 2.2 this example removes and which it leaves standing.

2.22 *Curvature destroyed, inequality retained*

Take $V_0(x) = \frac{1}{2}x^2$ and $V = V_0 + u$ with u a smooth bump of height ε and width δ. Show that $\mathrm{CD}(K,\infty)$ fails for every $K > 0$ once $\varepsilon/\delta^2 > 1$, however small ε is, while (2.38) still gives $\mathrm{LSI}(e^{-\varepsilon})$. Explain what this shows about which hypothesis to state a theorem under.

2.23 *The Dirichlet form cannot see the rotation*

For $V(x) = \frac{1}{2}|x|^2$ in $d = 2$ and $c(x) = \gamma Jx$ with J antisymmetric, verify directly that $\mathcal{E}(f,f)$ is independent of γ. Then compute the decay of $\mathrm{Var}_\pi(P_t f)$ for $f(x) = x_1$ as a function of γ and observe that it does depend on γ. Say precisely which step of the proof of Proposition 2.25(i) fails in the non-reversible case.

Tier IV Computational.

2.24 *Estimating the constant, and watching it collapse*

Take $V_\varepsilon(x) = \frac{1}{4\varepsilon}(x^2-1)^2$ on $\mathbb{R}$ for $\varepsilon \in \{1, \frac{1}{2}, \frac{1}{4}, \frac{1}{8}\}$.

(a) Discretise $\mathcal{L}$ on a grid and compute the spectral gap λ numerically. Plot $\log \lambda$ against the barrier height Δ and fit the slope.

(b) Estimate α by minimising the ratio $\mathcal{I}(\nu \,\|\, \pi)/(2\mathrm{KL}(\nu \,\|\, \pi))$ over a parametric family of ν, and compare with λ. Report the ratio α/λ.

(c) Simulate the reversible dynamics and measure the time for $\mathrm{KL}(\rho_t \,\|\, \pi)$ to halve. Compare with $\log 2/(2\alpha)$.

(d) State which of your three numbers is an upper bound on the true constant, which is a lower bound, and which is neither, and say what that implies about reporting a measured rate as a verification of Corollary 2.14.

Acceptance criteria: the fitted slope in (a) is reported with its uncertainty; (d) classifies all three numbers.

CHAPTER 3

Diffusion, Score, and Flow-Based Generative Models

A generative model is a way of turning noise into data. This chapter treats the family that does so by running a diffusion backwards, and the reason it belongs in this volume is that every step of the construction is an instance of something already proved: the reverse process is a diffusion by time-reversal theory, its drift is the score of the forward marginals, the deterministic sampler is a continuity equation, and the training objective is a least-squares regression whose minimiser is a conditional expectation.

What has to be resisted is the impression that this makes generative modelling a corollary of Chapter 1 and Chapter 2. Three things are genuinely new and none of them is supplied by the stochastic analysis. The forward process is time inhomogeneous and at no time stationary, so nothing from Section 2.1.2 about detailed balance is available. The score is not known but learned, so every statement about the exact reverse process is a statement about an object nobody has. And the interval on which the construction runs is closed at both ends, while the reverse drift is singular at one of them. Section 3.1 states the time-reversal theorem with the hypotheses that make these three points visible; Section 3.2 fixes the forward processes and shows that the degeneracy of the score–velocity dictionary of Section 3.4 is exactly the vanishing of the forward diffusion coefficient.

3.1 Time Reversal and the Probability-Flow ODE

3.1.1 The reverse-time equation

Throughout, X solves the forward equation

$$dX_t = f(t, X_t)\,dt + \sigma(t)\,dW_t, \qquad t \in [0, T], \qquad X_0 \sim p_{\text{data}}, \tag{3.1}$$

with $a(t) = \sigma(t)\sigma(t)^{\top}$ and $p_t = \text{Law}(X_t)$; the diffusion coefficient is taken state independent, which is the case used in practice and which Proposition 3.2 relaxes.

Theorem 3.1 (Time reversal of a diffusion) ESTABLISHED. *Assume*

(A1) f is locally Lipschitz with linear growth, and σ is continuous with $a(t) \succeq \varepsilon I$ for some $\varepsilon > 0$ on every $[\delta, T]$ with $\delta > 0$;

(A2) p_t has a density for each $t \in (0, T]$, and

$$\int_{\delta}^{T}\int_{\mathbb{R}^d} |\nabla \log p_t(x)|^2\, p_t(x)\,dx\,dt < \infty \qquad \text{for every } \delta > 0. \tag{3.2}$$

Then the reversed process $Y_s := X_{T-s}$, $s \in [0, T-\delta]$, is a weak solution of

$$dY_s = \Big[-f(T-s, Y_s) + a(T-s)\,\nabla \log p_{T-s}(Y_s)\Big]\,ds + \sigma(T-s)\,d\bar{W}_s, \tag{3.3}$$

for a Brownian motion $\bar{W}$ with respect to the reversed filtration, with $Y_0 \sim p_T$. The result is due to Anderson [2]; the general state-dependent form is Proposition 3.2.

Proof roadmap. The statement is proved by showing that the law of the reversed path solves the martingale problem of Definition 1.45 for the generator associated with (3.3). Concretely, for $f \in C_c^\infty$ one computes $\mathbb{E}\left[\phi(Y_{s+h}) - \phi(Y_s) \mid \mathcal{F}_s^Y\right]$ by conditioning the forward process on its later value; the conditional law of X_{t-h} given X_t has a density proportional to the forward transition density times p_{t-h}, and expanding the logarithm of that ratio to first order in h produces the extra drift $a\nabla \log p_t$. Hypothesis (A2) is what makes the expansion legitimate in L^2, and (A1) supplies the regularity of the transition density needed to differentiate it. The identification of $\bar{W}$ as a Brownian motion for the reversed filtration is then the martingale representation theorem, as in Theorem 1.46.

This is a roadmap. The full argument is a substantial piece of stochastic analysis and is quoted; what this volume uses is the statement together with the fact — visible in the roadmap and invisible in the formula — that (3.2) is doing real work. □

Scope of Theorem 3.1 and Proposition 3.2

Assumed. Non-degeneracy of a away from $t = 0$, and the path-integrated Fisher information bound (3.2). Neither is a formality. Proposition 3.2 assumes in addition that a is smooth in x and uniformly elliptic, and — this is the substantive part — the state-dependent time-reversal theorem itself, which is what discards the divergence-free field. The Fokker–Planck computation in its proof cannot do that step; see Remark 3.3.

Where it fails. Condition (3.2) fails as $\delta \to 0$ whenever p_{data} is not absolutely continuous — in particular whenever it is supported on a lower-dimensional manifold, which is the standing assumption of most of the applied literature. The relative Fisher information of p_t with respect to Lebesgue measure then blows up as $t \downarrow 0$, at a rate governed by the codimension, and (3.3) is not valid at $s = T$. Every practical construction stops at $t = \delta > 0$, and that truncation is one of the three error terms of Section 3.5. Degeneracy of a likewise removes the theorem: for a noise schedule with $\sigma(t) \to 0$ at an interior time, the reverse drift is undefined there, which Proposition 3.8 identifies with the degeneracy of the score–velocity dictionary.

What it does not say. It is a statement about the law of the reversed path, obtained from the exact score. It says nothing about a process built with an approximate score — that process is a different diffusion, and the distance between the two is not controlled by anything here — and nothing about pathwise agreement: the reversal is an identity in law, and there is no sense in which (3.3) run on a given Brownian path reconstructs the forward trajectory that produced the terminal point.

What would falsify it. A forward process satisfying (A1)–(A2) whose reversed law fails to solve the martingale problem for (3.3).

Proposition 3.2 (State-dependent diffusion: the divergence correction) ESTABLISHED. *Assume the state-dependent version of Theorem 3.1 — the reversed path law is that of a diffusion with the same covariance characteristic a — which for $a = a(t, x)$ smooth and uniformly elliptic is quoted here and is not a consequence of Theorem 3.1, whose statement fixes a state-independent a. The state-dependent theorem is due to Haussmann and Pardoux [10], whose coefficient regularity and integrability hypotheses are the ones assumed here. Under that theorem the reverse drift in (3.3)* *becomes*

$$-f + a\nabla\log p_t + \nabla\cdot a, \qquad (\nabla\cdot a)_i = \sum_j \partial_j a_{ij}, \tag{3.4}$$

the last term being absent from (3.3) only because a is constant in x there.

Remark 3.3 (What the computation below does and does not establish). The proof that follows is a Fokker–Planck matching: it shows that (3.4) is *a* drift whose marginals are the reversed ones, and it is complete as such. It does not show that (3.4) is *the* reversed path law, because the matching determines the drift only up to a divergence-free field — exactly the ambiguity Proposition 3.4 exploits. Selecting $J = 0$ is the work of the time-reversal theorem, which is a statement about path measures; that is why this proposition is tagged ESTABLISHED and quotes the state-dependent theorem rather than deriving its conclusion from the state-independent one. Presenting a marginal computation as though it settled a path-law question is the substitution Table 11.2 calls *path law against marginal*.

Proof. Write the forward Fokker–Planck equation (2.4) in flux form, $\partial_t p = -\nabla\cdot j$ with $j = fp - \frac{1}{2}\nabla\cdot(ap)$, using the notation of (2.6). The reversed marginals $q_s = p_{T-s}$ satisfy $\partial_s q = +\nabla\cdot j|_{T-s}$. A diffusion with drift $\bar{f}$ and the same diffusion matrix has $\partial_s q = -\nabla\cdot(\bar{f}q) + \frac{1}{2}\sum\partial_{ij}(a_{ij}q)$, so matching the two requires

$$-\nabla\cdot(\bar{f}q) + \tfrac{1}{2}\nabla\cdot(\nabla\cdot(aq)) = \nabla\cdot\Big(fq - \tfrac{1}{2}\nabla\cdot(aq)\Big),$$

whence $\bar{f}q = -fq + \nabla\cdot(aq) + J$ for some field J with $\nabla\cdot J = 0$. Matching Fokker–Planck equations does *not* force $J = 0$: in $d \geq 2$ there are non-zero, rapidly decaying divergence-free fields, and adding one produces a different diffusion with the same marginals — the phenomenon of Proposition 3.4, seen from the other side. What selects $J = 0$ is not the marginal equation but the time-reversal theorem, which identifies the law of the reversed *path* and not merely its marginals; Theorem 3.1 is that statement for constant a and its general form gives (3.4). Dividing by q and using $\nabla\cdot(aq) = a\nabla q + q\nabla\cdot a$ gives the displayed drift. ∎

3.1.2 The probability-flow ODE

The same marginals are produced by a deterministic flow, and by a whole family of stochastic ones interpolating between the two.

Proposition 3.4 (A one-parameter family with identical marginals) DERIVED. *Under the hypotheses of Theorem 3.1, for every $\lambda \geq 0$ the reverse-time equation*

$$\mathrm{d}Y_s = \Big[-f + \tfrac{1+\lambda}{2}\, a\nabla\log p\Big]_{T-s}(Y_s)\,\mathrm{d}s + \sqrt{\lambda}\,\sigma(T-s)\,\mathrm{d}\bar{W}_s \tag{3.5}$$

has marginals $\mathrm{Law}(Y_s) = p_{T-s}$ *for all s, provided* $\mathrm{Law}(Y_0) = p_T$ *and the displayed equation is well posed with marginals uniquely determined by the associated forward equation: for $\lambda > 0$ it suffices that the martingale problem for the displayed coefficients be well posed, and for $\lambda = 0$ that the velocity field be locally Lipschitz in y with the linear-growth bound needed for $\int_0^T\int|v_s|\,\mathrm{d}p_{T-s}\mathrm{d}s < \infty$, which by Theorem 1.2 and the uniqueness theory for the continuity equation makes the flow and its* *marginals unique. Both requirements can fail near the endpoints, where $\nabla \log p_t$ is unbounded for data on a manifold; the statement is then made on $[\delta, T]$. The two extreme members are*

$$\lambda = 0: \quad \dot{y} = -f(T-s, y) + \tfrac{1}{2}a(T-s)\nabla \log p_{T-s}(y) \qquad \textit{(the probability-flow ODE)}, \tag{3.6}$$

which is deterministic, and $\lambda = 1$, which is (3.3).

Proof. Write $q_s = p_{T-s}$, so $\partial_s q = -\partial_t p\big|_{T-s}$. By (2.4) with constant a, $\partial_t p = -\nabla\cdot(fp) + \frac{1}{2}\operatorname{tr}(a\nabla^2 p)$, so $\partial_s q = \nabla\cdot(fq) - \frac{1}{2}\operatorname{tr}(a\nabla^2 q)$. A process with drift $\bar{f}$ and diffusion λa has $\partial_s q = -\nabla\cdot(\bar{f}q) + \frac{\lambda}{2}\operatorname{tr}(a\nabla^2 q)$. Equating and using $\operatorname{tr}(a\nabla^2 q) = \nabla\cdot(a\nabla q)$ for constant a,

$$\nabla\cdot\left(\bar{f}q + fq - \tfrac{1+\lambda}{2}a\nabla q\right) = 0,$$

which is satisfied by the choice that makes the bracket vanish identically, $\bar{f} = -f + \frac{1+\lambda}{2}a\nabla \log q$: that is (3.5). This is all that is needed: the proposition asserts that these drifts *have* the stated marginals, not that they are the only ones that do. Adding any J/q with $\nabla\cdot J = 0$ gives a further equation with the same marginals and a different path law. Setting $\lambda = 0$ removes the noise and gives (3.6); setting $\lambda = 1$ gives (3.3). ∎

Convention

The factor of $\frac{1}{2}$ promised in Section 1.5.4 appears here, and it is worth seeing it cancel. In the Itô convention of that box the generator of (3.1) carries $\frac{1}{2}\operatorname{tr}(a\nabla^2)$. In Proposition 3.4 the deterministic member acquires $\frac{1}{2}a\nabla \log p$ and the stochastic member $\lambda = 1$ acquires $a\nabla \log p$ — with no $\frac{1}{2}$ — because the coefficient is $\frac{1+\lambda}{2}$ and $\lambda = 1$ doubles it. Writing $a = g^2 I$, these are the two formulas that circulate as $f - \frac{1}{2}g^2\nabla \log p$ and $f - g^2\nabla \log p$; they are not two conventions, they are two different equations, and (3.5) is the one-parameter family that contains both. Every occurrence of either in this volume is an instance of (3.5) with λ named.

Proposition 3.5 (Exact log-likelihood along the deterministic flow) DERIVED. *Let $v_s(y)$ denote the right-hand side of (3.6) and let $y(\cdot)$ solve $\dot{y} = v_s(y)$ with $y(0)$ given. Then, along the trajectory,*

$$\frac{\mathrm{d}}{\mathrm{d}s}\log q_s(y(s)) = -\nabla\cdot v_s(y(s)), \qquad q_s = p_{T-s}, \tag{3.7}$$

so that, for every $\delta > 0$ for which the hypotheses hold on $[\delta, T]$,

$$\log p_\delta(y(T-\delta)) = \log p_T(y(0)) - \int_0^{T-\delta} \nabla\cdot v_s(y(s))\,\mathrm{d}s. \tag{3.8}$$

Corollary 3.6 (The endpoint, and what it costs) DERIVED. *Suppose in addition that p_{data} has a positive density, that the score and hence v_s extend continuously to $s = T$, and that $\int_0^T |\nabla\cdot v_s(y(s))|\,\mathrm{d}s < \infty$. Then $\delta \downarrow 0$ in (3.8) is legitimate and*

$$\log p_{\text{data}}(y(T)) = \log p_T(y(0)) - \int_0^T \nabla\cdot v_s(y(s))\,\mathrm{d}s. \tag{3.9}$$

None of the three hypotheses is available for data supported on a lower-dimensional set: there p_{data} has no Lebesgue density, so the left-hand side of (3.9) is not merely hard to control but undefined, while $|\nabla \log p_t|$ blows up as $t \downarrow 0$ and the integral diverges. The truncated statement (3.8) survives that case unchanged, which is why it, and not (3.9), is what this volume uses.

Proof. By Proposition 3.4 with $\lambda = 0$, q solves the continuity equation $\partial_s q + \nabla \cdot (q v_s) = 0$. Hence along the flow

$$\frac{\mathrm{d}}{\mathrm{d}s} \log q_s(y(s)) = \frac{\partial_s q_s + \langle \nabla q_s, v_s \rangle}{q_s} = \frac{-\nabla \cdot (q_s v_s) + \langle \nabla q_s, v_s \rangle}{q_s} = -\nabla \cdot v_s,$$

using $\nabla \cdot (qv) = q \nabla \cdot v + \langle \nabla q, v \rangle$. Integrating from 0 to $T - \delta$ and using $q_{T-\delta} = p_\delta$ gives (3.8). For the corollary, the integrand is dominated by hypothesis and $\log p_\delta(y(T - \delta)) \to \log p_{\text{data}}(y(T))$ by continuity of the extended flow and of the density, so both sides converge. ∎

Interpretation

One family, two extremes, and the same marginals throughout. Setting $\lambda = 1$ gives a stochastic reverse process; $\lambda = 0$ gives a deterministic flow. Every member transports p_T back to p_{data} through the same sequence of laws, so any statement about the distribution of a sample transfers freely between them. What does not transfer is anything about a trajectory: the members differ as path measures, so couplings, likelihoods along a path, and the propagation of a solver's error are all λ-dependent. This is the volume's basic distinction — law against path law — in the place where it is easiest to lose.

Notes on Proposition 3.4 and Proposition 3.5

Assumed. The hypotheses of Theorem 3.1, the exact score, and well-posedness of each displayed equation together with uniqueness for the corresponding forward equation — without which the computation establishes an identity about a PDE and not a statement about a process. No decay hypothesis is needed for Proposition 3.4, because the proposition asserts only that the displayed drifts have the stated marginals: the proof exhibits one solution of the divergence equation and makes no claim that it is the only one. Proposition 3.5 additionally needs v_s locally Lipschitz so that the flow exists, which by Theorem 1.2 requires $\nabla \log q_s$ to be C^1 — and this is available on $[\delta, T]$ for $\delta > 0$, which is why the proposition is stated there. Corollary 3.6 needs three further hypotheses at the data endpoint, and each of them fails for a data law on a lower-dimensional set: an exact likelihood at $t = 0$ is a claim about a density that then does not exist. Reported likelihoods for a probability-flow sampler are therefore statements about p_δ for whatever δ the implementation stops at, and the comparison of two such numbers is meaningful only at a common δ.

These are not the only equations with these marginals. Adding to any member of (3.5) a field J/q with $\nabla \cdot J = 0$ leaves every time marginal unchanged and changes the path law, and in $d \geq 2$ such J exist in abundance with any prescribed decay. What singles out the one-parameter family is not the marginal equation — which does not determine a drift — but the time-reversal theorem, which identifies the law of the reversed path; Proposition 3.2 makes the same point from the other side. A derivation that reads uniqueness of the drift out of the Fokker–Planck equation has proved something weaker than it states.

Where the family stops being a family. All members agree in law at each time *when the score is exact*. They do not agree as processes: for $\lambda > 0$ the trajectory is random given Y_0, for $\lambda = 0$ it is not, and their couplings, path measures and error propagation differ. With an approximate score the marginals no longer agree either, and the discrepancy grows differently in λ; that comparison is Section 3.6 and nothing here bears on it.

What the likelihood formula does and does not give. (3.7) is exact for the exact velocity field. For a learned v_θ it computes the exact log-likelihood of the model *defined by* v_θ, which is a different distribution from p_{data}; the number obtained is a likelihood under the model and not a bound on any distance to the data. It also requires $\nabla \cdot v_\theta$, whose unbiased estimation by a Hutchinson trace estimator introduces a variance that is not accounted for in the identity.

What does not follow

Reversibility plays no part in any of this. The forward process (3.1) is not assumed stationary, and detailed balance is therefore unavailable as a route to the reverse drift; by Proposition 2.6 the vanishing of the probability flux is a property of an equilibrium law, not something the noising process has at a generic time. (This is weaker than saying the flux vanishes nowhere: a time-inhomogeneous law may well have zero flux at particular times or on particular regions, and nothing below needs to exclude that.) Nothing in Theorem 3.1 uses detailed balance, stationarity, or a Gibbs form for p_t; the theorem holds for forward processes that have no stationary measure at all. Arguments that derive the reverse drift by setting the flux to zero are deriving a different statement, and the coincidence of the resulting formula in the special case of an Ornstein–Uhlenbeck forward process is what makes the error hard to see.

Same marginals is not same process. Proposition 3.4 produces a continuum of processes agreeing in law at every fixed time. They differ in everything else. A claim about a sampler — its error under an approximate score, its behaviour under a coarse discretisation, the diversity of its outputs — is a claim about the process and is not settled by the marginal identity.

The score is not a modelling choice. In (3.3) the drift is $\nabla \log p_t$ for the *true* forward marginals, which are determined by p_{data} and the schedule. Learning enters only as an approximation to a fixed target, and every statement in this section is about the target. The moment $\nabla \log p_t$ is replaced by s_θ, one leaves this section and enters Section 3.5.

3.2 Forward Processes and Noise Schedules

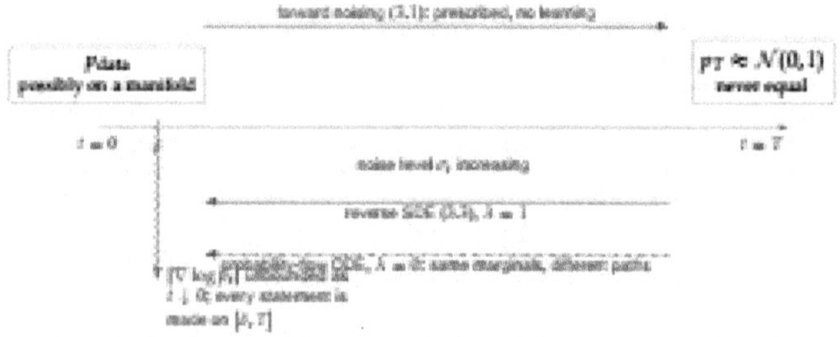

Figure 3.1: The forward process and the two constructions that reverse it. The forward direction is prescribed and involves no learning; both reverse constructions require the score. The two share *every time marginal* and differ as path measures (Proposition 3.4), so a statement about samples transfers between them and a statement about trajectories does not. What this figure does not imply: that p_T equals its Gaussian limit at any finite T (Proposition 3.10), or that the reverse constructions are defined at $t = 0$; the truncation at δ is a hypothesis, not a numerical convenience.

3.2.1 Linear forward processes

Proposition 3.7 (Solution of a linear forward process) DERIVED. *Let $f(t,x) = \mathfrak{f}(t)x$ with $\mathfrak{f}$ continuous and $\sigma(t) = g(t)\mathrm{I}$ with g continuous. Then the solution of (3.1) is*

$$X_t = a_t X_0 + \sigma_t \varepsilon, \qquad a_t = \exp \int_0^t \mathfrak{f}, \qquad \sigma_t^2 = a_t^2 \int_0^t \frac{g_s^2}{a_s^2}\,\mathrm{d}s, \tag{3.10}$$

with $\varepsilon \sim \mathcal{N}(0,\mathrm{I})$ independent of X_0; that is, $p_t(\cdot \mid x_0) = \mathcal{N}(a_t x_0, \sigma_t^2 \mathrm{I})$, which is exactly the Gaussian conditional path (3.18) of Section 3.4.

Proof. Apply Theorem 1.33 to $a_t^{-1} X_t$: since a is deterministic and C^1 with $\dot{a} = \mathfrak{f}a$, $\mathrm{d}(a_t^{-1} X_t) = -a_t^{-1}\mathfrak{f}X_t\mathrm{d}t + a_t^{-1}(\mathfrak{f}X_t\mathrm{d}t + g_t\mathrm{d}W_t) = a_t^{-1} g_t \mathrm{d}W_t$. Integrating, $X_t = a_t(X_0 + \int_0^t a_s^{-1} g_s \mathrm{d}W_s)$; the stochastic integral is a Wiener integral of a deterministic integrand, hence centred Gaussian with variance $\int_0^t g_s^2/a_s^2\,\mathrm{d}s$ per coordinate by (1.22). Multiplying by a_t^2 gives σ_t^2. ■

Proposition 3.8 (The forward diffusion coefficient is the dictionary coefficient) DERIVED. *With a, σ as in (3.10) and c_t as in Proposition 3.17,*

$$g_t^2 = -2c_t = -2\sigma_t^2 \frac{\mathrm{d}}{\mathrm{d}t} \log \mathrm{snr}_t. \tag{3.11}$$

Consequently the score–velocity dictionary is invertible at t if and only if $g_t \neq 0$: the dictionary degenerates exactly where the forward process stops injecting noise.

Proof. From (3.10), $\sigma_t^2/a_t^2 = \int_0^t g_s^2/a_s^2$, so differentiating, $g_t^2/a_t^2 = \frac{\mathrm{d}}{\mathrm{d}t}(\sigma_t^2/a_t^2)$. Expanding the right-hand side, $g_t^2 = a_t^2 \cdot \frac{2\sigma_t\dot{\sigma}_t a_t^2 - 2\sigma_t^2 a_t \dot{a}_t}{a_t^4} = 2\sigma_t\dot{\sigma}_t - 2\sigma_t^2 \frac{\dot{a}_t}{a_t} = -2\sigma_t^2\left(\frac{\dot{a}_t}{a_t} - \frac{\dot{\sigma}_t}{\sigma_t}\right)$, which is $-2c_t$ by the definition of c_t. Since $\sigma_t > 0$ on $(0,1)$, $c_t \neq 0$ if and only if $g_t \neq 0$. ■

Example 3.9 (Three schedules). The variance-preserving family takes $a_t^2 + \sigma_t^2 = 1$, so $\mathfrak{f} = \dot{a}/a$ and, by (3.11), $g_t^2 = -2\sigma_t^2 \frac{\mathrm{d}}{\mathrm{d}t} \log(a_t/\sigma_t) = -2\dot{a}_t a_t/1 = -2a_t\dot{a}_t$, positive because a decreases; with $a_t = \cos(\frac{\pi}{2}t)$ this is $\frac{\pi}{2}\sin(\pi t)$. The variance-exploding family takes $a_t \equiv 1$, so $\mathfrak{f} \equiv 0$ and $g_t^2 = 2\sigma_t\dot{\sigma}_t$; the process is a time-changed Brownian motion and has no stationary distribution. The linear interpolant $a_t = 1 - t$, $\sigma_t = t$ gives $\mathfrak{f} = -1/(1-t)$ and $g_t^2 = 2t/(1-t)$, which diverges as $t \uparrow 1$ — so the linear interpolant, natural as a probability path, is a forward *diffusion* only up to a blow-up at the noise end.

3.2.2 The terminal mismatch

Generation starts from a tractable reference law, usually $\mathcal{N}(0,\mathrm{I})$, whereas Theorem 3.1 requires starting from p_T. The gap is one of the three error sources of Section 3.5, and for linear forward processes it can be bounded exactly.

Proposition 3.10 (Terminal mismatch of a linear forward process) DERIVED. *Let $p_T = \mathrm{Law}(a_T X_0 + \sigma_T \varepsilon)$ with $m_2 := \mathbb{E}\,|X_0|^2 < \infty$. Then*

$$\mathrm{KL}(p_T \,\|\, \mathcal{N}(0,\mathrm{I})) \le \tfrac{1}{2} a_T^2 m_2 + \tfrac{d}{2}(\sigma_T^2 - 1 - \log \sigma_T^2). \tag{3.12}$$

In the variance-preserving case $\sigma_T^2 = 1 - \alpha_T^2$ the second term is $\frac{d}{4}\alpha_T^4 + O(\alpha_T^6)$, so the bound is $\frac{1}{2}\alpha_T^2 m_2 + O(d\alpha_T^4)$ and decays with the schedule.

Proof. Relative entropy is jointly convex, so applying the conditioning inequality to the mixture over X_0,

$$\mathrm{KL}(p_T \,\|\, \mathcal{N}(0, I)) \;\le\; \mathbb{E}_{X_0}\Big[\mathrm{KL}(\mathcal{N}(\alpha_T X_0, \sigma_T^2 I) \,\|\, \mathcal{N}(0, I))\Big].$$

The Gaussian relative entropy is $\frac{1}{2}(\alpha_T^2 |x_0|^2 + d\sigma_T^2 - d - d\log\sigma_T^2)$; taking the expectation over X_0 gives (3.12). For the expansion, put $\sigma_T^2 = 1 - u$ with $u = \alpha_T^2$: then $\sigma_T^2 - 1 - \log\sigma_T^2 = -u - \log(1-u) = \frac{u^2}{2} + O(u^3)$. ■

Where the schedule bites. By (3.11) a schedule is a forward diffusion only where snr_t is strictly decreasing, and by Example 3.9 the linear interpolant fails this at the noise end, where g_t^2 diverges. The consequences are not cosmetic: a diverging g means the reverse drift of (3.3) carries a diverging coefficient, and every discretisation error is multiplied by it. Conversely, at the data end $\sigma_t \to 0$ and $\nabla \log p_t$ diverges under the manifold hypothesis, by the scope note on Theorem 3.1. Both ends are singular, for different reasons, and every implementation truncates both.

What the mismatch bound does not include. Proposition 3.10 bounds the distance between p_T and the reference law only. It says nothing about how that distance propagates through the reverse process: the reverse dynamics may contract it, and for a strongly log-concave family it does, but that contraction is a separate estimate and is not supplied here. Quoting (3.12) as "the terminal error of the sampler" skips that step.

Operational reading

Read (3.11) as saying that there is one object, not two. A noise schedule can be specified by the pair (α_t, σ_t), by the pair (f, g), or by the single function $\log \mathrm{snr}_t$, and the three are related by (3.10) and (3.11). Papers specify whichever is convenient, so comparing two schedules means converting them to a common one first — most usefully $\log \mathrm{snr}_t$, since it is invariant under the rescaling $x \mapsto \lambda x$ that relates the variance-preserving and variance-exploding families.

That is one step beyond the proposition, and this box stops there. Which schedule produces better samples at a given number of function evaluations depends on where the learned score is accurate, on the solver, and on the data, and is the subject of Section 3.6.

3.3 Score Matching and Latent-Variable Readings

3.3.1 Three objectives with one minimiser

The reverse-time equation (3.3) needs $\nabla \log p_t$, which is not available: p_t is a convolution of the unknown p_{data} with a Gaussian. What is available is samples. The following identity is what makes the gap crossable.

Proposition 3.11 (Explicit and implicit score matching) CLASSICAL. *Let q be a density on $\mathbb{R}^d$,*

C^1 and vanishing at infinity together with $|s|\,q$, and let $s : \mathbb{R}^d \to \mathbb{R}^d$ be C^1 with $\mathbb{E}_q\,|s|^2 < \infty$. Then

$$\mathbb{E}_q\|s - \nabla \log q\|^2 = \mathbb{E}_q\left[2\,\nabla\cdot s + \|s\|^2\right] + \mathbb{E}_q\|\nabla \log q\|^2, \tag{3.13}$$

and the last term does not depend on s.

Proof. Expand the square: the cross term is $-2\int \langle s, \nabla \log q\rangle\, q = -2\int \langle s, \nabla q\rangle$, and integrating by parts, with the boundary term vanishing by hypothesis, gives $+2\int (\nabla\cdot s)\, q$. Collecting terms gives (3.13). ∎

Proposition 3.12 (Denoising score matching) DERIVED. *Let $X_t = a_t X_0 + \sigma_t \varepsilon$ as in (3.10), with $\sigma_t > 0$. Then for every measurable s with $\mathbb{E}\,\|s(X_t)\|^2 < \infty$,*

$$\mathbb{E}\left\|s(X_t) + \frac{\varepsilon}{\sigma_t}\right\|^2 = \mathbb{E}\|s(X_t) - \nabla \log p_t(X_t)\|^2 + C_t, \tag{3.14}$$

with C_t independent of s. Hence the three objectives — explicit score matching, written $\mathcal{J}_{\mathrm{exp}}(s) = \mathbb{E}\|s - \nabla \log p_t\|^2$, its implicit form $\mathcal{J}_{\mathrm{imp}}$ of (3.13), and the denoising form $\mathcal{J}_{\mathrm{DSM}}$ of (3.14) — differ pairwise by constants that do not depend on s, and therefore have the same minimiser $\nabla \log p_t$ over all measurable fields and the same ordering of any two admissible fields.

Proof. By Lemma 3.19 with $X = X_t$ and $Y = -\varepsilon/\sigma_t$, the left-hand side equals $\mathbb{E}\,\|s(X_t) - \mathbb{E}[-\varepsilon/\sigma_t \mid X_t]\|^2 + \mathbb{E}\,\|Y - \mathbb{E}[Y \mid X_t]\|^2$, and $\mathbb{E}[-\varepsilon/\sigma_t \mid X_t] = \nabla \log p_t(X_t)$ by (3.23). The second term is C_t. ∎

The three are more equivalent than their estimators are, and the distinction is worth stating precisely because it is easy to overshoot in either direction. Both (3.13) and (3.14) hold for *every* admissible field s, not merely at the optimum, and the constants they differ by do not depend on s. Substituting a parametric family $s = s_\theta$ whose members are all admissible therefore gives

$$\mathcal{J}_{\mathrm{exp}}(\theta) = \mathcal{J}_{\mathrm{imp}}(\theta) + C_1, \qquad \mathcal{J}_{\mathrm{DSM}}(\theta) = \mathcal{J}_{\mathrm{exp}}(\theta) + C_2, \tag{3.15}$$

with C_1, C_2 independent of θ. The three population objectives consequently have the same minimisers over every restricted class and, wherever differentiable, the same population gradient. What differs is everything else: only the denoising form is estimable from samples without knowing q; in the additive-Gaussian path used here its target is explicit as $-\varepsilon/\sigma_t$, whereas for a more general corruption kernel a denoising formulation needs that kernel's conditional score, which may or may not be tractable — so the restriction is on tractability, not on principle; and the three have different sample estimators, different stochastic-gradient variances, different boundary-regularity requirements for the integration by parts, and different conditioning near $\sigma_t = 0$. The choice among them is not neutral, and it is not neutral for those reasons rather than because the population problems differ.

This should not be confused with Theorem 3.20, which compares a flow-matching objective with a score-matching one across two parameterisations and two time weightings; there the two objectives are genuinely different functions of θ unless the tie and the weight relation both hold, and Example 3.21 exhibits the gap. Two parameterisations of one field is a different situation from three identities for one field.

3.3.2 The variational bound, and the latent-variable reading

Proposition 3.13 (Likelihood weighting) ESTABLISHED. *Let $\hat{p}_\theta$ be the law at time $\delta > 0$ of the reverse process (3.3) with $\nabla \log p_t$ replaced by s_θ and started from $\mathcal{N}(0, I)$ at $t = T$. Then, under the hypotheses of Theorem 3.1 and with g_t^2 from Proposition 3.8,*

$$\mathrm{KL}(p_\delta \,\|\, \hat{p}_\theta) \;\le\; \mathrm{KL}(p_T \,\|\, \mathcal{N}(0, I)) \;+\; \frac{1}{2}\int_\delta^T g_t^2 \, \mathbb{E}_{p_t}\big\| s_\theta(t,\cdot) - \nabla \log p_t \big\|^2 \, dt. \tag{3.16}$$

In particular the weighting $w(t) = \frac{1}{2} g_t^2$ is the one under which the score-matching objective is a bound on a divergence, and it is not the uniform weighting. The likelihood-weighting identity in this form is due to Song et al. [25].

Proof sketch. Let Q be the path law of the exact reverse process started from p_T and $\hat{\mathrm{Q}}$ that of the learned one started from $\mathcal{N}(0, I)$, both on $C([\delta, T]; \mathbb{R}^d)$ run backwards. The two have the same diffusion coefficient g, so by the path-space relative entropy formula (3.17) below,

$$\mathrm{KL}(\mathrm{Q} \,\|\, \hat{\mathrm{Q}}) = \mathrm{KL}(p_T \,\|\, \mathcal{N}(0, I)) + \frac{1}{2}\int_\delta^T \frac{1}{g_t^2}\, \mathbb{E}\big\| g_t^2 (\nabla \log p_t - s_\theta) \big\|^2 dt,$$

the first term being the relative entropy of the initial laws by the chain rule for relative entropy on path space, and the second the Girsanov term with drift difference $g_t^2(\nabla \log p_t - s_\theta)$. Simplifying the coefficient gives $\frac{1}{2} g_t^2 \|\cdot\|^2$. The data-processing inequality applied to the marginal at time δ gives (3.16).

This is a sketch: the Girsanov step needs the finiteness of the path-space relative entropy, which is Theorem 3.14's hypothesis and is not verified here for a general s_θ. ■

Theorem 3.14 (Path-space relative entropy) DERIVED. *Let $\mathrm{Q}^1, \mathrm{Q}^2$ be the laws on $C([0, T]; \mathbb{R}^d)$ of two diffusions with the same invertible diffusion coefficient $\sigma(t)$ and drifts b^1, b^2, with initial laws ν^1, ν^2. Assume*

(a) $\nu^1 \ll \nu^2$;

(b) for ν^1-almost every x, the stochastic exponential implementing the drift change is a true martingale under the conditional law $\mathrm{Q}^2(\cdot \mid X_0 = x)$ — Novikov's condition for $\theta = \sigma^{-1}(b^1 - b^2)$, imposed conditionally, is one sufficient criterion, Kazamaki's is another. The condition is conditional and not merely Q^2-averaged, because the density factors as $(d\nu^1/d\nu^2)(X_0)$ times a drift factor, and it is the second factor that must have conditional mean one for that product to be a probability density;

(c) $\mathbb{E}_{\mathrm{Q}^1} \int_0^T \|\sigma^{-1}(b^1 - b^2)\|^2 dt < \infty$;

(d) the martingale problem with initial law ν^1 and characteristics (b^1, a), $a = \sigma\sigma^\top$, is well posed, so that the law produced by the change of measure is the law Q^1 named above and not merely some law with drift b^1.

Then

$$\mathrm{KL}\big(\mathrm{Q}^1 \,\|\, \mathrm{Q}^2\big) = \mathrm{KL}\big(\nu^1 \,\|\, \nu^2\big) + \frac{1}{2}\mathbb{E}_{\mathrm{Q}^1}\int_0^T \big\|\sigma(t)^{-1}(b^1 - b^2)(t, X_t)\big\|^2 dt. \tag{3.17}$$

Hypothesis (d) is easy to overlook and is what makes the statement about the two named diffusions rather than about a measure the construction happens to produce; the alternative — defining Q^1 to be the transformed measure — costs nothing.

This is a restatement, for use in this chapter, of Proposition 10.2, which carries the same four hypotheses (a)–(d) in the same conditional form and is proved there; the badge records that, and the historical source is Girsanov's theorem (Theorem 10.1) together with the chain rule for relative entropy. It is restated here because Proposition 3.13 and Proposition 3.24 both rest on it, and every later use — Proposition 7.5 in Chapter 7 and the Schrödinger constructions of Chapter 10 — refers to Proposition 10.2 rather than re-deriving it.

Notes on Proposition 3.12, Proposition 3.13 and Theorem 3.14

Which object the common coefficient really is. As stated, Theorem 3.14 fixes a common invertible σ, which is the convenient setting for the generative constructions of this chapter. The load-bearing object, however, is the *covariance characteristic* $a = \sigma\sigma^{\top}$ and not the square root: by Proposition 2.6 and the remark after Definition 1.32, two square roots of the same a give the same generator, the same martingale problem and the same law, so replacing σ by $-\sigma$ or by σO is no obstruction at all. And the failure, when the characteristics do differ, is *oriented*: by Proposition 10.2, if $a^1 \neq a^2$ on a set of positive $dt \otimes Q^1$ occupation measure then $Q^1 \not\ll Q^2$ and $\mathrm{KL}(Q^1 \,\|\, Q^2) = \infty$, while *mutual* singularity needs a separating condition both measures see. Proposition 10.2 is the single source of truth for all of this and should be quoted in preference to any restatement here. Finiteness of $\mathbb{E}_{Q^1} \int \|\theta\|^2$ is an L^2 statement under Q^1; it neither implies that the stochastic exponential is a true Q^2-martingale nor, by itself, that $Q^1 \ll Q^2$, and calling it "Novikov in disguise" — which requires an exponential moment under Q^2 — would be a substitution of exactly the kind this volume is written to prevent. Where the drift energy alone is wanted as a hypothesis, the correct reference is a finite-energy Girsanov theorem, whose absolute-continuity conclusion must then be quoted with its own assumptions. Proposition 3.13: the hypotheses of Theorem 3.1, finiteness of the path-space relative entropy, and $\delta > 0$.

Where the weighting comes from and what it costs. The factor $\frac{1}{2}g_t^2$ is not a modelling choice; it is what (3.17) produces. Training under the uniform weighting $w \equiv 1$ is legitimate and common, and it optimises a different functional that is not a bound on (3.16); by Theorem 3.20(iii) the two have different minimisers over a restricted class. A model reported as achieving a given likelihood must therefore have been trained, or at least evaluated, under the weighting of (3.16).

Direction of the divergence. Equation (3.16) bounds $\mathrm{KL}(p_0 \,\|\, \hat{p}_0)$ and needs $\mathrm{KL}(p_T \,\|\, \mathcal{N}(0, I))$, which is the direction bounded by Proposition 3.10. Proposition 3.24 below consumes the same direction, so the two compose without an extra hypothesis. What is *not* available is the opposite orientation, $\mathrm{KL}(\mathcal{N}(0, I) \,\|\, p_T)$: it does not follow from Proposition 3.10, and the notes there say what it would cost. Every relative entropy in this chapter is written as an ordered pair and none is silently reversed.

What it does not say. It bounds a divergence between the exact reverse process and the learned one, both simulated exactly. Discretisation is absent, δ is arbitrary and its effect is not bounded here, and nothing is claimed about sample quality under any perceptual criterion.

What does not follow

A diffusion model is not an infinitely deep variational autoencoder. The two share a hierarchical latent-variable structure and a variational bound, and (3.16) is the continuous-time form of that bound. They are not the same model class. The encoder of a diffusion model is fixed by the schedule and has no parameters; the encoder of a variational autoencoder is learned, and the bound is optimised jointly over it. The continuum limit carries the weight $\frac{1}{2}g_T^2$ and a terminal term, neither of which appears in the finite-depth statement. And the discrete-time bound converges to (3.16) only under the schedule regularity of Proposition 3.7. The correspondence is worth stating and is stated; the identity is not available and is not used anywhere in this volume.

One population problem is not one algorithm. By Proposition 3.11 and Proposition 3.12 the explicit, implicit and denoising forms differ by θ-independent constants over any admissible parametric class, so they share their minimisers and their population gradients there — this is (3.15), and it is stronger than an agreement at the optimum. What they do not share is their behaviour as *estimators*: three different sample objectives, three different gradient variances, three different requirements for the integration by parts that produced them, and three different behaviours near $\sigma_T = 0$. Reporting that two methods optimise "the same objective" is therefore correct here and says nothing about which will train.

And this is not the situation of Theorem 3.20. There the comparison is across two parameterisations with two weightings, the identity requires both (3.28) and $\hat{v} = c_T^2 w$, and without them the objectives are genuinely different functions of θ with different minimisers over a restricted class. An identity that holds for every admissible field and an identity that holds under a dictionary are different kinds of statement, and the second does not license the first's conclusion or the reverse.

3.4 Score, Velocity, and Flow Matching

Section 3.3 trained a network to estimate the score of a noised data distribution and used it to run a stochastic differential equation backwards. This section trains a network to estimate a velocity field and uses it to run an ordinary differential equation backwards. The two procedures are usually described as equivalent. They are related by an exact and completely explicit dictionary — but the dictionary holds for a restricted family of interpolating paths, it relates objects and not objectives, and the further statement that the two *training objectives* coincide is true only under two additional conditions that published practice routinely violates.

The section is organised so that each level of the relationship is stated separately and its conditions are visible: Section 3.4.1 constructs the marginal velocity field; Section 3.4.2 proves the exact score-to-velocity dictionary and identifies where it degenerates; Section 3.4.3 proves the equivalence of objectives under two named conditions and exhibits a computable failure when they are dropped; Section 3.4.4 treats optimal-transport couplings and states the one-step sampling result as what it is — a theorem about an oracle.

3.4.1 Conditional paths and the marginal velocity field

Convention

Time direction. In this volume $t = 0$ carries the data and $t = 1$ the reference law, for every construction: the forward process adds noise as t increases, and generation integrates backwards from $t = 1$ to $t = 0$. Much of the flow-matching literature runs t the other way, with data at $t = 1$. Both are correct and the translation is $t \mapsto 1 - t$ together with a sign change on every velocity field; what is not correct is to import a formula written in one convention into a derivation written in the other, which changes the sign of the drift and leaves every marginal law unchanged, so that the error is invisible until samples are drawn.

Path parameters. A *Gaussian conditional path* is specified by $\alpha, \sigma \in C^1([0,1])$ with $\alpha_t > 0$ and $\sigma_t > 0$ on $(0,1)$, $\alpha_0 = 1$, $\sigma_0 = 0$, and $(\alpha_1, \sigma_1) = (0,1)$ in the limit. We write

$$\mathrm{snr}_t := \frac{\alpha_t}{\sigma_t}$$

for the signal-to-noise ratio, which is the quantity every statement below is really about. Variance-preserving, variance-exploding and linear-interpolant paths are the choices $\alpha_t^2 + \sigma_t^2 = 1$, $\alpha_t \equiv 1$, and $(\alpha_t, \sigma_t) = (1-t, t)$ respectively; nothing in this section is particular to any of them.

Fix a Gaussian conditional path and let $Z \sim p_{\mathrm{data}} \in \mathcal{P}_2(\mathbb{R}^d)$ and $\varepsilon \sim \mathcal{N}(0, I)$ be independent. Define

$$X_t := \alpha_t Z + \sigma_t \varepsilon, \qquad t \in [0,1], \tag{3.18}$$

so that the conditional law is $p_t(\cdot \mid z) = \mathcal{N}(\alpha_t z, \sigma_t^2 I)$ and the marginal is $p_t(x) = \int p_t(x \mid z)\, p_{\mathrm{data}}(\mathrm{d}z)$, with $p_0 = p_{\mathrm{data}}$ and $p_1 = \mathcal{N}(0, I)$.

Differentiating (3.18) in t with (Z, ε) held fixed gives the *conditional velocity*

$$u_t(x \mid z, \varepsilon) = \dot{\alpha}_t z + \dot{\sigma}_t \varepsilon, \tag{3.19}$$

which is a random variable, not a field: it is the velocity of one sample path, and it is known in closed form precisely because the path was chosen rather than learned. The object that generates the marginal family is its conditional expectation.

Proposition 3.15 (Marginalisation of the velocity field) ESTABLISHED. *Let p_t and X_t be as above and suppose $p_t \in C^1$ in t with $\int_0^1 \mathbb{E}\,|\dot{\alpha}_t Z + \dot{\sigma}_t \varepsilon|\, \mathrm{d}t < \infty$. Define*

$$u_t(x) := \mathbb{E}\big[\dot{\alpha}_t Z + \dot{\sigma}_t \varepsilon \mid X_t = x\big]. \tag{3.20}$$

Then (p_t, u_t) solves the continuity equation $\partial_t p_t + \nabla \cdot (p_t u_t) = 0$ on $(0,1) \times \mathbb{R}^d$.

If, in addition, u_t is regular enough in x for the ordinary differential equation $\dot{x} = u_t(x)$ to generate a unique flow — local Lipschitz continuity with linear growth suffices, and so does the weaker Sobolev-plus-bounded-divergence condition under which a unique regular Lagrangian flow exists — then that flow transports p_1 to p_0 when integrated from $t = 1$ to $t = 0$. The transport clause is a statement about characteristics and needs spatial regularity; the continuity equation is a statement about distributions and needs none beyond the stated integrability, which is why the two are separated here.

Proof. For each fixed z, the conditional density $p_t(\cdot \mid z)$ is the law of the deterministic-in-(z,ε) curve (3.18) averaged over ε, and satisfies the continuity equation with the field $u_t(x \mid z) := \mathbb{E}[\dot a_t z + \dot\sigma_t \varepsilon \mid X_t = x, Z = z]$ — a direct computation for Gaussians, or the transport of $\mathcal{N}(0, I)$ by $\varepsilon \mapsto a_t z + \sigma_t \varepsilon$. Integrating $\partial_t p_t(x \mid z) + \nabla\cdot(p_t(x \mid z) u_t(x \mid z)) = 0$ against $p_{\text{data}}(dz)$ and exchanging derivative and integral, justified by the stated integrability,

$$\partial_t p_t(x) + \nabla\cdot\int p_t(x \mid z)\, u_t(x \mid z)\, p_{\text{data}}(dz) = 0.$$

By Bayes' rule the integral equals $p_t(x)\,\mathbb{E}[u_t(x \mid Z) \mid X_t = x] = p_t(x) u_t(x)$, and the tower property turns the inner conditioning into (3.20). ∎

Remark 3.16 (Attribution). Proposition 3.15 is the construction that makes flow matching trainable, and it belongs to the modern research literature rather than to this volume; it is tagged ESTABLISHED accordingly, and the source is Lipman et al. [17]; the proof is reproduced because the argument is three lines and the reader should see where the independence of Z and ε is used.

The content of Proposition 3.15 is that an intractable object — a field whose definition involves a conditional expectation under p_{data} — is the regression target of a tractable one. Nothing in it says that this field is a good field to learn, that its integral curves are short, or that the resulting sampler is efficient. Those are separate questions, and Section 3.4.4 shows that the answer to the first two depends on a choice not yet made: the coupling between Z and ε.

3.4.2 The dictionary between score and velocity

Proposition 3.17 (Score–velocity dictionary) DERIVED. *Let $s_t(x) := \nabla \log p_t(x)$ and let u_t be as in (3.20). For every $t \in (0,1)$ and every $x \in \mathbb{R}^d$,*

$$u_t(x) = \lambda_t x + c_t s_t(x), \qquad \lambda_t := \frac{\dot a_t}{a_t}, \qquad c_t = \sigma_t^2 \frac{d}{dt} \log \mathrm{snr}_t. \tag{3.21}$$

The relation is invertible in s_t at time t if and only if $c_t \neq 0$, that is, if and only if the signal-to-noise ratio is strictly monotone at t. The same affine map relates the conditional targets:

$$\dot a_t z + \dot\sigma_t \varepsilon = \lambda_t x + c_t\left(-\frac{\varepsilon}{\sigma_t}\right) \qquad \text{whenever } x = a_t z + \sigma_t \varepsilon. \tag{3.22}$$

Proof. Since $p_t(x \mid z) = \mathcal{N}(a_t z, \sigma_t^2 I)$ we have $\nabla_x \log p_t(x \mid z) = -(x - a_t z)/\sigma_t^2 = -\varepsilon/\sigma_t$, and differentiating $p_t(x) = \int p_t(x \mid z) p_{\text{data}}(dz)$ under the integral sign gives the standard identity

$$s_t(x) = \mathbb{E}\big[\nabla_x \log p_t(x \mid Z) \mid X_t = x\big] = -\frac{1}{\sigma_t}\mathbb{E}[\varepsilon \mid X_t = x]. \tag{3.23}$$

From $x = a_t Z + \sigma_t \varepsilon$, $\mathbb{E}[Z \mid X_t = x] = a_t^{-1}(x - \sigma_t \mathbb{E}[\varepsilon \mid X_t = x]) = a_t^{-1}(x + \sigma_t^2 s_t(x))$. Substituting both conditional expectations into (3.20),

$$u_t(x) = \dot a_t \frac{x + \sigma_t^2 s_t(x)}{a_t} - \dot\sigma_t \sigma_t s_t(x) = \lambda_t x + (\lambda_t \sigma_t^2 - \sigma_t \dot\sigma_t) s_t(x),$$

and $\lambda_t\sigma_t^2 - \sigma_t\dot{\sigma}_t = \sigma_t^2(\dot{a}_t/a_t - \dot{\sigma}_t/\sigma_t) = \sigma_t^2 \frac{d}{dt} \log \mathrm{snr}_t$, which is c_t. Invertibility in s_t is equivalent to $c_t \neq 0$, and since $\sigma_t > 0$ on $(0,1)$ this holds exactly when $\frac{d}{dt} \log \mathrm{snr}_t \neq 0$. For (3.22), substitute $x = a_t z + \sigma_t \varepsilon$ into the right-hand side and use $c_t/\sigma_t = \lambda_t\sigma_t - \dot{\sigma}_t$ to obtain $\dot{a}_t z + \dot{\sigma}_t \varepsilon$. ■

Example 3.18 (The two standard paths). For the Ornstein–Uhlenbeck path $a_t = e^{-t/2}$, $\sigma_t^2 = 1 - e^{-t}$ on a rescaled time interval, one finds $\lambda_t = -\frac{1}{2}$ and $c_t = -\frac{1}{2}$, so $u_t(x) = -\frac{1}{2}x - \frac{1}{2}s_t(x)$ — exactly the probability-flow ODE of Equation (3.6) for $dX = -\frac{1}{2}X dt + dW$, as it must be. For the linear interpolant $a_t = 1 - t$, $\sigma_t = t$ one finds $\lambda_t = -1/(1-t)$ and $c_t = -t/(1-t)$. In both cases snr_t is strictly decreasing on $(0,1)$ and the dictionary is invertible throughout; it degenerates only in the limits $t \downarrow 0$, where $\sigma_t \to 0$ and the score blows up, and $t \uparrow 1$, where $a_t \to 0$.

Notes on Proposition 3.15 and Proposition 3.17

Where it fails. Everything in (3.21) is a consequence of the conditional density being Gaussian with mean linear in z. For a non-Gaussian conditional path — a Brownian bridge with a non-Gaussian endpoint law, a diffusion bridge, a stochastic interpolant with a non-Gaussian latent, a path on a manifold, or a discrete-state process — the score is in general not an affine function of the velocity, and no dictionary of this kind exists. The correspondence between the two objects survives in a weaker form, as two parameterisations of the same continuity equation; the closed-form conversion does not. Section 3.7 treats the general interpolant, and Section 3.8 the discrete case, where the score is replaced by a ratio of rates and the analogy with (3.21) is only structural.

Within the Gaussian family, the dictionary degenerates wherever snr_t is stationary. A path with a_t/σ_t constant on an interval is a pure rescaling on that interval, and on it the velocity field carries no information about the score at all. This is not a pathology to be dismissed: it is the reason noise schedules are chosen to make $\log \mathrm{snr}_t$ strictly monotone, and it is what makes the conversion numerically ill-conditioned near the endpoints, where $c_t \to 0$ or $c_t \to \infty$.

3.4.3 Three levels of equivalence between the objectives

Proposition 3.17 relates two *fields*. Training relates two *objectives*, and the step from one to the other is where the usual claim of equivalence is made too quickly. We need one standard lemma.

Lemma 3.19 (Least squares identifies the conditional expectation) CLASSICAL. *Let (X, Y) be square-integrable with values in $\mathbb{R}^d \times \mathbb{R}^d$. For every measurable g with $\mathbb{E}\,|g(X)|^2 < \infty$,*

$$\mathbb{E}\,\|g(X) - Y\|^2 = \mathbb{E}\,\|g(X) - \mathbb{E}[Y \mid X]\|^2 + \mathbb{E}\,\|Y - \mathbb{E}[Y \mid X]\|^2, \tag{3.24}$$

so the minimiser over all measurable g is $g^*(x) = \mathbb{E}[Y \mid X = x]$, p_X-almost surely, and the second term does not depend on g.

Proof. Expand the square and use $\mathbb{E}\big[\langle g(X) - \mathbb{E}[Y \mid X], Y - \mathbb{E}[Y \mid X]\rangle\big] = 0$, which holds by the tower property because the first factor is $\sigma(X)$-measurable. ■

Define, for positive weights $w, \tilde{w} \in C([0,1])$, the conditional flow matching and denoising score matching objectives

$$\mathcal{L}^{w}_{\mathrm{CFM}}(\theta) := \int_0^1 w(t)\,\mathbb{E}\left\| v_\theta(t, X_t) - \dot{\alpha}_t Z - \dot{\sigma}_t \varepsilon \right\|^2 \mathrm{d}t, \tag{3.25}$$

$$\mathcal{L}^{\tilde{w}}_{\mathrm{DSM}}(\theta) := \int_0^1 \tilde{w}(t)\,\mathbb{E}\left\| s_\theta(t, X_t) + \frac{\varepsilon}{\sigma_t} \right\|^2 \mathrm{d}t, \tag{3.26}$$

the expectations being over $Z \sim p_{\text{data}}$ and $\varepsilon \sim \mathcal{N}(0, 1)$ independent, with X_t as in (3.18).

Theorem 3.20 (Three levels of equivalence) [illegible]. *Fix a Gaussian conditional path with $c_t \neq 0$ on $(0, 1)$, and assume the time-integrated targets are square integrable:*

$$\int w(t)\,\mathbb{E}\left\| \dot{\alpha}_t Z + \dot{\sigma}_t \varepsilon \right\|^2 \mathrm{d}t < \infty, \qquad \int \tilde{w}(t)\,\mathbb{E}\left\| \varepsilon / \sigma_t \right\|^2 \mathrm{d}t < \infty. \tag{3.27}$$

The second condition is the binding one and is not implied by positivity and continuity of $\tilde{w}$: for the standard paths $\sigma_0 = 0$, so $\mathbb{E}\|\varepsilon/\sigma_t\|^2 = d/\sigma_t^2$ blows up at the endpoint and $\int \tilde{w}/\sigma_t^2$ may diverge. Equivalently, and this is what a practical statement should do, restrict all four objectives to a truncated interval $[\delta, 1 - \delta]$ away from the singular endpoints.

(i) (Same object.) *For any positive weights, the minimisers of (3.25) and (3.26) over all measurable fields are $v_t^* = u_t$ and $s_t^* = s_t$ respectively, and these satisfy the dictionary (3.21). The two objectives therefore identify the same object in two coordinate systems.*

(ii) (Same function of θ.) *Suppose in addition that the two parameterisations are tied by*

$$v_\theta(t, x) = \lambda_t x + c_t s_\theta(t, x) \qquad \text{for all } \theta, \tag{3.28}$$

and that the weights satisfy $\tilde{w}(t) = c_t^2 w(t)$. Then

$$\mathcal{L}^{w}_{\mathrm{CFM}}(\theta) = \mathcal{L}^{\tilde{w}}_{\mathrm{DSM}}(\theta) \qquad \text{for every } \theta, \tag{3.29}$$

identically — hence the same gradients, the same optimisation trajectory, and the same minimiser over any restricted class.

(iii) (Otherwise, no guarantee.) *If either (3.28) or $\tilde{w} = c_t^2 w$ fails, (3.29) is not guaranteed, and there exist restricted classes and admissible weightings for which the two objectives, and their minimisers, differ. Two admissible weightings applied to the same objective over the same restricted class $\{v_\theta\}$ may likewise have different minimisers; Example 3.21 computes a case in which they differ by a factor of more than three. If in addition the class $\{s_\theta\}$ is rich enough to separate two times at which $\tilde{w} \neq c_t^2 w$ — so that the map $\theta \mapsto \mathbb{E}\|s_\theta(t, X_t) + \varepsilon/\sigma_t\|^2$ is not carried by a single time — then failure of the weight condition forces (3.29) to fail. No such separation is available for a degenerate class, and for a singleton class the two objectives can agree by coincidence.*

Proof. (i) Apply Lemma 3.19 at each t with $X = X_t$ and Y the conditional target. For (3.25) the target is $\dot{\alpha}_t Z + \dot{\sigma}_t \varepsilon$ and its conditional expectation is u_t by (3.20); for (3.26) the target is $-\varepsilon/\sigma_t$ and its conditional expectation is s_t by (3.23). The positive weight $w(t)$ does not affect the pointwise-in-t minimiser. Proposition 3.17 relates the two.

(ii) By (3.28) and (3.22), for every θ, t and every realisation of (Z, ε),

$$v_\theta(t, X_t) - (\dot{a}_t Z + \dot{\sigma}_t \varepsilon) = c_t\left(s_\theta(t, X_t) + \frac{\varepsilon}{\sigma_t}\right),$$

the two affine maps cancelling term by term. Squaring gives $\|\cdot\|^2 = c_t^2\|\cdot\|^2$ pointwise; multiplying by $w(t)$, integrating, and using $\tilde{w} = c_t^2 w$ gives (3.29).

(iii) With the tie (3.28) in force, the computation in (ii) shows $\mathcal{L}^{w}_{\mathrm{CFM}}(\theta) = \mathcal{L}^{c_t^2 w}_{\mathrm{DSM}}(\theta)$, so (3.29) with a different $\tilde{w}$ would force $\int(\tilde{w} - c_t^2 w)\,\mathbb{E}\|s_\theta + \varepsilon/\sigma_t\|^2 \mathrm{d}t = 0$ for all θ. Under the stated separation hypothesis this is impossible, which is the conditional necessity claim; without it the integral can vanish accidentally — a class containing a single θ, or one whose error profile in t happens to be orthogonal to $\tilde{w} - c_t^2 w$, produces no contradiction — so no unconditional necessity is asserted. Without the tie, the two objectives are functions of different maps and (3.29) has no meaning as an identity in θ; the existence claim is then witnessed by Example 3.21. ∎

Interpretation

Read the theorem as a dictionary with conditions, not as the slogan that flow matching and diffusion are the same method. Three separate things are in play. The population target *fields* are related pointwise, always: a velocity and a score determine each other through (3.28). Two *parameterised objectives* become the same function of θ only when the parameterisations are tied *and* the time weights match. And *over a restricted class*, without both conditions, the two can have different minimisers — Example 3.21 computes a case where they differ by a factor of 3.6.

The practical reading is that a change of parameterisation is free and a change of weighting is not. What a weighting does is decide where in t the approximation error is allowed to sit, and two schemes that agree on the field can disagree on that.

Scope of Theorem 3.20

Assumed. A Gaussian conditional path with $c_t \neq 0$ on $(0, 1)$; positive continuous weights; square-integrability of the conditional targets at each t, so that Lemma 3.19 applies there; and, for the statements about the time-integrated objectives, (3.27) or a truncation of the time interval. The last is a real hypothesis and not a formality: without it both sides of (3.29) can equal $+\infty$ for every θ, in which case "the same function of θ" and "the same gradients" assert nothing. An alternative repair, used when the divergent part is θ-independent, is to compare *excess* risks, subtracting the conditional-variance constant of Proposition 3.12 from each objective before integrating; that version needs no truncation and is what an implementation with a σ_t^{-2} weight is implicitly optimising.

Where each part fails. Part (i) is insensitive to the weights; the two least-squares problems identify u_t and s_t separately whatever c_t does, and what fails where $c_t = 0$ — that is, where snr_t is stationary — is the invertible *dictionary* between them, not the identification. Part (ii) fails if either condition is dropped, and the two are independent: matching the weights without tying the parameterisations gives two functions of different maps, and tying the parameterisations without matching the weights gives two different functions of the same θ. Part (iii) is an existence statement about restricted classes, plus a necessity claim that is *conditional on a separation hypothesis about the class*: it is not asserted that every failure of the two conditions produces a failure of (3.29), because a sufficiently poor class can make the two objectives agree for reasons that have nothing to do with the dictionary. Part (iii) also says nothing about how large the discrepancy is in any particular case; Example 3.21 supplies one case and no general bound is claimed.

What this does not say. Nothing here compares the two estimators at finite sample size, or their variance, or their behaviour near $t = 0$ and $t = 1$, where the dictionary is ill-conditioned in opposite directions.

What would falsify it. For (iii), a path, a tied parameterisation and weights with $\tilde{w} = c_1^2 w$ for which the two objectives differ at some θ; the proof is an identity, so such an example would locate an error in the cancellation.

Example 3.21 (Two admissible weightings, one restricted class, two answers). Let $d = 1$, $p_{\text{data}} = \mathcal{N}(0, \tau^2)$ and take the linear interpolant $a_t = 1 - t$, $\sigma_t = t$. Then $p_t = \mathcal{N}(0, m_t)$ with $m_t = (1-t)^2\tau^2 + t^2$, and by (3.20) and joint Gaussianity the exact marginal velocity is linear,

$$u_t(x) = \beta_t x, \qquad \beta_t = \frac{\dot{a}_t a_t \tau^2 + \dot{\sigma}_t \sigma_t}{m_t} = \frac{\dot{m}_t}{2m_t}.$$

Restrict the model class to time-independent linear fields $v_\theta(t, x) = \theta x$ — a caricature, chosen because it can be solved in closed form, and because every real network class is restricted in some way. Minimising (3.25) over θ gives, by Lemma 3.19 and $\mathbb{E}[X_t u_t(X_t)] = \beta_t m_t$, $\mathbb{E}[X_t^2] = m_t$,

$$\theta_w^* = \frac{\int_0^1 w(t)\,\beta_t m_t \,dt}{\int_0^1 w(t)\, m_t \,dt} = \frac{\frac{1}{2}\int_0^1 w(t)\, \dot{m}_t \,dt}{\int_0^1 w(t)\, m_t \,dt}; \tag{3.30}$$

a $w m_t$-weighted average of β_t. Take $\tau^2 = 4$. For the uniform weighting $w \equiv 1$,

$$\theta_1^* = \frac{\frac{1}{2}(m_1 - m_0)}{\int_0^1 m_t \,dt} = \frac{\frac{1}{2}(1-4)}{\frac{1}{3}(4+1)} = -\frac{9}{10},$$

while for the equally standard choice $w(t) = \sigma_t^2 = t^2$, using $\int_0^1 t^2 \dot{m}_t \,dt = \frac{1}{2} - \frac{\tau^2}{6}$ and $\int_0^1 t^2 m_t \,dt = \frac{\tau^2}{30} + \frac{1}{5}$,

$$\theta_{t^2}^* = \frac{\frac{1}{4} - \frac{4}{12}}{\frac{4}{30} + \frac{1}{5}} = \frac{-\frac{1}{12}}{\frac{1}{3}} = -\frac{1}{4}.$$

The two minimisers differ by a factor of 3.6, and the induced samplers differ correspondingly. Both weightings are admissible, both objectives have the same population minimiser over the unrestricted class by Theorem 3.20(i), and neither of those facts constrains the answer once the class is restricted.

What does not follow

"Flow matching and diffusion models are equivalent" is not a statement that can be checked. Three different statements travel under that sentence and they have different truth values. *The fields are related* by (3.21): exact, for Gaussian conditional paths, whenever snr_t is strictly monotone. *The objectives are the same function of θ* under (3.28) and matched weights: exact, under two conditions that must be checked and are frequently not met. *The trained models are two different estimators*, differing in their parameterisation, their weighting, their variance, and their behaviour near $t = 0$ and $t = 1$; no result here compares them. Any claim of equivalence in this volume names which of the three it means.

Same marginals is not same sampler. The probability-flow ODE and the reverse-time stochastic differential equation have identical time marginals when the score is exact. They do not have identical behaviour when it is not: the stochastic sampler re-injects noise at every step and contracts certain errors, while the deterministic one transports them. Statements that the two "give the same samples" hold for the exact field only, which is the one case that never occurs in practice.

A weighting is not a detail. By (3.30) the weight selects which times the fitted field is accurate at, and under a restricted class that choice moves the answer. Comparisons of published results across different weightings therefore measure the weighting as much as the method, and a reported improvement accompanied by a change of weighting is not attributable.

The dictionary is not a change of variables in the loss landscape. (3.28) ties the parameterisations by an explicit t-dependent affine map. If the network instead predicts the velocity directly, with its own architecture, no such tie exists, and (ii) of Theorem 3.20 does not apply — even if the weights happen to match.

3.4.4 Optimal-transport couplings and one-step sampling

Nothing so far constrained the joint law of (Z, ε): they were taken independent, and that choice is what makes u_t curved. Replacing the independent coupling by another coupling of p_{data} and $\mathcal{N}(0, I)$ changes the conditional paths, changes the marginal velocity field, and leaves the endpoint marginals untouched. The extreme choice is the optimal one.

Definition 3.22 (Displacement interpolant). Let $\mu, \nu \in \mathcal{P}_2(\mathbb{R}^d)$ with μ absolutely continuous, and let $T = \nabla\phi$ with ϕ convex be the Brenier map pushing μ to ν. The *displacement interpolant* is the family $\mu_t := (T_t)_{\#}\mu$ with $T_t(z) := (1-t)z + tT(z)$, $t \in [0, 1]$.

Proposition 3.23 (Exact one-step sampling under an optimal-transport oracle) MODEL-DEPENDENT. *In the setting of Definition 3.22, the maps T_t are injective for $t \in [0, 1)$, the family (μ_t) solves the continuity equation with the marginal velocity field*

$$u_t(x) = T(z) - z \qquad \text{for the unique } z \text{ with } x = T_t(z), \tag{3.31}$$

and this field is constant along its own integral curves. Consequently, for every starting time $t \in [0, 1)$ and every $t' \in [0, 1]$, the single explicit Euler step $x \mapsto x + (t' - t)\, u_t(x)$ maps μ_t exactly onto $\mu_{t'}$.

No claim is made for a step starting at $t = 1$. There $T_1 = T$ need not be injective — ν may carry atoms, and a Brenier map may collapse sets of positive measure onto them — so (3.31) need not define a field at all. A one-step map from $t = 1$ to $t = 0$ requires a separately available inverse transport map, which exists when ν too is absolutely continuous and is then the Brenier map from ν to μ; that is a second oracle, not a consequence of the first.

Proof. $T_t = \nabla\phi_t$ with $\phi_t(z) = (1-t)\frac{1}{2}|z|^2 + t\phi(z)$, which is strictly convex for $t \in [0, 1)$ because the first summand is; the gradient of a strictly convex function is injective, giving the first claim and the well-posedness of (3.31). Since $t \mapsto T_t(z)$ is affine with velocity $T(z) - z$ independent of t, the curve $t \mapsto T_t(z)$ is an integral curve of the field defined by (3.31), and the family (μ_t) is its push-forward, hence solves the continuity equation. Along that integral curve the velocity is constant, so for $t < 1$ — where u_t is defined — the exact flow map between times t and t' is $x \mapsto x + (t' - t)u_t(x)$, which is the Euler step. At $t = 1$ the argument has no starting point: strict convexity of ψ_t was used for $t < 1$ only, and nothing in the hypotheses makes T injective. ∎

The model behind Proposition 3.23

The model. An oracle supplies (a) the *exact* optimal-transport coupling of the two full marginals, with a *forward* Brenier map T from μ that exists and is unique — which requires μ absolutely continuous and finite second moments — and (b) the *exact* marginal velocity field (3.31), which is available at $t \in [0, 1)$. A step launched from the ν endpoint needs a *third* oracle, the inverse map, and with it the additional hypothesis that ν is absolutely continuous. No estimation, no minibatching, no parametric approximation, and exact arithmetic.

Exact inside it. Everything stated: injectivity of T_t for $t < 1$, constancy of the velocity along integral curves, and exactness of the single Euler step from any $t \in [0, 1)$ to any $t' \in [0, 1]$, with no error term. The proof is two lines and there is no inequality in it. The generation direction — a single step launched at the reference endpoint — is not inside the model as stated.

What does not transfer. Four things, each of which alone destroys exactness. *(1) Minibatch couplings.* The optimal coupling of two empirical measures of size B is not an unbiased estimator of the optimal coupling of the marginals, and the paths it induces are not displacement interpolants of p_{data} and $\mathcal{N}(0, I)$; the induced marginal field is curved, with curvature that decreases in B at a rate that is itself dimension-dependent. *(2) A learned field.* $v_\theta \neq u_t$, and a one-step scheme has no mechanism for correcting the difference: the sampling error is the approximation error of the field, undamped. Straightness removes the discretisation error and nothing else. *(3) Straightness is a property of the exact field only.* A field fitted under a restricted class and a weighting is not straight even when its target is (Example 3.21 is the one-dimensional shadow of this), so the premise of the one-step argument is not available at inference time. *(4) Non-uniqueness, and the endpoint.* Without absolute continuity of μ the Brenier map need not exist, the interpolant is defined only through a plan, and (3.31) is not a function of x. Symmetrically, at $t = 1$ the field is undefined unless ν is absolutely continuous and the inverse map is supplied: the oracle that makes the *forward* step exact does not by itself make the *generation* step exact, and a one-step sampler claimed on this basis is quoting the proposition outside its domain.

Operational reading

The dictionary (3.21) is worth using at inference. A model trained on either objective can be evaluated in the other coordinate system at no cost, which means the choice between a stochastic sampler — which needs s_t — and a deterministic one — which needs u_t — may be made after training rather than before it, for any Gaussian conditional path with $c_t \neq 0$.

That is one step beyond the theorem, and this box stops there. Which sampler produces better samples at a given number of function evaluations depends on how the field's error is distributed over t, on the solver, and on the data, and is settled in Section 3.6 against a table of assumptions rather than here. In particular, nothing in this section supports converting a published step count obtained under one weighting into a claim about another.

The section contributes three rows to the Equivalence Atlas of Chapter 11. They are written now, in the form in which they will be audited later.

Atlas entries generated by this section

R3. Marginal velocity field = affine image of the score. EXACT

Exact conditions. Gaussian conditional path (3.18); t with $\sigma_t > 0$ and $\frac{d}{dt}\log \mathrm{snr}_t \neq 0$.

What goes wrong outside them. For non-Gaussian interpolants, bridges, manifold-valued and discrete-state paths there is no affine dictionary; where snr_t is stationary the velocity carries no information about the score.

R4. Conditional flow matching objective = denoising score matching objective. EXACT

Exact conditions. The parameterisations tied by (3.28) *and* the weights related by $\tilde{w} = c_t^2 w$. Both, not either. Also the time-integrated square-integrability (3.27), or a truncation away from the endpoints.

What goes wrong outside them. Equality is no longer guaranteed: there *exist* restricted classes for which the two become different functions of θ with different minimisers, and Example 3.21 is one explicit witness, where two admissible weightings give answers differing by a factor of 3.6. The gap is not asymptotically negligible. A degenerate class — a singleton, or one whose error profile in t happens to be orthogonal to $\tilde{w} - c_t^2 w$ — may still give equal objectives by coincidence, so no universal separation is claimed.

R5. Optimal-transport flow matching = one-step exact sampling. ANALOGY

Exact conditions. None available in practice. The exact statement is Proposition 3.23, which is tagged MODEL-DEPENDENT and assumes an oracle for both the coupling and the field.

What goes wrong outside them. Minibatch coupling error, field approximation error, and loss of straightness under a restricted class; only the discretisation error is removed by straightness, and it is typically not the dominant one.

3.5 End-to-End Error

What a user runs is not (3.3). It is a discretisation of (3.5) with a learned score, started from a reference law rather than from p_T, and stopped at $\delta > 0$. Three errors separate the output from p_{data}, and they have three different sources, three different sets of hypotheses, and three different literatures. This section states them separately and gives the one decomposition that is exact.

3.5.1 The decomposition

Proposition 3.24 (Exact two-term decomposition, exact simulation) DERIVED. *Let $\hat{\mathbb{Q}}$ be the path law of (3.3) with s_θ in place of $\nabla \log p_t$, started from $\mathcal{N}(0, I)$, simulated exactly on $[\delta, T]$, and let $\hat{p}_\delta$ be its law at time δ. Write*

$$\varepsilon_{\mathrm{score}}^2 := \int_\delta^T g_t^2\, \mathbb{E}_{p_t} \left\| s_\theta(t, \cdot) - \nabla \log p_t \right\|^2 dt. \tag{3.32}$$

Then, under the hypotheses of Theorem 3.14,

$$\mathrm{KL}(p_\delta \,\|\, \hat{p}_\delta) \;\le\; \underbrace{\mathrm{KL}(p_T \,\|\, \mathcal{N}(0,I))}_{\text{terminal mismatch}} + \underbrace{\tfrac{1}{2}\,\varepsilon_{\text{score}}^2}_{\text{score error}}, \tag{3.33}$$

and consequently, by Pinsker's inequality, $\mathrm{TV}(p_\delta, \hat{p}_\delta) \le \left(\tfrac{1}{2}\mathrm{KL}(p_\delta \,\|\, \hat{p}_\delta)\right)^{1/2}$.

Proof. This is Proposition 3.13 with the notation (3.32); the two terms are the two on the right of (3.17), and the passage from path laws to time-δ marginals is the data-processing inequality for relative entropy. Pinsker is standard. ∎

Two things about (3.33) deserve emphasis before the third term is added. The score error enters through the weight g_{T-t}^2, the same weight as in Proposition 3.13, so the quantity a bound needs is exactly the quantity the likelihood-weighted objective trains. And no convexity, log-concavity or isoperimetric hypothesis on p_{data} appears anywhere: the *path-space bound* does not assume a log-Sobolev inequality or log-concavity, which is the structural reason it does not inherit the guarantee-degradation of Example 2.29. That is a statement about which hypotheses (3.33) consumes, and not a claim that multimodality is harmless for generation: a barrier can still enter through the score-estimation error $\varepsilon_{\text{score}}$, through the Hessian constants of Remark 3.25, through the terminal mismatch, and through the stiffness of the reverse solver. What is removed is one specific dependence, not the difficulty.

Remark 3.25 (Literature note: discretisation of the reverse process). *This is a literature note, not a claim of this volume.* It carries no status badge, and nothing downstream may cite it as a theorem; the reason is given after the statement and is worth reading before the statement.

For $\hat{p}_\delta$ the law produced by an exponential-integrator discretisation of (3.3) with N steps on $[\delta, T]$, under smoothness hypotheses on s_θ and a schedule with log snr Lipschitz, results of the form

$$\mathrm{KL}(\hat{p}_\delta \,\|\, \tilde{p}_\delta) \;\lesssim\; \frac{\Theta}{N}, \qquad \Theta = \Theta\big(d,\ \text{schedule},\ \|\nabla^2 \log p_t\|\big), \tag{3.34}$$

are available in the literature, typically with Θ linear in d.

Why this is not stated as a theorem here. Equation (3.34) as displayed is schematic: it is the shape shared by several results rather than any one of them, and the shared shape is not the intersection of their hypotheses. A statement of theorem strength must fix the integrator and its grid; the truncation δ; the norm in which the score error and the score's regularity are measured; the regularity assumed of p_{data}; the divergence in which the error is reported, since $\mathrm{KL}(\cdot\,\|\,\cdot)$, TV and W_2 results are not interchangeable; and the exact dimension dependence with its constant. Different sources control different ones of these. Writing the common shape and attaching a badge to it would assert a theorem no single source proves — which is the synthesis error this volume's apparatus exists to prevent, and it would be incoherent to commit it while displaying a badge that promises one primary source per claim.

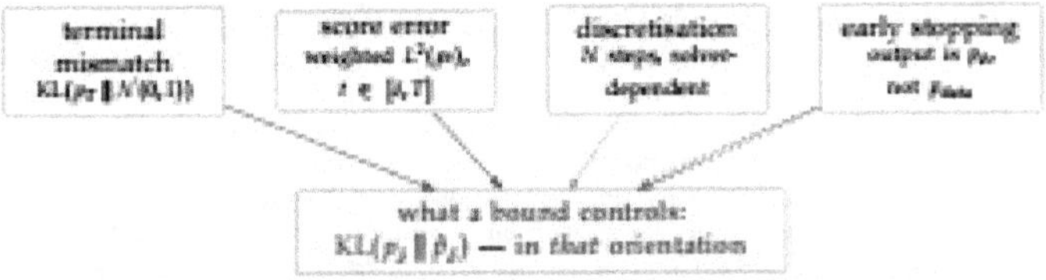

Figure 3.2: The four sources of error in score-based generation, and what a bound on them controls. Solid arrows mark the two terms combined exactly by (3.33); the plain arrow marks a term quoted from the literature (Remark 3.25); the jagged arrow marks the one that no bound removes, because early stopping changes the target rather than approximating it. What this figure does not imply: that the four add up. They have different hypotheses and different orientations, and this volume states no end-to-end theorem combining all four (Table 3.1).

Table 3.1: The three error terms. They are three results, not one: each row has its own hypotheses and its own source, and the last column is what the row does not cover. A single end-to-end statement would have to merge three hypothesis sets, and no such merge is performed in this volume.

Term	Controlled by	Needs	Does not cover
Terminal mismatch	The schedule: $\frac{1}{2}\alpha_T^2 m_2 + O(d\alpha_T^4)$ (Proposition 3.10)	Finite second moment of p_{data}	KL($\mathcal{N}(0, I)$ ‖ p_T). That opposite orientation is what a reverse-oriented marginal comparison would need; it requires a lower bound on the density of p_T and is not supplied here. The orientation (3.33) actually consumes is the one in this row
Score error	$\varepsilon_{\text{score}}$, weighted by g_t^2 (Proposition 3.24)	Finite path-space relative entropy; no convexity	How $\varepsilon_{\text{score}}$ is achieved: that is a statistical question about estimation, not an analytic one
Discretisation	N, the solver, and $\nabla^2 \log p_t$ (Remark 3.25, a literature note)	Smoothness of s_θ and a regular schedule	Any statement at N small, where the asymptotic constant is not the operative one
Early stopping	δ, through the tail of $\mathbb{E}_{p_\delta} \lvert\nabla \log p_\delta\rvert^2$	Nothing beyond (3.2) for $t \ge \delta$	The manifold case, where $\delta \downarrow 0$ is not available at all (Theorem 3.1, scope note)

Notes on Proposition 3.24, Remark 3.25 and Table 3.1

The direction gap. Equation (3.33) consumes KL(p_T ‖ $\mathcal{N}(0, I)$) and Proposition 3.10 supplies exactly that, so the two compose. The reverse composition — bounding KL($\mathcal{N}(0, I)$ ‖ p_T), which is what a bound on KL($\hat{p}_\delta$ ‖ p_δ) would need — does not follow from Proposition 3.10 and requires a lower bound on the density of p_T. Which direction a downstream result needs is stated wherever one is used, and the two are never silently exchanged.

> **What the three terms are not.** They are not additive contributions to one theorem. Adding them requires that the hypotheses of the three rows hold simultaneously, and the third row's smoothness hypothesis is in tension with the first row's tolerance of a rough p_{data}: a data distribution on a manifold makes $\nabla^2 \log p_t$ blow up as $t \downarrow 0$, so Θ in (3.34) degrades exactly where ε_{score} is hardest to make small. The three terms interact through δ, and optimising over δ is a trade-off, not a sum.
>
> **What "optimal" can and cannot refer to here.** Minimax optimality statements in this area concern the statistical problem of estimating $\nabla \log p_t$ from n samples, under smoothness or manifold assumptions on p_{data}. They are results about ε_{score} — the second row alone — and they do not become statements about the sampler by substitution into (3.33), because that substitution requires the estimator to be the one being run and the other two rows to hold at the same time. This volume states no end-to-end optimality claim.

The manifold hypothesis does not help uniformly. It reduces the effective dimension of the estimation problem, improving the achievable ε_{score}, and it destroys (3.2) at $t = 0$, forcing $\delta > 0$ and inflating the constant in (3.34). Reporting the first effect without the second is reporting half of a trade-off.

A bound in KL(‖) is not a bound on sample quality. Equation (3.33) and Pinsker give total variation. Total variation is not Fréchet inception distance, is not perceptual quality, and is not controlled by any metric reported in the applied literature. Nothing in this section licenses a numerical comparison with a reported benchmark.

ε_{score} is not the training loss. It is the weighted L^2 error against the *true* score, in expectation under the *true* p_t. The training loss is an empirical average of a different functional, differing by the constant C_t of (3.14), by the choice of weighting, and by the gap between empirical and population risk. A reported training loss bounds none of the three terms of Table 3.1.

3.6 Samplers and Solvers

A sampler is a discretisation of a member of the family (3.5). The design space has three axes and the literature's headline numbers move along all three at once, which is why they are not comparable as stated.

Proposition 3.26 (Exponential integrators, and why they are the natural scheme) DERIVED. *Let the reverse equation be written in semi-linear form* $dY_s = \left[A(s)Y_s + h(s, Y_s)\right]ds + \ldots$, *where* A *collects the part linear in* Y *coming from* $f(t,x) = f(t)x$. *Then over a step* $[s, s']$ *the linear part can be integrated exactly:*

$$Y_{s'} = \frac{a_{t'}}{a_t} Y_s + \int_s^{s'} \frac{a_{t'}}{a_r} h(r, Y_r)\, dr + \ldots, \qquad t = T - s,\ t' = T - s', \tag{3.35}$$

and freezing h *at the left endpoint gives a scheme whose error involves only the variation of* h, *not of the stiff linear part. For* $\lambda = 0$ *and* h *frozen this is the deterministic scheme usually written in terms of* $\log \mathrm{snr}$ *as the integration variable.*

Proof. Multiply the equation by the integrating factor a_t^{-1} and apply Theorem 1.33 exactly as in the proof of Proposition 3.7; the linear term cancels and integrating the result over $[s, s']$ gives (3.35). Freezing h leaves an error $\int_s^{s'} \frac{a_{t'}}{a_r}\left(h(r, Y_r) - h(s, Y_s)\right)dr$, which is controlled by the modulus of continuity of h alone. ∎

What it buys. Exactness on the linear part, hence a scheme whose step-size restriction is set by the score and not by the schedule. That is why step counts fall sharply when such a scheme replaces Euler–Maruyama, and it is a statement about the solver, obtained without any change to the model.

What it does not buy. Nothing about the score's own smoothness, which sets the remaining error; nothing at $\lambda > 0$ beyond the same treatment of the drift; and nothing about the regime in which h varies fast, near $t = \delta$, where the frozen-endpoint approximation is worst exactly where the score is largest.

Table 3.2: Solvers on three axes. The three columns move independently, and a reported reduction in the first is uninformative without the other two. "Bias" means the error that does not vanish as the number of steps grows at fixed model.

Family	**Evaluations**	**Bias at fixed model**	**Extra training cost**
Euler–Maruyama on (3.3)	Many; step limited by the schedule	Vanishes with N at the order recorded in Remark 3.25	None
Exponential integrator, $\lambda \in [0, 1]$	Fewer; step limited by the score (Proposition 3.26)	Vanishes with N; smaller constant	None
Higher-order and predictor–corrector	Fewer still, at more evaluations per step	Vanishes faster, under stronger smoothness of s_θ	None
Distillation and flow maps	Very few, by construction	Does not vanish with N: the student is a different model	A second training run, comparable to the first

What does not follow

A step count is not a property of a method. Reported reductions from several hundred evaluations to fewer than ten are obtained by moving along different columns of Table 3.2 — some by a better solver, some by distilling into a different model, some by changing the guidance scale, which changes the target. The three are not comparable, and a comparison that holds the model fixed is comparing solvers while a comparison that permits distillation is comparing models. Any figure of the form "N steps" carries an implicit choice of all three columns and of an error metric, and this volume states none.

Distillation moves the bias, it does not remove it. The last row's bias does not vanish with N because the object being simulated is no longer the reverse process of (3.5). Its error is the distance between the student and the teacher, which is a training error and belongs in the second row of Table 3.1, not the third. Reporting it as a solver improvement puts it in the wrong row.

3.7 General Interpolants

Section 3.4 took the conditional path Gaussian. The construction survives without that, and the dictionary does not.

Definition 3.27 (Stochastic interpolant). Let (Z_0, Z_1) be a coupling of p_{data} and a reference law, let $I : [0,1] \times \mathbb{R}^d \times \mathbb{R}^d \to \mathbb{R}^d$ be C^1 in t with $I(0, z_0, z_1) = z_0$ and $I(1, z_0, z_1) = z_1$, and let $\gamma \in C^1([0,1])$ with $\gamma_0 = \gamma_1 = 0$, $\gamma > 0$ on $(0,1)$. The *stochastic interpolant* is $X_t = I(t, Z_0, Z_1) + \gamma_t \varepsilon$ with $\varepsilon \sim \mathcal{N}(0, I)$ independent.

Proposition 3.28 (What survives and what does not) DERIVED. *For a stochastic interpolant with* $\mathbb{E}\,|\partial_t I| < \infty$*:*

(i) the marginalisation of Proposition 3.15 holds verbatim, with $u_t(x) = \mathbb{E}[\partial_t I(t, Z_0, Z_1) + \dot{\gamma}_t \varepsilon \mid X_t = x]$ *generating* p_t*;*

(ii) if $\gamma > 0$ *then* $\nabla \log p_t(x) = -\gamma_t^{-1} \mathbb{E}[\varepsilon \mid X_t = x]$*, so the score is still a conditional expectation of the noise;*

(iii) there is in general no function Φ *with* $u_t(x) = \Phi(t, x, \nabla \log p_t(x))$ *affine in its last argument. The affine dictionary (3.21) is a consequence of* I *being linear in* (z_0, z_1) *with deterministic coefficients, and fails as soon as it is not.*

Proof. (i) The proof of Proposition 3.15 used only that the conditional law solves a continuity equation with a known conditional velocity and that the tower property applies; neither uses Gaussianity of the conditional law in x, only that X_t is a known function of (Z_0, Z_1, ε). (ii) is (3.23) with $\dot{a}_t z_0$ replaced by $I(t, Z_0, Z_1)$, which changes nothing in the computation. (iii) By (i) and (ii), $u_t(x)$ is determined by $\mathbb{E}[\partial_t I \mid X_t = x]$ and $\mathbb{E}[\varepsilon \mid X_t = x]$, and in the Gaussian case of Section 3.4 the first is an affine function of the second because $\partial_t I = \dot{a}_t Z_0$ and Z_0 is recovered from x and ε by a fixed linear map. For nonlinear I no such recovery exists: take $d = 1$, Z_1 independent of Z_0, and $I(t, z_0, z_1) = (1-t)z_0 + tz_1 + t(1-t)z_0 z_1$; then $\partial_t I$ contains the product $Z_0 Z_1$, whose conditional expectation given X_t is not a function of $\mathbb{E}[\varepsilon \mid X_t]$ and x alone. ∎

What this means for the atlas. Row R3 of Chapter 11 is tagged EXACT with the Gaussian conditional path in its conditions column. Part (iii) is the entry in its failure column, made precise: the relation is exact under a hypothesis that is a property of the interpolant, not of the framework, and outside it the two objects remain two parameterisations of one continuity equation and nothing more.

Not addressed here. Bridges with both endpoints constrained — where I is not chosen but determined by a conditioning — are Chapter 10; nothing in this section applies to them, because Definition 3.27 takes the coupling as given.

3.8 Discrete State Spaces

The constructions above are written for $\mathbb{R}^d$, and the objects a language model produces are not in $\mathbb{R}^d$. It is sometimes said that the continuous-time picture has no discrete counterpart. It has an exact one, and this section states it, because the claim to the contrary is the single largest gap between this volume's subject and the systems it is about.

Definition 3.29 (Forward chain). Let $\mathcal{X}$ be finite. A continuous-time Markov chain with rate matrix $R_t(x, y) \geq 0$ for $x \neq y$ and $R_t(x, x) = -\sum_{y \neq x} R_t(x, y)$ has marginals solving the master equation

$$\partial_t p_t(y) = \sum_x p_t(x) R_t(x, y), \tag{3.36}$$

the discrete counterpart of (2.4).

Proposition 3.30 (Exact time reversal of a chain) [illegible]. *Let $p_t(x) > 0$ for all x and t. Then $Y_s = X_{T-s}$ is a continuous-time Markov chain with rates*

$$\bar{R}_s(y, x) = R_{T-s}(x, y) \frac{p_{T-s}(x)}{p_{T-s}(y)}, \qquad x \neq y. \tag{3.37}$$

Proof. Write $t = T - s$ and $q_s = p_t$. For $x \neq y$ the reversed chain must satisfy $q_s(y)\bar{R}_s(y, x) = p_t(x)R_t(x, y)$, because both sides are the instantaneous rate of the transition $y \to x$ in reversed time, equal to the rate of $x \to y$ in forward time weighted by the forward occupancy. Dividing by $q_s(y) = p_t(y)$ gives (3.37); that the resulting rates define a chain whose marginals are q_s is the master equation (3.36) run backwards, which holds by construction. ■

Where positivity bites. If $p_t(y) = 0$ the reverse rate (3.37) is undefined, and this is not an edge case: an absorbing forward chain drives most states to zero probability at large t, so the reversal is stated on the support and the learned object is a conditional law rather than a rate. On a countably infinite state space the same formula holds pointwise but the resulting chain may explode, and positivity no longer suffices.

What it does not say. It gives the reverse rates in terms of the *true* marginals, exactly as (3.3) gives the reverse drift in terms of the true score. Everything about learning those ratios — the objective, its variance, the error it propagates — is outside this proposition, and no analogue of Proposition 3.24 is stated in this volume for the discrete case.

The ratio $p_t(x)/p_t(y)$ plays the role that $\nabla \log p_t$ plays in (3.3), and it is what is learned: the objective is a regression onto these ratios, in the same way that Proposition 3.12 is a regression onto the score. Absorbing or masking forward chains, in which one state is a sink, make many ratios degenerate and reduce the learned object to a conditional distribution over the unmasked coordinates, which is why they connect to autoregressive models.

What does not follow

Discrete time and discrete space both have exact time reversal. Proposition 3.30 is exact, elementary, and requires nothing beyond positivity of the marginals. The often-repeated claim that the reverse-time construction is a specifically continuous-space phenomenon is false, and the correct statement of what continuous time supplies is narrower: a uniform differential formalism, a probability-flow counterpart, and the calculus of Chapter 1. None of those is the reversal itself.

The score has no counterpart, but the ratio does. There is no gradient on a finite set, so $\nabla \log p_t$ does not survive; what survives is the ratio $p_t(x)/p_t(y)$ between neighbouring states. The structural similarity between the two is worth noticing and is not a dictionary: no result of Section 3.4 transfers, and Proposition 3.17 has no discrete analogue.

A token sequence is not a diffusion. Writing an autoregressive model as a continuous-time process is a modelling step with hypotheses of its own, and (3.37) does not supply them. Section 7.2 treats the discrete result and the continuous-time analogue separately, and names the gap between them.

3.9 Chapter Summary

Chapter 3 builds the generative constructions on the reversal theorem of Theorem 3.1 and keeps four error sources apart. The reverse-time SDE and the probability-flow ODE are the two ends of a one-parameter family sharing every time marginal and differing as path measures, so a statement about samples transfers between them and a statement about trajectories does not. The three population score-matching objectives differ by constants independent of the field, hence share their minimisers over any admissible restricted class; what does *not* transfer without both a parameterisation tie and a weight relation is the comparison between the flow-matching and denoising objectives (Theorem 3.20).

The chapter states no end-to-end theorem. Terminal mismatch, score error, discretisation and early stopping have different hypotheses and different orientations of the divergence, and Table 3.1 records them as three results plus one change of target rather than as a sum. Every relative entropy here is written as an ordered pair, and none is silently reversed.

Exercises

§3.1 and §3.2

Tier I Direct application.

3.1 *Reversing an Ornstein–Uhlenbeck process*

Take $f \equiv -\frac{1}{2}$, $g \equiv 1$, and p_{data} Gaussian. Compute p_t in closed form from Proposition 3.7, write out (3.3) and (3.6) explicitly, and verify Proposition 3.4 by solving both and comparing the marginals.

3.2 *The two formulas, reconciled*

Show that the drift of (3.6) for $a = g^2 I$ is $f - \frac{1}{2} g^2 \nabla \log p_t$ and that of (3.3) is $f - g^2 \nabla \log p_t$ after the change of time direction, and identify the value of λ in (3.5) for each. State in one sentence why these are not two conventions.

Tier II Completing the derivation.

3.3 *Where the divergence term comes from*

Derive (3.4) by repeating the proof of Proposition 3.4 with $a = a(t, x)$, and check that the extra term $\nabla \cdot a$ vanishes exactly when a is constant in x. Then verify that (3.4) agrees with the reversal of a Brownian motion on a Riemannian manifold written in local coordinates, where $\nabla \cdot a$ carries the metric derivative.

3.4 *The mismatch bound, sharpened*

Show that (3.12) is an equality when p_{data} is a point mass, and compute both sides when $p_{\text{data}} = \mathcal{N}(0, \tau^2 I)$. Report the ratio of the two as a function of τ and α_T, and say for which τ the bound is loose.

Tier III Probing the boundary.

3.5 *The manifold hypothesis breaks (A2)*

Let p_{data} be the uniform measure on the unit circle in $\mathbb{R}^2$. Compute p_t for the variance-exploding schedule and show that $\int |\nabla \log p_t|^2 \, p_t$ diverges like σ_t^{-2} as $t \downarrow 0$. Deduce the largest δ for which (3.2) is finite, and state what this implies for the starting time of any implementation.

3.6 *A schedule that is not a diffusion*

For the linear interpolant $\alpha_t = 1 - t$, $\sigma_t = t$, compute f and g from Proposition 3.8 and confirm that g_t^2 diverges at $t = 1$. Then show that the probability path is nevertheless perfectly well defined for all $t \in [0, 1]$, and explain the discrepancy: which object exists, which does not, and which of the constructions in Section 3.4 needs only the first.

3.7 *Marginals agree, samplers do not*

Fix $d = 1$ and $p_{\text{data}} = \mathcal{N}(0, 1)$, and perturb the exact score by a constant ε. Compute, in closed form, the law produced by (3.5) for $\lambda \in \{0, 1\}$ with the perturbed score, and report the two errors as functions of ε. State which is larger and identify the mechanism, referring to the notes on Proposition 3.4.

Tier IV Computational.

3.8 *Verifying the family*

Take p_{data} a mixture of three Gaussians in $\mathbb{R}^2$ and the variance-preserving schedule with $\alpha_t = \cos(\frac{\pi}{2} t)$.

(a) Compute $\nabla \log p_t$ in closed form and simulate (3.5) for $\lambda \in \{0, \frac{1}{2}, 1, 2\}$ from 10^5 samples of p_T. Report a distributional distance to p_{data} for each, at three step sizes.

(b) Confirm numerically that the four marginals agree at λ fixed and step size small, and report the step size at which they cease to.

(c) Replace the exact score by a perturbed one, $s + \varepsilon w$ with w a fixed smooth field, and repeat (a). Report how the error depends on λ.

(d) State which of (a)–(c) is testing Proposition 3.4 and which is testing something the proposition does not claim.

Acceptance criteria: (b) reports the step size at which agreement fails; (d) is answered in two sentences.

§3.4

Tier I Direct application.

3.9 *The dictionary for three schedules*

Compute λ_t and c_t for the variance-preserving path $\alpha_t^2 + \sigma_t^2 = 1$ with $\alpha_t = \cos(\frac{\pi}{2} t)$, for the variance-exploding path $\alpha_t \equiv 1$, $\sigma_t = t$, and for the linear interpolant. In each case identify the times at which the dictionary degenerates and say what happens to u_t and to s_t there.

Hint. $c_t = \sigma_t^2 \frac{d}{dt} \log \mathrm{snr}_t$ is quicker than the unsimplified expression.

3.10 *Recovering the probability-flow ODE*

Show that for any Gaussian conditional path the field (3.21) coincides with the probability-flow ODE velocity $f_t(x) - \frac{1}{2}g_t^2 s_t(x)$ of Proposition 3.4, and identify f and g in terms of α, σ. State which direction of time each formula was written in and check that the signs agree.

Tier II Completing the derivation.

3.11 *The marginalisation identity without Gaussianity*

Prove Proposition 3.15 for a general conditional path $p_t(\cdot \mid z)$ satisfying the continuity equation with a known conditional field $u_t(\cdot \mid z)$, and identify exactly which step of the proof used the Gaussian form and which did not. Then state which of Proposition 3.15 and Proposition 3.17 survives.

3.12 *Necessity of both conditions in Theorem 3.20(ii)*

Give an explicit pair $(w, \tilde{w})$ with $\tilde{w} \neq c_t^2 w$ and a two-parameter class $\{s_\phi\}$ for which $\mathcal{L}^{w}_{\mathrm{CFM}}$ and $\mathcal{L}^{\tilde{w}}_{\mathrm{DSM}}$ have different minimisers, even with the tie (3.28) in force. Then give a class for which they coincide despite $\tilde{w} \neq c_t^2 w$, and explain why this does not contradict part (iii).

Tier III Probing the boundary. Each of these asks where a statement stops being true.

3.13 *Reproducing and generalising Example 3.21*

Verify the two numbers $-\frac{9}{10}$ and $-\frac{1}{4}$ by direct computation. Then treat τ^2 as a parameter and plot θ_1^*/θ_2^* against τ^2. Identify the value of τ^2 at which the two coincide, and explain, from (3.30), why they must coincide there and only there. Finally, explain in one sentence why enlarging the class to $v_\theta(t, x) = \theta(t)x$ with θ an arbitrary continuous function makes the discrepancy vanish, and what that shows about which claims of equivalence are safe.

3.14 *A path on which the score is unrecoverable*

Construct a Gaussian conditional path with snr_t constant on $[\frac{1}{3}, \frac{2}{3}]$ and strictly decreasing elsewhere. Show that on the middle interval the marginal velocity field is $\lambda_t x$ independently of p_{data}. Explain what a network trained by (3.25) learns on that interval and what happens if one attempts to invert (3.21) numerically there.

3.15 *Straight is not the same as short, and neither is exact*

Let p_{data} be a mixture of two well-separated Gaussians in $\mathbb{R}$ and $\mu = \mathcal{N}(0, 1)$. Compute the Brenier map T numerically and confirm Proposition 3.23 to machine precision. Then replace T by the optimal coupling of two size-B empirical samples, for $B \in \{16, 128, 1024\}$, and measure the curvature of the resulting marginal field — for instance by $\int_0^1 \mathbb{E}\,\|u_t(X_t) - u_0(X_0)\|^2 \,dt$ along its own flow. Report the one-step sampling error against B, and say which of the four items in the mini-ledger of Proposition 3.23 your experiment isolates and which it does not.

3.16 *Where the deterministic and stochastic samplers part*

Take the exact score for a one-dimensional Gaussian target, then perturb it by a fixed smooth error $\delta(t, x)$ of size ε. Compute, to first order in ε, the resulting error in the law produced by the probability-flow ODE and by the reverse-time stochastic differential equation. Explain why the two differ even though the exact fields have identical marginals, and state which of the two is being described whenever a paper reports that "the ODE and SDE samplers are equivalent".

Tier IV Computational.

3.17 *Weighting, parameterisation, and what is actually being compared*

On CIFAR-scale data is out of scope; use a two-dimensional target with structure, for example a mixture of eight Gaussians on a circle.

(a) Train four models: $\{\text{velocity}, \text{score}\}$ parameterisation $\times$ $\{w \equiv 1, w = \sigma_t^2\}$, with the same architecture, the same optimiser, and the same number of steps.

(b) Convert all four to a common coordinate system using (3.21) and report, for each, the error $\mathbb{E}\,\|v_\theta(t, X_t) - u_t(X_t)\|^2$ as a function of t. The four curves should differ in where they are accurate, not uniformly in level.

(c) Sample from all four with the same deterministic solver at NFE $\in \{2, 5, 10, 50\}$, and report a distributional error.

(d) Identify which of the differences in (c) are attributable to the weighting, which to the parameterisation, and which to the solver. State what would have to be held fixed for a comparison between two published methods to be attributable at all.

Acceptance criteria: (b) shows the crossing of the error curves, not just their levels; (d) is answered with the evidence that fixes each attribution, and.

§3.3 to §3.8

Tier I Direct application.

3.18 *The constant in denoising score matching*

Compute C_t in (3.14) explicitly for $p_{\text{data}} = \mathcal{N}(0, \tau^2)$ in $d = 1$, and verify that it diverges as $\sigma_t \downarrow 0$. State what this implies for the numerical stability of the objective near the data end.

3.19 *Likelihood weighting for two schedules*

Compute $w(t) = \frac{1}{2}g_t^2$ for the variance-preserving schedule $\alpha_t = \cos(\frac{\pi}{2}t)$ and for the variance-exploding schedule with $\sigma_t = t$. Plot both against t and against $\log \mathrm{snr}_t$, and say which times each weighting emphasises.

Tier II Completing the derivation.

3.20 *The decomposition, step by step*

Derive (3.33) from (3.17) in full, including the chain rule for relative entropy on path space and the data-processing step. Identify the exact point at which the direction of the divergence is fixed, and state what would have to change to obtain a bound on $\mathrm{KL}(\hat p_\delta \,\|\, p_\delta)$ instead.

3.21 *Reversal of a two-state chain*

For $\mathcal{X} = \{0,1\}$ and constant rates, solve (3.36) explicitly and verify (3.37) by direct computation. Then let state 1 be absorbing and describe what happens to the reverse rates.

Tier III Probing the boundary.

3.22 *Where the three terms fight*

Take p_{data} uniform on a circle in $\mathbb{R}^2$. Using the notes on Theorem 3.1, estimate the growth of $\mathbb{E}_{p_\delta}|\nabla \log p_\delta|^2$ and of $\|\nabla^2 \log p_\delta\|$ as $\delta \downarrow 0$. Then write the sum of the three terms of Table 3.1 as a function of δ and N, and find the δ that minimises it. State which two terms are in tension and which is indifferent.

3.23 *An interpolant with no dictionary*

Verify part (iii) of Proposition 3.28 for the stated example by computing u_t and $\nabla \log p_t$ numerically for Z_0, Z_1 standard Gaussian, and showing that the map from the second to the first is not affine. Report a quantitative measure of the departure from affinity.

3.24 *Distillation is not a solver*

Design an experiment that distinguishes a solver improvement from a distillation improvement, holding the error metric fixed. State what each would predict as the number of steps is increased beyond the value used at training time, and identify the row of Table 3.2 each prediction tests.

Tier IV Computational.

3.25 *Measuring the bound*

Take p_{data} a mixture of eight Gaussians in $\mathbb{R}^2$, for which $\nabla \log p_t$ is available in closed form.

(a) Train s_θ under the weighting $w \equiv 1$ and under $w = \tfrac{1}{2} g_t^2$, and measure $\varepsilon^2_{\mathrm{score}}$ from (3.32) for both.

(b) Sample from both with an exact-score reference and with each learned score, at $N \in \{10, 50, 500\}$, and estimate $\mathrm{KL}(p_\delta \,\|\, \hat p_\delta)$ on a grid.

(c) Plot the measured divergence against the right-hand side of (3.33) plus (3.34), and report the ratio.

(d) Say which of your measurements tests Proposition 3.24, which tests Remark 3.25, and which tests neither because it varies two rows of Table 3.1 at once.

Acceptance criteria: the ratio in (c) is a number at each N; (d) classifies every measurement.

CHAPTER 4

Sampling

Chapter 2 established that the law of a reversible diffusion descends the free energy, and Section 2.4 established when it does so exponentially. Neither statement is about an algorithm. This chapter is, and the distance between the two is larger than it looks.

Three things separate the continuous-time picture from a program. The trajectory must be discretised, which introduces a bias that does not vanish as the run lengthens. The dynamics may be chosen — reversible or not, first order or second — and the choice changes the rate without changing the target. And whatever is proved must be proved under hypotheses on the target that, by Table 2.2, are rarely verifiable. The chapter takes these in order, and its last section is about what can honestly be said when none of the hypotheses can be checked.

One organising decision runs through it: every rate stated below has its hypotheses in the first column of the table it appears in, not in a footnote. A rate quoted without its assumptions is not a weaker statement than one with them; it is a different statement, usually a false one.

4.1 Discretisation

4.1.1 The unadjusted algorithm and its bias

Definition 4.1 (Unadjusted Langevin algorithm). For $V \in C^1(\mathbb{R}^d)$, $\pi \propto e^{-V}$ and step size $\eta > 0$, the *unadjusted Langevin algorithm* is

$$x_{k+1} = x_k - \eta \nabla V(x_k) + \sqrt{2\eta}\,\xi_k, \qquad \xi_k \sim \mathcal{N}(0, I) \text{ independent}, \tag{4.1}$$

that is, Euler–Maruyama (Definition 1.47) applied to (2.10).

The scheme has a stationary law, and for a Gaussian target it can be computed exactly. The computation is worth doing because it exhibits both the size of the bias and the reason for the step-size restriction, without any approximation.

Proposition 4.2 (Exact stationary law of ULA on a Gaussian target). *Let $V(x) = \frac{1}{2}\langle x, Hx\rangle$ with $H \succ 0$, so $\pi = \mathcal{N}(0, H^{-1})$, and suppose $\eta\lambda < 2$ for every eigenvalue λ of H. Then (4.1) has the unique stationary law $\pi_\eta = \mathcal{N}(0, \Sigma_\eta)$ with*

$$\Sigma_\eta = \left(I - \tfrac{\eta}{2}H\right)^{-1} H^{-1}, \tag{4.2}$$

so in the eigendirection of λ the variance is inflated by the factor $(1 - \eta\lambda/2)^{-1}$, a relative error

$$\frac{\eta\lambda}{2 - \eta\lambda} = \frac{\eta\lambda}{2} + O(\eta^2\lambda^2). \tag{4.3}$$

which does not decrease as the number of steps grows and diverges as $\eta \uparrow 2/\lambda$. If $\eta\lambda_{\max}(H) \geq 2$ the iteration diverges.

Proof. The iteration is linear: $x_{k+1} = (I - \eta H)x_k + \sqrt{2\eta}\,\xi_k$. A centred Gaussian law with covariance Σ is stationary if and only if $\Sigma = (I - \eta H)\Sigma(I - \eta H)^\top + 2\eta I$. Since H is symmetric, Σ may be taken to commute with H, and in an eigenbasis the equation is scalar: $s = (1 - \eta\lambda)^2 s + 2\eta$, so $s\big(1 - (1 - \eta\lambda)^2\big) = 2\eta$, that is $s\big(2\eta\lambda - \eta^2\lambda^2\big) = 2\eta$ and $s = \big(\lambda(1 - \eta\lambda/2)\big)^{-1}$, which is (4.2). Uniqueness and convergence hold precisely when $|1 - \eta\lambda| < 1$ for every λ, that is $\eta\lambda < 2$; otherwise the corresponding component is multiplied by a factor of modulus at least one at every step. Dividing s by the exact variance $1/\lambda$ gives (4.3). ■

The step-size restriction $\eta < 2/\lambda_{\max}$ is the same threshold as for gradient descent in Proposition 6.9(ii), and for the same reason: the noise does not affect the stability of the linear map. What the noise adds is the bias (4.3), and the two are in tension — a larger step converges faster and lands further away.

Proposition 4.3 (Non-asymptotic bound for ULA) ESTABLISHED. *Assume $\pi \propto e^{-V}$ satisfies $\mathrm{LSI}(\alpha)$ in the normalisation (2.14) and ∇V is β-Lipschitz. Let ρ_K be the law of x_K from (4.1) with $\eta \leq \alpha/(4\beta^2)$. Then*

$$\mathrm{KL}(\rho_K \,\|\, \pi) \;\leq\; e^{-\alpha\eta K}\,\mathrm{KL}(\rho_0 \,\|\, \pi) \;+\; \frac{8\,\eta\, d\,\beta^2}{\alpha}, \tag{4.4}$$

so that $\mathrm{KL}(\rho_K \,\|\, \pi) \leq \varepsilon$ is reached with $\eta \asymp \alpha\varepsilon/(d\beta^2)$ and $K = \tilde{O}\big(d\beta^2/(\alpha^2\varepsilon)\big)$. The bound in this form — under a log-Sobolev inequality and a bounded Hessian, rather than under convexity — is due to Vempala and Wibisono [27].

Interpretation

A discretised Langevin diffusion does not sample π; it samples a nearby law, and the Gaussian case shows exactly which one. The step size enters twice — once as a discretisation error and once as an inflation of the stationary variance — so running longer cannot remove the bias, only the transient. That is the structural reason the next section pays for a Metropolis test: rejection buys back the exact invariant law at the cost of throwing work away.

Notes on Proposition 4.2 and Proposition 4.3

Two limits that do not commute. Equation (4.4) has two terms: one vanishing as $K \to \infty$ at fixed η, one vanishing as $\eta \to 0$ at fixed $K\eta$. Running longer at a fixed step size drives the first to zero and leaves the second, which is the bias (4.3); taking $\eta \to 0$ at fixed K leaves the first. Only the joint limit with η tied to ε produces a complexity, and that is what the second display records.

Where the dimension enters. The continuous-time rate of Corollary 2.14 is 2α and contains no d. Every factor of d in (4.4) comes from the discretisation term, through the variance $2\eta d$ injected per step. The dimension dependence of sampling is therefore not a property of the Langevin dynamics; it is a property of simulating it. Proposition 4.17 and the notes below say what happens when this is forgotten.

What it does not say. Nothing usable about targets without a usable log-Sobolev constant. The distinction matters: by Example 2.29 a multimodal target with a nontrivial barrier may well satisfy $\mathrm{LSI}(\alpha)$ — a one-dimensional double well does — with α exponentially small in the barrier height, so (4.4) is true, applies, and is useless. The failure is in the constant, not in the hypothesis, and saying "no log-Sobolev inequality" misidentifies which of the two has gone wrong. Nothing about a target known only through a learned potential. And nothing about the law of the iterates at finite K beyond the divergence bound: samples produced by ULA are correlated, and the effective sample size is a separate quantity.

A mixing time is not a continuous-time rate. Corollary 2.14 gives $\mathrm{KL}(\rho_t \,\|\, \pi) \le e^{-2\alpha t}\mathrm{KL}(\rho_0 \,\|\, \pi)$ for the exact diffusion — no dimension, no smoothness constant, no step size. Proposition 4.3 gives an iteration count $\tilde{O}(d\beta^2/\alpha^2\varepsilon)$ for an algorithm. Statements of the form "the mixing time is proportional to d/λ_{gap}" merge the two: they attach the dimension dependence of a discretisation to the spectral quantity of a semigroup. The two live in different rows of Table 4.1 and are never quoted for one another in this volume.

The bias is not an error that shrinks with effort. By Proposition 4.2 the stationary law of ULA is a different distribution from π, at distance of order η. Running for longer moves the iterates closer to π_η and no closer to π. This is the third row of Table 1.2 appearing where it matters most: neither the strong nor the weak order of Definition 1.47 sees it, because both are defined at a fixed horizon.

Smoothness is a hypothesis about the potential, not about the target. The constant β in (4.4) is the Lipschitz constant of ∇V on all of $\mathbb{R}^d$. Two potentials differing by a bounded function have the same π up to a bounded factor and can have wildly different β; the bound is not a functional of π alone.

4.1.2 Metropolis adjustment

The bias of (4.1) can be removed exactly, at the cost of a rejection step and of a stricter relationship between η and d.

Definition 4.4 (Metropolis-adjusted Langevin algorithm). Given x_k, propose y from the density $q(x_k,\cdot) = \mathcal{N}(x_k - \eta\nabla V(x_k),\, 2\eta I)$, accept it with probability

$$\min\left\{1,\ \frac{\pi(y)\,q(y,x_k)}{\pi(x_k)\,q(x_k,y)}\right\},$$

and otherwise set $x_{k+1} = x_k$.

Proposition 4.5 (Exactness of the adjustment) [illegible]. *The chain of Definition 4.4 is reversible with respect to π, hence π is stationary and the bias (4.3) is removed, for every $\eta > 0$ for which the kernel is defined. Irreducibility is not used and is not needed: detailed balance alone gives the invariant law. What irreducibility (with aperiodicity) is needed for is the separate statement that the chain converges to π from an arbitrary start, which is not asserted here.*

Proof. Write $P(x, dy)$ for the transition kernel. For $x \neq y$ the density of P is $q(x,y)\min\{1, r(x,y)\}$ with $r(x,y) = \pi(y)q(y,x)/(\pi(x)q(x,y))$, so

$$\pi(x)\,q(x,y)\,\min\{1, r(x,y)\} = \min\{\pi(x)q(x,y),\ \pi(y)q(y,x)\},$$

which is symmetric in (x,y); this is detailed balance in the sense of (2.5) for the discrete kernel. Stationarity follows by integrating in x. ∎

Interpretation

The Metropolis test converts a biased proposal into an exact one, and the currency is rejection. Detailed balance is arranged by construction, so the invariant law is π for *any* step size — nothing about the accuracy of the proposal enters. What the step size controls is how often the proposal is accepted, and hence how much work is wasted. The trade is therefore not bias against variance but bias against cost, and Proposition 4.6 prices it.

Notes on Proposition 4.5 and Proposition 4.6

Exactness is not usefulness. For a positive smooth target and a Gaussian proposal, no finite η makes the acceptance probability vanish identically — the proposal density is everywhere positive and $r(x, y) > 0$ — so the failure mode at large η is an acceptance rate close to zero, not a chain frozen at its starting point. The consequence is the same in practice and different in kind: the invariant law remains exactly π while the effective sample size per unit cost collapses, which is a statement about mixing and belongs with Proposition 4.6, not with the exactness above.

What exactness costs. The adjustment removes the bias for every η at which the chain moves, and the scaling result says that the range of such η shrinks with d. The two statements are about different things — one about the invariant law, one about the rate — and neither implies the other; a chain can be exact and useless.

What it does not say. Nothing about the target class of Proposition 4.3: the two rows of Table 4.1 they occupy have different first columns. Nothing about the cost of evaluating π, which for a target defined by an unnormalised model may dominate the gradient cost.

Proposition 4.6 (The price of exactness) ESTABLISHED. *For product targets and under regularity conditions, the acceptance probability of Definition 4.4 degenerates unless $\eta = O(d^{-1/3})$, the optimal asymptotic acceptance rate being a constant strictly between zero and one; the resulting mixing time scales as $\tilde{O}(d^{1/3})$ steps for that class, against $\tilde{O}(d)$ for the random-walk proposal. The scaling statements are quoted; the exponents depend on the target class and on the stationarity or warmness of the start, and should be read with those.*

Proposition 4.5 and Proposition 4.6 together are the honest statement of the trade: the adjustment removes the bias exactly and costs a step-size restriction that is more severe than the stability restriction of Proposition 4.2. Which is preferable depends on the tolerance ε; Table 4.1 records the comparison and Section 4.4.2 says what it takes for such a comparison to mean anything.

4.2 Kinetic Dynamics

4.2.1 Underdamped Langevin and degenerate dissipation

Definition 4.7 (Underdamped Langevin dynamics). For $U \in C^2(\mathbb{R}^d)$ and friction $\gamma > 0$,

$$\mathrm{d}X_t = V_t\,\mathrm{d}t, \qquad \mathrm{d}V_t = -\nabla U(X_t)\,\mathrm{d}t - \gamma V_t\,\mathrm{d}t + \sqrt{2\gamma}\,\mathrm{d}W_t, \tag{4.5}$$

on $\mathbb{R}^d \times \mathbb{R}^d$, with generator $\mathcal{L} = \underbrace{\langle v, \nabla_x\rangle - \langle \nabla U, \nabla_v\rangle}_{\mathcal{L}_{\mathrm{ham}}} + \underbrace{\gamma\left(-\langle v, \nabla_v\rangle + \Delta_v\right)}_{\mathcal{L}_{\mathrm{ou}}}$.

Proposition 4.8 (Structure of the generator, and degenerate dissipation) DERIVED. *Let $\mathcal{M}(x, v) \propto e^{-U(x) - |v|^2/2}$. Then:*

(i) $\mathcal{M}$ is stationary for $\mathcal{L}$; moreover $\mathcal{L}_{\mathrm{ham}}$ is antisymmetric and $\mathcal{L}_{\mathrm{ou}}$ symmetric in $L^2(\mathcal{M})$, so the Dirichlet form is

$$\mathcal{E}(f,f) = \gamma \int |\nabla_v f|^2 \, \mathrm{d}\mathcal{M}, \tag{4.6}$$

which involves no x-derivative;

(ii) along the law f_t of (4.5),

$$\frac{\mathrm{d}}{\mathrm{d}t}\mathrm{KL}(f_t \,\|\, \mathcal{M}) = -\gamma \int f_t \left|\nabla_v \log \tfrac{f_t}{\mathcal{M}}\right|^2, \tag{4.7}$$

which vanishes on every f with $f/\mathcal{M}$ depending on x alone — in particular on every product density $f = \rho \otimes \mathcal{M}_v$.

Consequently no inequality of the form $\mathrm{KL}(f \,\|\, \mathcal{M}) \le C \int f \,|\nabla_v \log(f/\mathcal{M})|^2$ can hold, and the Grönwall argument of Proposition 2.25 does not close.

Proof. (i) A direct computation as in Proposition 2.5: for $f, g \in C_c^\infty$, integrating by parts in v turns $\int f \mathcal{L}_{\mathrm{ou}} g \, \mathrm{d}\mathcal{M}$ into $-\gamma \int \langle \nabla_v f, \nabla_v g \rangle \, \mathrm{d}\mathcal{M}$, symmetric; and $\int f \mathcal{L}_{\mathrm{ham}} g \, \mathrm{d}\mathcal{M} = -\int g \mathcal{L}_{\mathrm{ham}} f \, \mathrm{d}\mathcal{M}$ because $\mathcal{L}_{\mathrm{ham}}$ is a divergence-free transport field for $\mathcal{M}$, being the Hamiltonian vector field of $H(x,v) = U(x) + |v|^2/2$ and $\mathcal{M} \propto e^{-H}$. Taking $f \equiv 1$ gives stationarity, and (4.6) is the symmetric part. (ii) Repeat the proof of Proposition 2.12: the Hamiltonian part contributes nothing because it is antisymmetric — exactly as the non-gradient part did in Proposition 2.16(ii) — and the Ornstein–Uhlenbeck part contributes only v-derivatives. For the final assertion take $f(x,v) = \rho(x)\,\mathcal{M}_v(v)$ with ρ a probability density on $\mathbb{R}^d$ and $\mathcal{M}_v \propto e^{-|v|^2/2}$ the velocity marginal of $\mathcal{M}$: this is a genuine probability density on $\mathbb{R}^d \times \mathbb{R}^d$, the ratio $f/\mathcal{M} \propto \rho(x)e^{U(x)}$ depends on x alone, so $\nabla_v \log(f/\mathcal{M}) = 0$ and the right-hand side of (4.7) vanishes, while $\mathrm{KL}(f \,\|\, \mathcal{M}) = \mathrm{KL}(\rho \,\|\, \pi)$ is positive whenever $\rho \neq \pi$. A function of x alone is not itself a density on the phase space, and writing one is the quickest way to make this argument look illegitimate. ∎

> Interpretation
>
> Kinetic dynamics splits into a conservative part that transports and a dissipative part that acts only on velocity. Nothing damps position directly, so the free-energy argument of Chapter 2 has no purchase: the dissipation is degenerate, and a rate has to come from the *interaction* of the two parts rather than from either alone. That is what hypocoercivity supplies, and it is why the friction has to be tuned rather than maximised — too much friction and the conservative transport is suppressed along with the noise.

What the degeneracy means. Equation (4.7) is a genuine identity and a genuine obstruction: the dissipation controls only $\nabla_v \log(f/\mathcal{M})$, so a functional inequality bounding $\mathrm{KL}(f \,\|\, \mathcal{M})$ by it cannot exist, as the product densities $f = \rho \otimes \mathcal{M}_v$ show. The missing control is recovered by working with a modified functional that mixes x- and v-derivatives, which is what Proposition 4.9 does and what this proposition does not.

What it does not say. It does not say the kinetic dynamics converges more slowly — it converges faster, by Proposition 4.9 — only that the argument of Proposition 2.25 is unavailable. The pattern is the one of Proposition 2.16: an antisymmetric part of the generator is invisible to the Dirichlet form and changes the behaviour anyway.

Proposition 4.9 (Hypocoercivity) ESTABLISHED. *Assume U is μ-strongly convex with β-Lipschitz gradient. Then there is a functional $\mathcal{H}$ equivalent to* $\mathrm{KL}(\cdot \,\|\, \mathcal{M})$ *up to constants depending only on γ, μ, β, and a rate $\varkappa > 0$, such that $\mathcal{H}(f_t) \le e^{-\varkappa t}\mathcal{H}(f_0)$; with the friction chosen as $\gamma \asymp \sqrt{\mu}$ the rate satisfies $\varkappa \asymp \sqrt{\mu}$, against $\varkappa = 2\mu$ for the overdamped dynamics in its normalisation. A non-asymptotic analysis of the discretised dynamics, with the $\sqrt{d}/\varepsilon$ step count that motivates the tuning above, is given by Cheng, Chatterji, Bartlett and Jordan [7]. Their normalisation of (4.5), time scaling and mass matrix should be read together with the rate: it is not invariant under them.*

Remark 4.10 (The quadratic test, which fixes the friction). The scaling $\gamma \asymp \sqrt{\mu}$ is not free: it is forced by the exactly solvable case, and any general rate statement must agree with it. Take $U(x) = \frac{\lambda}{2}x^2$ in one dimension. The mean of (4.5) obeys $\ddot{x} = -\lambda x - \gamma\dot{x}$, whose characteristic roots solve

$$r^2 + \gamma r + \lambda = 0, \tag{4.8}$$

so the decay rate of that mode is $\gamma/2$ when $\gamma^2 \le 4\lambda$ (oscillatory, both roots with real part $-\gamma/2$) and is asymptotically λ/γ when $\gamma^2 \gg 4\lambda$ (overdamped, the slow root). In short it is $\min\{\gamma, \lambda/\gamma\}$ up to constants. A single γ must serve every curvature in $[\mu, \beta]$, and the worst mode is the flattest one, $\lambda = \mu$; maximising $\min\{\gamma/2, \mu/\gamma\}$ over γ gives $\gamma = \sqrt{2\mu}$ and rate $\asymp \sqrt{\mu}$, with *every* stiffer mode then decaying at exactly $\gamma/2$ as well. Choosing $\gamma \asymp \sqrt{\beta}$ instead would over-damp the flat directions and give $\mu/\sqrt{\beta} \ll \sqrt{\mu}$: the friction must be matched to the smallest curvature, not the largest.

The improvement is the square root of the condition number, and the units in which that is true must be said. Rescale time so that $\beta = 1$; then $\mu = 1/\kappa$, the overdamped rate 2μ corresponds to a time of order κ, and the kinetic rate $\sqrt{\mu}$ to a time of order $\sqrt{\kappa}$. Without that normalisation the two rates are not comparable at all: the overdamped equation is the $\gamma \to \infty$ limit of (4.5) *after* a time rescaling by γ, so quoting $\sqrt{\mu}$ against 2μ across the two conventions is a normalisation substitution of the kind Table 6.2 catalogues. Even stated correctly, this is a claim about the *continuous-time* dynamics; the discretisation is a separate factor, and the $\sqrt{\kappa}$ improvement usually quoted for kinetic samplers is an algorithmic statement about admissible step size, not this one.

4.2.2 Hamiltonian Monte Carlo

Definition 4.11 (Leapfrog and HMC). One leapfrog step of size h for $H(x, v) = U(x) + \frac{1}{2}|v|^2$ is

$$v^- = v - \tfrac{h}{2}\nabla U(x), \qquad x' = x + hv^-, \qquad v' = v^- - \tfrac{h}{2}\nabla U(x'). \tag{4.9}$$

HMC draws $v \sim \mathcal{N}(0, 1)$, applies L leapfrog steps, and accepts the resulting x' with probability $\min\{1, e^{-\Delta H}\}$ where $\Delta H = H(x', v') - H(x, v)$.

Proposition 4.12 (Why the acceptance ratio is $e^{-\Delta H}$) DERIVED. *The leapfrog map $\Psi_h : (x, v) \mapsto (x', v')$ is volume preserving and reversible in the sense that $S \circ \Psi_h \circ S \circ \Psi_h = \mathrm{id}$, where $S(x, v) = (x, -v)$. Consequently the HMC kernel is reversible with respect to $\mathcal{M}$ and the Metropolis ratio reduces to $e^{-\Delta H}$, with no Jacobian factor.*

Proof. Each of the three substeps in (4.9) is a shear — one variable is updated by a function of the other — so each has unit Jacobian determinant, and so does the composition. For reversibility, applying S then Ψ_h then S recovers the inverse map, as is checked by running (4.9) with v negated. Volume preservation kills the Jacobian in the Metropolis–Hastings ratio and reversibility makes the proposal symmetric, leaving $\mathcal{M}(x',v')/\mathcal{M}(x,v) = e^{-\Delta H}$. ■

Interpretation

The acceptance ratio is simple because of what the integrator is, not because of how accurate it is. Each half-kick and each drift is a shear: a map that changes one variable by a function of the other. A shear has unit Jacobian, so the composition does too, and reversing the momentum reverses the composition. Those two geometric facts — unit Jacobian, momentum-flip symmetry — are exactly what cancels in the Metropolis–Hastings ratio, leaving $e^{-\Delta H}$. Local truncation accuracy plays no part in that cancellation; it affects only how large ΔH is, hence the acceptance rate, and not whether the chain has the right invariant law.

A useful consequence: the force used inside the integrator need *not* be the exact gradient of the U appearing in the target. If the force is any fixed deterministic function of position, the map is still a composition of shears, still volume preserving and still reversible; and if the acceptance test is computed with the intended Hamiltonian, the correction still targets $\mathcal{M}$ exactly. An inaccurate force costs acceptance rate, not correctness.

What does break the argument is randomness or mismatch that the acceptance ratio does not see: a proposal using a fresh stochastic gradient at each step, so that the map is not a fixed deterministic function and the reverse proposal has a different density; or an acceptance test evaluated with an approximate or learned energy different from the target's. Either requires the extra randomness to be carried in an augmented state, or a corrected proposal ratio. In floating point, finally, volume is preserved only up to rounding, and the accumulated drift over a long trajectory biases ΔH — small, but one of the reasons long trajectories are not free. The proposition says nothing about the *size* of ΔH, which is what determines the acceptance rate and hence the usable step size; that is Proposition 4.13, and it needs a third-derivative hypothesis.

Proposition 4.13 (Dimension dependence of HMC) ESTABLISHED. *For targets in the class of μ-strongly log-concave densities with β-Lipschitz gradient and bounded third derivatives, and from a warm start, HMC with an appropriately tuned step size and trajectory length reaches total-variation accuracy ε in $\tilde{O}(\kappa d^{1/4})$ gradient evaluations, where $\kappa = \beta/\mu$, against $\tilde{O}(\kappa d^{1/2})$ for the corresponding Metropolis-adjusted first-order method. The exponents are those of one comparison within one target class; the warmness of the start and the normalisation of the third-derivative bound both move them, and a statement quoted without those is not comparable with another.*

Scope of Proposition 4.9 and Proposition 4.13

Assumed, and stated first. Strong log-concavity with modulus μ; Lipschitz gradient with constant β; for Proposition 4.9 the friction tuned as in Remark 4.10 and the normalisation of (4.5) exactly as written, with unit mass and no preconditioning — a mass matrix changes the rate and is the usual source of mismatched quotations; for Proposition 4.13 additionally bounded third derivatives and a warm start whose definition enters the suppressed logarithmic factors. Every one of these is a hypothesis on the target; none is verifiable for a target given by a learned potential.

Source-sensitive comparisons. Both Proposition 4.13 and Remark 4.16 are quoted results whose exponents and quantifiers depend on choices made in the statement one reads. For Proposition 4.13: the target class, the warmness of the start, and the normalisation of the third-derivative bound, each of which moves the exponents in κ and d. For Remark 4.16: which of the three quantities called a gap is meant, and under what spectral hypotheses the strictness clause holds. Read either against the conditions in the row of Table 4.1 it is being compared with, and not as a ranking of methods.

What the exponent is about. The $d^{1/4}$ compares HMC with a first-order adjusted method *within that class*. It is not a statement about arbitrary targets, and it is not a statement about the same target sampled by an unadjusted method, whose complexity has a different form (Proposition 4.3). The factor κ in front is at least as important as the exponent in d and is routinely dropped when the result is quoted.

Where it gives nothing. For a multimodal target the barrier is a property of π, not of the sampler: by Proposition 4.18 any chain whose one-step proposal has a *hard* bound on displacement has conductance bounded by the mass of the bottleneck region, so for such a chain no tuning removes an exponential factor. Standard HMC is *not* in that class: the momentum is refreshed from an unbounded Gaussian, so even at bounded trajectory length Lh the position proposal has no deterministic radius, and Proposition 4.18 does not apply to it. The expectation that momentum improves conditioning and not metastability is well founded, but establishing it for HMC needs its own energy-barrier or conductance argument — or a truncated-momentum variant, for which one obtains a high-probability localisation rather than a hard range — and this volume proves neither.

What would falsify them. A target in the stated class and a correctly tuned implementation whose measured gradient count exceeds the stated order by a growing factor in d.

4.3 Non-Reversible Perturbations

Proposition 2.16 showed that an admissible non-reversible perturbation dissipates the free energy at exactly the same instantaneous rate and is not the W_2 gradient flow of it. What it did not do is say whether the perturbation helps. It does, and the Gaussian case shows by how much.

Proposition 4.14 (A computable improvement) [illegible]. *Let $\pi = \mathcal{N}(0,\Sigma)$ with $\Sigma \succ 0$ and, for S antisymmetric, consider*

$$dX_t = -AX_t\,dt + \sqrt{2}\,dW_t, \qquad A = (I + S)\Sigma^{-1}. \tag{4.10}$$

Then π is stationary for every antisymmetric S. The mean and the centred covariance obey the closed linear equations

$$m_t = e^{-tA}m_0, \qquad \Sigma_t - \Sigma = e^{-tA}(\Sigma_0 - \Sigma)e^{-tA^T}, \tag{4.11}$$

and

$$\lim_{t\to\infty} \frac{1}{t} \log \left\| e^{-tA} \right\| = -\min \Re\,\mathrm{spec}(A), \tag{4.12}$$

so (4.12) is the asymptotic operator-norm exponent of e^{-tA}, hence the worst case over initial directions: the mean deviation decays at that exponent for a generic m_0 — faster for an m_0 with no component in the slow generalised eigenspace, and not at all for $m_0 = 0$ — and the induced map $B \mapsto e^{-tA}Be^{-tA^T}$ on covariance deviations has worst-case exponent twice it. No metric-level rate for $\mathrm{Law}(X_t)$ is asserted here; see *Remark 4.15. For $d = 2$, $\Sigma = \mathrm{diag}(1,\kappa)$ with $\kappa > 1$, and*

$S = s\left(\begin{smallmatrix}0 & -1\\ 1 & 0\end{smallmatrix}\right)$,

$$\min \Re\,\mathrm{spec}(A) = \begin{cases}\kappa^{-1} + O(s^2) & s \text{ small},\\ \frac{1}{2}(1+\kappa^{-1}) & s^2 \geq \frac{1}{4}(1-\kappa^{-1})^2\kappa,\end{cases} \tag{4.13}$$

so a sufficiently large rotation improves this exponent from the slow eigenvalue κ^{-1} of Σ^{-1} to the arithmetic mean $\frac{1}{2}(1+\kappa^{-1})$ of its two eigenvalues, a factor of order κ.

Remark 4.15 (Why this is a matrix statement and not a law statement). It is tempting to read (4.12) as "$\mathrm{Law}(X_t)$ approaches π at rate $\min \Re\,\mathrm{spec}(A)$", and that sentence has no referent until three things are fixed: the initial law, the metric, and which functional of the law is being measured. Each choice changes the answer. If $X_0 \sim \pi$ then $\mathrm{Law}(X_t) = \pi$ for every t and there is no decay rate at all, of any size. Within Gaussian initial laws, the mean deviation decays at the exponent (4.12) and the covariance deviation at *twice* it, by (4.11); a metric that weights the two differently — W_2, $\mathrm{KL}(\,\|\,)$, χ^2 — inherits whichever dominates for the given initial law, and the constants differ. What this proposition supplies is a finite-dimensional, fully computable illustration that an admissible non-reversible perturbation can move a spectral exponent by a factor of order κ. The semigroup-level statement is Remark 4.16, which is recorded as a literature note rather than proved here; the finite-time guarantee that transfers to *every* initial law is Corollary 2.26, which is unchanged by c. Three objects, three statements, and this is the first of them.

Proof. Stationarity of $\mathcal{N}(0,\Sigma)$ for $\mathrm{d}X = -AX\mathrm{d}t + \sqrt{2}\mathrm{d}W$ requires $A\Sigma + \Sigma A^{\top} = 2I$; with $A = (I+S)\Sigma^{-1}$ this is $(I+S) + (I+S^{\top}) = 2I$, which holds for every antisymmetric S. The mean solves $\dot{m} = -Am$, giving the first half of (4.11); the covariance solves $\dot{\Sigma}_t = -A\Sigma_t - \Sigma_t A^{\top} + 2I$, and subtracting the stationary equation $A\Sigma + \Sigma A^{\top} = 2I$ leaves a homogeneous Lyapunov equation for $\Sigma_t - \Sigma$ with the stated solution. Then (4.12) is Proposition 1.6. In the stated 2×2 case, $A = \left(\begin{smallmatrix}1 & -s\kappa^{-1}\\ s & \kappa^{-1}\end{smallmatrix}\right)$, with $\mathrm{tr}\,A = 1 + \kappa^{-1}$ and $\det A = \kappa^{-1}(1+s^2)$. The eigenvalues are real and distinct while $(\mathrm{tr}\,A)^2 > 4\det A$, that is $s^2 < \frac{1}{4}\kappa(1-\kappa^{-1})^2$, and the smaller one is $\kappa^{-1} + O(s^2)$; beyond that threshold they are complex conjugate with real part $\frac{1}{2}\,\mathrm{tr}\,A$, which is the second line. ∎

Scope of Proposition 4.14 and Remark 4.16

Assumed. A Gaussian target and a linear drift, so that everything is a 2×2 eigenvalue computation, and — the point of Remark 4.15 — an object that is the matrix exponential e^{-tA} rather than a divergence between laws. The corresponding general statement is recorded, and deliberately not proved or asserted here, in Remark 4.16.

Three quantities called "the gap", and only one of them improves. For a non-selfadjoint generator the word covers (a) the Poincaré coercivity constant, defined through the Dirichlet form; (b) the spectral bound $\inf \Re\,\mathrm{spec}(-\mathcal{L}^{c})$; and (c) the asymptotic decay rate (4.14) of the semigroup. They coincide in the reversible case and need not otherwise. Quantity (a) is *unchanged* by an admissible c — that is Corollary 2.26, and it is why the finite-time guarantee neither improves nor degrades. What the literature note Remark 4.16 concerns is (b) or (c) depending on the source; the two agree when the semigroup is eventually norm-continuous with isolated spectrum, and not in general. Neither says anything about the size of the improvement, which is what Proposition 4.14 supplies in one computable case.

Asymptotic is not finite time, and the transient is not where one might look for it. Equation (4.13) is a statement about $\min \Re \operatorname{spec}(A)$, hence about $t \to \infty$. For large s the matrix A is strongly non-normal and, by the box following Proposition 1.6, $\|e^{-tA}\|$ in the Euclidean norm on $\mathbb{R}^d$ can grow substantially before it decays — so a functional of the mean or of the covariance, measured in that norm, can move away from equilibrium first.

That transient does not occur in the semigroup on $L^2(\pi)$. By Corollary 2.26 the Dirichlet form is unchanged by an admissible perturbation, so the variance of an observable and the relative entropy of the law decay monotonically at the reversible rates $e^{-2\lambda t}$ and $e^{-2\alpha t}$ for every t, perturbed or not. The two statements are about different objects in different norms, and the first is not evidence against the second. Any claim that a non-reversible sampler is "worse over a short run" must therefore name the observable and the norm; it is false for the variance in $L^2(\pi)$ and for the relative entropy.

What it does not say. It says nothing about a discretisation — non-reversible schemes have no Metropolis adjustment preserving π without extra construction — and nothing about multimodal targets, where by Proposition 4.18 the obstruction is again a property of π.

What would falsify it. A 2×2 instance in which the measured asymptotic decay rate disagrees with (4.13); the proof is a characteristic polynomial.

Remark 4.16 (Literature note: the general speed-up statement). The mechanism computed in Proposition 4.14 is known in greater generality: for $\mathcal{L}$ reversible with respect to π and c admissible in the sense of Definition 2.15, results in the literature compare the perturbed and unperturbed semigroups and conclude that the perturbation does not slow convergence, with strict improvement under further conditions.

This volume does not state that as a theorem, and the reason is worth giving. Three quantities travel under the name "rate" here — the Poincaré constant of the symmetric part, the spectral bound $-\sup \Re \operatorname{spec}(\mathcal{L}^c)$ on the mean-zero subspace, and the asymptotic decay exponent

$$e(\mathcal{L}^c) := -\lim_{t\to\infty} \frac{1}{t} \log \|P_t^c\|_{L^2_0(\pi)\to L^2_0(\pi)} \tag{4.14}$$

— and for a non-self-adjoint generator they need not agree. Which of the three a given source establishes, and what spectral hypotheses its strictness clause carries, decides the statement; quoting one exponent under another's name is exactly the substitution Table 11.2 catalogues. The load-bearing evidence in this chapter is therefore Proposition 4.14, where all three quantities are computed explicitly in the Gaussian case and the improvement is a factor of order κ.

What does not follow

Faster is not steeper. By Proposition 2.16(ii) the perturbed dynamics dissipates the free energy at exactly the same instantaneous rate at every law. Its advantage is entirely in which laws it visits, and by the notes on Proposition 2.20, the Dirichlet form cannot see the perturbation at all. Any explanation of the speed-up in terms of a larger dissipation, a steeper descent or a better geometry is describing something that does not happen.

Piecewise-deterministic samplers are not diffusions. The zig-zag and bouncy-particle processes are non-reversible by construction, move deterministically between random events, and have no diffusion limit at fixed target; nothing in Chapter 1 applies to them, and their analysis is by different tools. They belong in this section by their role and not by their mathematics, and this volume states no result about them.

4.4 What Can Be Compared

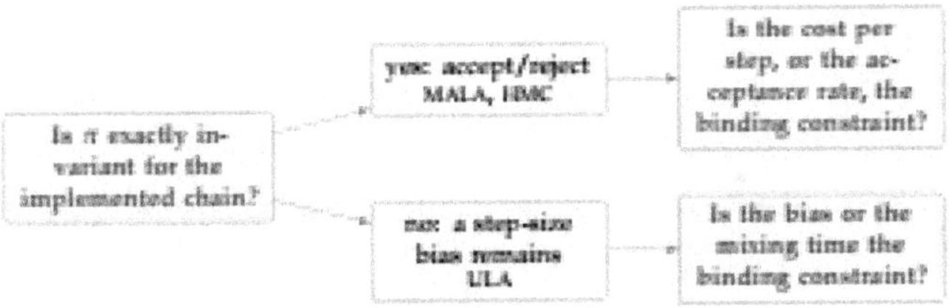

Figure 4.1: The samplers of this chapter, separated by the questions that decide which guarantees apply — not by performance. What this figure does not imply: any ordering of the methods. Two branches are comparable only when the hypotheses in the first column of Table 4.1 agree; where they do not, the complexity figures are not on the same axis and no ranking is defined.

4.4.1 What a lower bound can and cannot say

Proposition 4.17 (No universal lower bound follows from a spectral gap) DERIVED. *The spectral gap λ of a generator $\mathcal{L}$ bounds the decay of $\mathrm{Var}_\pi(P_t f)$ for that semigroup, by Proposition 2.25(i). It bounds nothing about a different algorithm targeting the same π: for the Gaussian of Proposition 4.14 the reversible generator has gap κ^{-1} while an admissible non-reversible generator with the same stationary law has gap of order 1, and an exact independent-sampling oracle is not governed by λ at all — viewed as the Markov kernel $P(x, dy) = \pi(dy)$ it annihilates the mean-zero subspace, so its discrete-time spectral gap is 1 and it mixes in one step, whatever λ is. Consequently no statement of the form "every sampling algorithm requires time at least $f(\lambda)$" can be derived from λ.*

Proof. Proposition 2.25(i) is a statement about P_t generated by $\mathcal{L}$, and its proof uses (2.31) for that semigroup and no other. Proposition 4.14 exhibits two generators with the same π and gaps differing by a factor of order κ. For the third example, $P(x, dy) = \pi(dy)$ satisfies $Pf = \int f \, d\pi$, so $Pf = 0$ for every mean-zero f: its gap is 1 and is unrelated to λ. Hence λ is not a function of π alone and cannot bound a quantity depending only on π and the algorithm class. ∎

What can be proved is a lower bound for a *named* class of algorithms, and the following is the version this volume uses. It is honest about its restriction, which is that the algorithm moves locally.

Proposition 4.18 (Conductance barrier for local chains) DERIVED. *Let P be a Markov kernel reversible with respect to π on $\mathbb{R}^d$ whose one-step proposal is supported in a ball of radius r: $P(x,B) = 0$ whenever $\mathrm{dist}(x,B) > r$ and $x \notin B$. Let A be measurable with $\pi(A) \le \frac{1}{2}$ and let $\partial_r A = \{x : \mathrm{dist}(x,\partial A) \le r\}$. Then the conductance satisfies*

$$\Phi := \inf_{\pi(B)\le 1/2} \frac{\int_B P(x,B^c)\,\pi(\mathrm{d}x)}{\pi(B)} \le \frac{\pi(\partial_r A)}{\pi(A)}, \tag{4.15}$$

and the relaxation time is at least $(2\Phi)^{-1}$.

Proof. Take $B = A$ in the infimum. If $x \in A$ and $\mathrm{dist}(x,\partial A) > r$ then every point reachable in one step lies in A, so $P(x,A^c) = 0$; hence the numerator is at most $\pi(\partial_r A)$ and (4.15) follows. The relaxation-time bound is the easy direction of Cheeger's inequality, $\lambda \le 2\Phi$, applied to the reversible kernel. ■

What the restriction excludes. A chain with unbounded proposal range — an independence sampler, a mode-jumping proposal, a normalising flow proposal, or a diffusion model used as a sampler — is not covered, and indeed such methods can cross a barrier in one step. That is the content of the bound, not a gap in it: the exponential cost of multimodality is a cost of *local* exploration.

Where the barrier comes from. For the double well of Example 2.29, $\pi(\partial_r A)/\pi(A) \asymp re^{-\Delta}$, so (4.15) reproduces the exponential dependence on the barrier height obtained from the logarithmic Sobolev constant, by an argument that needs no functional inequality at all.

What it does not say. It is a bound on the conductance, hence on the relaxation time of that chain. It is not a lower bound over all algorithms, is not information-theoretic, and does not follow from a spectral gap; those three phrasings are what Proposition 4.17 withdraws.

4.4.2 An accounting template for samplers

Table 4.1: Sampling methods, with hypotheses in the first column. Two rows are comparable only if their first columns agree; where they do not, the remaining columns are not on the same axis and no ordering between the methods is implied. This is the volume's accounting template of Table 11.4, specialised. *Every entry is a representative bound from the literature, not a claim of optimality*, and the conditions under which each is stated are in the row itself; complexity results in this area are revised frequently, and each row should be read together with the conditions recorded in its row.

Hypotheses on the target	Method	Cost, in gradient evaluations	What is not included
LSI(α), β-smooth	ULA	$\tilde{O}(d\beta^2/(\alpha^2\varepsilon))$ in KL($\|$) (Proposition 4.3)	A bias that is removed only by taking $\eta \to 0$; correlation between iterates

continued overleaf

Table 4.1: (continued)

Hypotheses on the target	Method	Cost, in gradient evaluations	What is not included
π computable pointwise up to a constant; irreducibility	MALA	$\tilde{O}(d^{1/3})$ steps for product-like targets (Proposition 4.6)	The cost of the accept step; the deterioration of the acceptance rate away from the analysed class
μ-strongly log-concave, β-smooth	Underdamped Langevin	Continuous-time rate $\asymp \sqrt{\mu}$ at friction $\gamma \asymp \sqrt{\mu}$, against 2μ for the overdamped dynamics (Proposition 4.9)	The discretisation entirely; this row is a statement about a flow. The two rates are comparable only after normalising $\beta = 1$ (Remark 4.10)
μ-strongly log-concave, β-smooth, bounded third derivatives, warm start	HMC	$\tilde{O}(\kappa d^{1/4})$ in TV (Proposition 4.13)	The tuning of h and L; behaviour outside the class; metastability, which is untouched
Reversible, admissible c	Non-reversible perturbation	At least the reversible rate for every t (Corollary 2.26); the mean/operator spectral exponent asymptotically better by up to a factor κ (Proposition 4.14) — an exponent for e^{-tA}, not a rate for $\mathrm{Law}(X_t)$ in a named metric (Remark 4.15)	The sharp finite-time constant; the transient of mean and covariance in the Euclidean norm; discretisation without an adjustment
Bounded proposal range r; reversible	*Any such chain*	Relaxation time $\geq \pi(A)/(2\pi(\partial_r A))$ (Proposition 4.18)	Methods with unbounded range, which the bound does not cover

What does not follow

There is no information-theoretic lower bound for "any MCMC algorithm". Proposition 4.17 says why: the spectral gap is a property of a particular semigroup, and two semigroups with the same stationary law can have gaps differing by an arbitrary factor. What exists is Proposition 4.18, whose restriction — bounded proposal range — is in its statement, and query-complexity bounds in specified oracle models, whose restrictions are likewise in their statements. A bound whose class restriction is not in the same sentence as its conclusion is not a bound.

Rows of Table 4.1 are not comparable across hypotheses. The first and fourth rows report costs of $\tilde{O}(d/\kappa)$ and $\tilde{O}(d^{1/4})$, which look like a comparison and are not: they use different hypotheses, different error metrics, and one of them counts the adjustment. Comparing them requires a target satisfying both hypothesis sets, at which point the constants matter more than the exponents.

Conditioning is not dimension. Every row above carries a factor in $\kappa = \beta/\mu$ as well as one in d, and for realistic targets κ is the larger. Reporting only the exponent in d is reporting the smaller half of the cost.

4.5 Chapter Summary

Chapter 4 separates four quantities that are often quoted for one another: a continuous-time contraction rate, a discretisation bias, an iteration complexity, and a cost per step. The unadjusted and adjusted Langevin algorithms, underdamped dynamics, and non-reversible perturbations are compared only where the hypotheses in the first column of Table 4.1 agree; where they do not, no ordering between methods is defined, and Figure 4.1 is an assumptions-first map rather than a ranking.

Two limits are structural. Proposition 4.17 shows that a spectral gap belongs to one semigroup and cannot yield a lower bound over all algorithms. And by Example 2.29 a barrier degrades the *constant* in a log-Sobolev inequality rather than removing the hypothesis, so (4.4) remains true and useless on any accessible time scale. The complexity statements here are the most source-sensitive in the volume; each is representative of the literature it is drawn from, and none is claimed to be sharpest.

Exercises

Tier I Direct application.

4.1 *The ULA bias, measured*

For $V(x) = \frac{1}{2}\lambda x^2$ in $d = 1$, verify (4.2) numerically at $\lambda = 1$ and $\eta \in \{0.1, 0.5, 1.0, 1.9\}$, and plot the relative error against (4.3). Report what happens at $\eta = 2.1$ and identify which statement of Proposition 4.2 that tests.

4.2 *Leapfrog is a shear composition*

Verify by direct computation that each substep of (4.9) has unit Jacobian, and that $S \circ \Psi_h \circ S \circ \Psi_h = \mathrm{id}$. Then exhibit a one-step integrator for the same Hamiltonian that is volume preserving and *not* reversible, and say what the acceptance ratio becomes.

Tier II Completing the derivation.

4.3 *Degenerate dissipation, in coordinates*

Prove Proposition 4.8(ii) in full, following the proof of Proposition 2.12, and identify the exact point at which the x-derivatives cancel. Then explain why the same cancellation occurred in Proposition 2.16(ii) and what the two have in common.

4.4 *Stationarity of the rotated Ornstein–Uhlenbeck process*

Verify that $A\Sigma + \Sigma A^\top = 2I$ for $A = (I + S)\Sigma^{-1}$ and any antisymmetric S, and show that this is exactly the condition $\nabla\cdot(\pi c) = 0$ of Proposition 2.5(iii) written for a linear drift.

Tier III Probing the boundary.

4.5 *Where the improvement is paid for*

In the setting of Proposition 4.14 with $\kappa = 100$, take s at the threshold of (4.13) and compute $\sup_{t\geq 0} \|e^{-tA}\|$ numerically as a function of s. Determine the time at which the non-reversible dynamics first beats the reversible one, and say which quantity of Proposition 1.6 predicts the transient.

4.6 *A chain the conductance bound does not cover*

Construct a Markov chain, reversible with respect to the double well of Example 2.29, that mixes in $O(1)$ steps. Identify which hypothesis of Proposition 4.18 it violates, and state what its cost per step must be for the construction to be honest.

4.7 *Two rows, one target*

Find a target satisfying the hypotheses of both the first and the fourth rows of Table 4.1. Write down both cost expressions with all constants, and determine for which (d, κ, ε) each is smaller. Report the crossover surface and say what it shows about comparing exponents.

Tier IV Computational.

4.8 *The four methods on one target*

Take $\pi = \mathcal{N}(0, \Sigma)$ in $d = 50$ with $\Sigma = \mathrm{diag}(1, \ldots, \kappa)$ logarithmically spaced, $\kappa \in \{10, 10^3\}$.

(a) Run ULA, MALA, underdamped Langevin and HMC, tuned as well as you can, and report the number of gradient evaluations to reach a fixed error in the slowest coordinate's variance.

(b) For ULA, report the residual bias at your chosen η and compare with (4.3).

(c) Add the non-reversible perturbation of Proposition 4.14 to the overdamped dynamics and repeat. Report the improvement and compare with (4.13).

(d) State, for each of the four methods, which row of Table 4.1 your measurement corresponds to, and which column of that row your experiment does *not* test.

Acceptance criteria: (b) reports the bias as a number against the prediction; (d) names a row and an untested column for each method.

CHAPTER 5

Particle and Variational Inference

Chapter 4 approximated π by running a dynamics long enough that its law arrives there. This chapter approximates π by optimisation: a family of candidate distributions is fixed, and one of them is chosen. The two strategies exchange one difficulty for another. An exact sampler targets π itself, so it carries no *variational-family* error — though as Chapter 4 showed it still carries a discretisation bias, a Monte-Carlo error, and a mixing time that may be exponential. Variational inference has a tractable optimisation and, in exchange, a family approximation error that no amount of running time removes.

Between them sits a family of methods that keeps a finite set of interacting particles and moves them deterministically. Those are the subject of Section 5.2 and Section 5.3, and they are the reason this chapter is in this volume rather than in a book on optimisation: their mean-field limits are the flows of Chapter 2, in a different geometry.

Which geometry is the question the chapter is organised around. There are four in play — Wasserstein, Wasserstein restricted to a family, Fisher–Rao restricted to a family, and Euclidean on the parameters — and they give four different trajectories for the same functional on the same family. Table 5.2 sets them side by side; the material before it exists to make that table checkable.

5.1 Variational Inference in a Parametric Family

5.1.1 The objective

Let $\pi = Z^{-1}e^{-V}$ with Z unknown, and let $\{q_\varphi\}_{\varphi\in\Phi}$ be a family of densities on $\mathbb{R}^d$ parameterised by φ in an open subset of $\mathbb{R}^p$.

Proposition 5.1 (The evidence decomposition) DERIVED. *For every φ with $\mathrm{KL}(q_\varphi \,\|\, \pi) < \infty$,*

$$\log Z = \underbrace{\mathbb{E}_{q_\varphi}\big[-V\big] + \mathcal{H}(q_\varphi)}_{=:\,\mathcal{L}(\varphi)} + \mathrm{KL}(q_\varphi \,\|\, \pi), \tag{5.1}$$

where $\mathcal{H}(q) = -\int q \log q$ is the differential entropy. Since $\log Z$ does not depend on φ, maximising $\mathcal{L}$ is exactly minimising $\mathrm{KL}(q_\varphi \,\|\, \pi)$, and $\mathcal{L}(\varphi) \le \log Z$ with equality if and only if $q_\varphi = \pi$.

Proof. $\mathrm{KL}(q \,\|\, \pi) = \int q \log q - \int q \log \pi = -\mathcal{H}(q) + \int qV + \log Z$; rearranging gives (5.1), and non-negativity of the relative entropy gives the inequality and the equality case. ■

Equation (5.1) is the same identity as (2.13): $\mathcal{L} = -\mathcal{F}$ with the free energy of Definition 2.8, and the unknown constant $\log Z$ is again exactly what an algorithm does not have. What is new here is the restriction to a family, and everything difficult about variational inference is in that restriction.

5.1.2 Three flows, and which one is which

Suppose the family is a smooth submanifold $\mathcal{Q} \subset \mathcal{P}_2(\mathbb{R}^d)$ and write $T_\varphi\mathcal{Q}$ for its tangent space at q_φ, a subspace of $\mathrm{Tan}_{q_\varphi}\mathcal{P}_2(\mathbb{R}^d)$. Three different dynamics are called "gradient descent on the variational objective" in the literature, and they are three different curves.

Definition 5.2 (The three flows). With $\mathcal{F}[\rho] = \mathrm{KL}(\rho \,\|\, \pi)$ up to the additive constant of (5.1):

- (F1) the *unconstrained* W_2 gradient flow of $\mathcal{F}$, which by Theorem 2.13 is the Fokker–Planck equation (2.11) and leaves $\mathcal{Q}$ immediately;
- (F2) the *constrained* W_2 gradient flow, the curve in $\mathcal{Q}$ whose velocity is the $L^2(q_\varphi)$-orthogonal projection onto $T_\varphi\mathcal{Q}$ of the unconstrained velocity $-\nabla \log(q_\varphi/\pi)$;
- (F3) the *parameter-space Euclidean* gradient flow $\dot\varphi = -\nabla_\varphi \mathrm{KL}(q_\varphi \,\|\, \pi)$.

Proposition 5.3 (The constrained flow is a preconditioned parameter flow) DERIVED. *Let $\partial_i q_\varphi$ denote the tangent vector generated by ∂_{φ_i}, represented as $-\nabla\cdot(q_\varphi \psi_i)$. The field ψ_i is not unique — adding any w with $\nabla\cdot(q_\varphi w) = 0$ changes nothing — so take throughout the minimal representative, that is the unique $\psi_i \in \mathrm{Tan}_{q_\varphi}\mathcal{P}_2(\mathbb{R}^d) = \overline{\{\nabla\phi : \phi \in C_c^\infty\}}^{L^2(q_\varphi)}$, which is the one of least $L^2(q_\varphi)$ norm. With that choice set*

$$\left(G_W(\varphi)\right)_{ij} := \int \langle \psi_i, \psi_j \rangle \, q_\varphi \, \mathrm{d}x. \tag{5.2}$$

Then (F2) is the flow

$$\dot\varphi = -G_W(\varphi)^{-1} \, \nabla_\varphi \mathrm{KL}(q_\varphi \,\|\, \pi), \tag{5.3}$$

and it is invariant under smooth reparameterisation of the family, whereas (F3) is (5.3) with G_W replaced by the identity and is not. Both statements presuppose that $G_W(\varphi)$ is invertible, equivalently that the chart has full rank at φ — that the minimal representatives $\psi_1, \ldots, \psi_k$ are linearly independent in $L^2(q_\varphi)$. At a parameter where the family is degenerate, G_W drops rank, (5.3) is to be read with a pseudoinverse, and the projection in (F2) is onto a smaller space than the parameter count suggests.

Proof. A curve in $\mathcal{Q}$ has velocity $\sum_i \dot\varphi_i \psi_i$, and by Proposition 2.12 the rate of decrease of $\mathcal{F}$ along it is $\langle \nabla \log(q_\varphi/\pi), \sum_i \dot\varphi_i \psi_i \rangle_{L^2(q_\varphi)} = \sum_i \dot\varphi_i \partial_{\varphi_i} \mathrm{KL}(q_\varphi \,\|\, \pi)$, the last equality being the chain rule.

Rather than maximise the decrease under a speed constraint — which fixes the *direction* of the motion but leaves its speed to a normalisation — take the gradient-flow definition directly: $\dot\varphi$ is the velocity in $\mathcal{Q}$ whose associated tangent vector is the W_2-orthogonal projection of $-\nabla \log(q_\varphi/\pi)$ onto the span of $\psi_1, \ldots, \psi_k$. The projection is characterised by the normal equations

$$\left\langle \sum_i \dot\varphi_i \psi_i + \nabla \log \frac{q_\varphi}{\pi}, \psi_j \right\rangle_{L^2(q_\varphi)} = 0 \qquad (j = 1, \ldots, k),$$

that is $\sum_i (G_W)_{ji} \dot\varphi_i = -\partial_{\varphi_j} \mathrm{KL}(q_\varphi \,\|\, \pi)$ by (5.2) and the chain rule above. With G_W invertible this is exactly (5.3), with no free constant left over. Invariance holds because G_W transforms as a metric tensor and ∇_φ as a covector, so (5.3) transforms as a vector; replacing G_W by the identity destroys that, since the identity is not a tensor. ■

Example 5.4 (The same family, two coordinates, two trajectories). Take $d = 1$, $\pi = \mathcal{N}(0,1)$ and the family $q_{m,\varsigma} = \mathcal{N}(m, \varsigma^2)$, so

$$\mathrm{KL}(q \,\|\, \pi) = \tfrac{1}{2}\left(m^2 + \varsigma^2 - 1\right) - \log \varsigma. \tag{5.4}$$

For Gaussians the W_2 distance is $W_2^2 = (m - m')^2 + (\varsigma - \varsigma')^2$, so in the coordinates (m, ς) the tensor (5.2) is the identity, and (F2) and (F3) coincide:

$$\dot m = -m, \qquad \dot\varsigma = -\left(\varsigma - \frac{1}{\varsigma}\right) = -\frac{\varsigma^2 - 1}{\varsigma}. \tag{5.5}$$

Reparameterise the same family by the variance $v = \varsigma^2$. Then $\mathrm{KL}(\,\|\,) = \frac{1}{2}(m^2 + v - 1 - \log v)$ and (F3) in these coordinates is $\dot v = -\frac{1}{2}(1 - v^{-1})$, that is

$$\dot\varsigma = \frac{\dot v}{2\varsigma} = -\frac{\varsigma^2 - 1}{4\varsigma^3}, \tag{5.6}$$

which differs from (5.5) by the factor $4\varsigma^2$ in the ς component alone — and that asymmetry, not the mere presence of a non-constant factor, is what matters. Written back in the (m, ς) coordinates the two vector fields are

$$\left(-m, -\frac{\varsigma^2 - 1}{\varsigma}\right) \quad \text{and} \quad \left(-m, -\frac{\varsigma^2 - 1}{4\varsigma^3}\right). \tag{5.7}$$

A time reparameterisation would multiply the *whole* field by a positive scalar, leaving the trajectories unchanged and only the clock different. Here the m component is identical in the two fields, forcing that scalar to be 1, while the ς component demands $1/(4\varsigma^2)$. For $m \neq 0$ and $\varsigma \neq 1$ the two are inconsistent, so the fields are not collinear and the *orbits* differ: the two parameterisations trace out different curves in $\mathcal{Q}$, not the same curve at different speeds. Only on the invariant set $\{m = 0\}$, where the first component vanishes and no constraint on the scalar survives, is the discrepancy a change of clock. Flow (F2) is unchanged by the reparameterisation, by Proposition 5.3.

Connection

One functional, four geometries. The object being minimised in this chapter is the same $\mathrm{KL}(\cdot \,\|\, \pi)$ that Chapter 2 descended in W_2 and that Chapter 4 attacked with a diffusion. What changes from method to method is not the objective but the norm in which "steepest" is measured: Euclidean in the parameters, Fisher–Rao, Wasserstein restricted to a family, or the Stein metric of Section 5.2. Figure 5.1 and Table 5.2 show the four together. Reporting a trajectory without naming the geometry is therefore not an omission of detail; it leaves the statement without a referent.

Notes on Proposition 5.1, Proposition 5.3 and Example 5.4

Assumed. Proposition 5.1: finiteness of $\mathrm{KL}(q_\varphi \,\|\, \pi)$, which fails whenever the family has lighter tails than π — the direction of the divergence is what makes a Gaussian family usable against a heavy-tailed target and unusable in the other direction, and that asymmetry is a property of the objective, not of the optimiser. For the rest: that $\mathcal{Q}$ is a smooth submanifold of $\mathcal{P}_2(\mathbb{R}^d)$ with tangent vectors representable in $\mathrm{Tan}_{q_\varphi}\mathcal{P}_2(\mathbb{R}^d)$, and that G_W is invertible — which fails exactly where the parameterisation is degenerate, for instance at $\varsigma = 0$ in Example 5.4.

What the example shows and what it does not. It shows that (F3) depends on the coordinates and (F2) does not, and that the discrepancy is not a reparameterisation of time — for the reason given after (5.7), namely that the two fields are not collinear off $\{m = 0\}$, so the orbits themselves differ. Note that a non-constant factor would not by itself establish this: a state-dependent positive scalar multiplying the entire field is exactly a time change, and leaves every orbit where it was. It does not show that either flow is better. Both converge to π here, because π lies in the family; the interesting case is the one in which it does not, and then the two flows converge to different points of $\mathcal{Q}$ in general.

Where the Gaussian case is misleading. The coincidence of (F2) and (F3) in the (m, ς) coordinates is special to the Gaussian family with that parameterisation, where the Wasserstein geometry happens to be flat. It does not persist for mixtures, for normalising flows, or for the Gaussian family with a full covariance in its Cholesky parameterisation. A method validated on (5.5) has been validated on the one case that cannot distinguish the two flows — the same warning as in the notes on Proposition 1.43.

What does not follow

Maximising the evidence lower bound is not the Wasserstein gradient flow of the relative entropy. The unconstrained W_2 gradient flow of $\mathrm{KL}(\cdot \,\|\, \pi)$ is the Fokker–Planck equation of Theorem 2.13, whose solution leaves any finite-dimensional family immediately and converges to π itself. Variational inference does not follow it, cannot follow it, and does not converge to π unless $\pi \in \mathcal{Q}$. What variational inference follows is (F2) or (F3), and Proposition 5.3 says these are two different equations, of which only the first is a gradient flow in W_2 — restricted to $\mathcal{Q}$, which is a further qualification and not a technicality.

Three flows, three limits. (F1) converges to π; (F2) and (F3) converge to stationary points of $\mathrm{KL}(\cdot \,\|\, \pi)$ on $\mathcal{Q}$, which differ from each other when the objective has several. Statements comparing variational inference with sampling by comparing their rates are comparing the speed of convergence to two different limits.

A gradient in parameter space is not a gradient on the family. The expression $\nabla_\varphi \mathrm{KL}(q_\varphi \,\|\, \pi)$ is a covector: it needs a metric to become a direction. Choosing the Euclidean metric on Φ is a choice, it is invisible in the notation, and by Example 5.4 it changes the trajectory. Section 5.4 makes the third choice explicit and Table 5.2 sets all four side by side.

5.2 The Geometry of Stein Variational Gradient Descent

Section 5.1 described variational inference as descent on $\mathrm{KL}(\cdot \,\|\, \pi)$ over a parametric family, and noted that the descent trajectory in parameter space is not the descent trajectory in the space of measures. This section treats an algorithm that removes the parametric family altogether and keeps a set of interacting particles instead. It has an exact gradient-flow description — and the description is not the one usually quoted.

The pattern of the section is deliberately the same as Section 2.2. There, the Fokker–Planck equation was the steepest descent of the free energy in the Wasserstein geometry, its dissipation rate was the relative Fisher information, and a rate required a functional inequality that was a hypothesis on the target. Here, the mean-field limit of Stein variational gradient descent is the steepest descent of the *same functional* in a *different geometry*, its dissipation rate is the squared kernel Stein discrepancy, and a rate again requires a functional inequality — one that implies the ordinary log-Sobolev inequality, hence is never weaker as a hypothesis on the pair (π, k). Once the parallel is set out in that form, the statement that circulates as "SVGD is the Wasserstein gradient flow of the squared kernel Stein discrepancy" can be seen to confuse two of the three slots at once: it names the wrong geometry and promotes a dissipation rate to a driving functional. Section 5.2.4 settles the geometry half outright — the SVGD velocity is the steepest-descent direction of $\mathrm{KL}(\cdot \,\|\, \cdot)$ in the Stein tangent norm, not in the W_2 one — and shows that the second half rests on a confusion between a dissipation rate and a functional; whether KSD^2 could serve as some other flow's driving functional is left open there, and stated as Open Problem 5.17.

5.2.1 The Stein operator, the discrepancy, and the field

Throughout, $\pi \propto e^{-V}$ is a positive C^1 density on $\mathbb{R}^d$ with $s_\pi := \nabla \log \pi = -\nabla V$, and $k : \mathbb{R}^d \times \mathbb{R}^d \to \mathbb{R}$ is a positive-definite kernel, symmetric and $C^{1,1}$, with reproducing kernel Hilbert space $\mathcal{H}$. We write $\mathcal{H}^d := \mathcal{H} \otimes \mathbb{R}^d$ with $\|\phi\|^2_{\mathcal{H}^d} = \sum_{i=1}^d \|\phi_i\|^2_{\mathcal{H}}$, and

$$\kappa := \sup_{x \in \mathbb{R}^d} k(x, x) < \infty. \tag{5.8}$$

For the Gaussian kernel $k(x, y) = \exp(-|x - y|^2 / 2h)$, $\kappa = 1$.

Definition 5.5 (Stein operator and kernel Stein discrepancy). For $\phi \in \mathcal{H}^d$ with ϕ and $\nabla \cdot \phi$ integrable, the *Stein operator* of π is

$$A_\pi \phi(x) := \langle s_\pi(x), \phi(x) \rangle + \nabla \cdot \phi(x), \tag{5.9}$$

and for $\rho \in \mathcal{P}_2(\mathbb{R}^d)$ the *kernel Stein discrepancy* is

$$\mathrm{KSD}(\rho \,\|\, \pi) := \sup_{\|\phi\|_{\mathcal{H}^d} \le 1} \mathbb{E}_\rho\big[A_\pi \phi\big]. \tag{5.10}$$

The Stein identity $\mathbb{E}_\pi[A_\pi \phi] = 0$, valid whenever $\pi\phi \to 0$ at infinity, is what makes (5.10) a discrepancy: it vanishes at $\rho = \pi$. Two facts organise everything below, and both are consequences of one integration by parts.

Lemma 5.6 (The Stein functional is the Wasserstein velocity in disguise) [illegible]. *Let ρ be absolutely continuous with ρ / π differentiable and $\rho\phi \to 0$ at infinity, and set*

$$f_\rho := \nabla \log \frac{\rho}{\pi} \in L^2(\rho; \mathbb{R}^d), \tag{5.11}$$

the field whose negative drives the Wasserstein gradient flow of Section 2.2. Then for every $\phi \in \mathcal{H}^d$

$$\mathbb{E}_\rho\big[A_\pi \phi\big] = -\langle f_\rho, \phi \rangle_{L^2(\rho)}. \tag{5.12}$$

Consequently, writing $P_\rho : L^2(\rho; \mathbb{R}^d) \to \mathcal{H}^d$ for the integral operator

$$(P_\rho g)(y) := \int_{\mathbb{R}^d} k(x, y) \, g(x) \, \rho(\mathrm{d}x), \tag{5.13}$$

the supremum in (5.10) equals $\left\|\phi_\rho^*\right\|_{\mathcal{H}^d} = \mathrm{KSD}(\rho\|\pi)$, *where*

$$\boxed{\phi_\rho^* = -P_\rho f_\rho} \tag{5.14}$$

If $\mathrm{KSD}(\rho\|\pi) > 0$ *the supremum is attained at* $\phi_\rho^* / \left\|\phi_\rho^*\right\|_{\mathcal{H}^d}$. *If* $\mathrm{KSD}(\rho\|\pi) = 0$ *then* $\phi_\rho^* = 0$, *the Stein functional vanishes identically on the unit ball of* $\mathcal{H}^d$, *the supremum is 0, and there is no distinguished normalised maximiser — the case* $\rho = \pi$, *at which the flow is already at rest.*

Proof. For (5.12), integrate the divergence term by parts:

$$\int (\langle s_\pi, \phi\rangle + \nabla\cdot\phi)\rho = \int \langle \nabla \log \pi - \nabla \log \rho, \phi\rangle\, \rho = -\int \langle \nabla \log(\rho/\pi), \phi\rangle\, \rho,$$

the boundary term vanishing by hypothesis. So $\phi \mapsto \mathbb{E}_\rho[\mathcal{A}_\pi \phi]$ is the bounded linear functional $\phi \mapsto -\langle f_\rho, \phi\rangle_{L^2(\rho)}$ on $\mathcal{H}^d$. Its Riesz representer is computed from the reproducing property: for each i, $\langle f_{\rho,i}, \phi_i\rangle_{L^2(\rho)} = \int f_{\rho,i}(x)\, \langle k(x,\cdot), \phi_i\rangle_{\mathcal{H}}\, \rho(\mathrm{d}x) = \left\langle P_\rho f_\rho|_i, \phi_i\right\rangle_{\mathcal{H}}$, so the representer is $-P_\rho f_\rho$. The supremum of a linear functional over the unit ball is the norm of its representer, attained at the normalised representer. ■

Two readings of (5.14) are worth stating separately, because the section turns on keeping them apart.

- **As a field.** ϕ_ρ^* is the Wasserstein velocity $-f_\rho$ of Section 2.2, *smoothed by the kernel*. SVGD does not use a different descent direction; it uses the same one, passed through P_ρ. Everything that distinguishes the two algorithms is in that operator.
- **As a number.** $\mathrm{KSD}(\rho\|\pi)$ is the $\mathcal{H}^d$-norm of that smoothed field. It is not a free-energy-like quantity: it is built from the velocity, not from the position.

Remark 5.7 (Attribution). Definition 5.5, the closed form of ϕ_ρ^*, and the identification of the SVGD update with it belong to the modern research literature; see Liu and Wang [18] and Liu [19]. The derivation is given in full because (5.12) — which is where the Wasserstein velocity enters — is usually presented in a form that hides it.

5.2.2 The mean-field flow and what it dissipates

The n-particle SVGD algorithm applies $x_i \mapsto x_i + \eta\, \phi_{\hat\rho_n}^*(x_i)$ with $\hat\rho_n$ the empirical measure; its mean-field limit as $n \to \infty$ and $\eta \downarrow 0$, treated in Section 5.3, is the nonlinear transport equation

$$\partial_t \rho_t + \nabla\cdot(\rho_t \phi_{\rho_t}^*) = 0, \qquad \phi_{\rho_t}^* = -P_{\rho_t} f_{\rho_t}. \tag{5.15}$$

Two limits are being taken there, $n \to \infty$ and $\eta \to 0$, and this volume does not establish that they may be interchanged; that this section studies (5.15) and not the algorithm is a restriction recorded in the notes on Proposition 5.8.

Proposition 5.8 (Dissipation identity) DERIVED. *Let (ρ_t) be a positive classical solution of (5.15) with enough decay for the integrations by parts of Lemma 5.6. Then*

$$\frac{\mathrm{d}}{\mathrm{d}t}\mathrm{KL}(\rho_t \,\|\, \pi) = -\langle f_{\rho_t}, P_{\rho_t} f_{\rho_t}\rangle_{L^2(\rho_t)} = -\left\|\phi^*_{\rho_t}\right\|^2_{\mathcal{H}^d} = -\,\mathrm{KSD}^2(\rho_t \,\|\, \pi). \tag{5.16}$$

In particular $\mathrm{KL}(\rho_t \,\|\, \pi)$ *is non-increasing, and it is strictly decreasing whenever* $\mathrm{KSD}(\rho_t \,\|\, \pi) > 0$.

Proof. As in Section 2.2, for any velocity field v generating the flow, $\frac{\mathrm{d}}{\mathrm{d}t}\mathrm{KL}(\rho_t \,\|\, \pi) = \int \rho_t \langle f_{\rho_t}, v\rangle$. Substituting $v = \phi^*_{\rho_t} = -P_{\rho_t} f_{\rho_t}$ gives the first equality. For the second, the reproducing property gives, for any $g \in L^2(\rho)$,

$$\|P_\rho g\|^2_{\mathcal{H}^d} = \int \langle k(x,\cdot)g(x), P_\rho g\rangle_{\mathcal{H}^d}\,\rho(\mathrm{d}x) = \int \langle g(x), (P_\rho g)(x)\rangle\,\rho(\mathrm{d}x) = \langle g, P_\rho g\rangle_{L^2(\rho)},$$

so the two expressions agree; note in passing that this shows $\langle g, P_\rho g\rangle_{L^2(\rho)} \geq 0$, i.e. P_ρ is a positive operator on $L^2(\rho)$. The third equality is $\mathrm{KSD} = \left\|\phi^*_\rho\right\|_{\mathcal{H}^d}$ from Lemma 5.6. ■

Equation (5.16) is the sentence this section exists to make precise, and it is worth setting beside its Wasserstein counterpart. The following table is the section's summary; it is also the one place a reader should look before writing any sentence containing the words "gradient flow" and "Stein".

Table 5.1: The same functional, descended in two geometries. Only the middle row differs between the columns in any essential way; every other difference follows from it. The convention of Section 2.2 applies: a gradient flow is undefined until its metric is named.

	Wasserstein geometry	Stein geometry
Norm on velocities	$\|\cdot\|_{L^2(\rho)}$	$\|\cdot\|_{\mathcal{H}^d}$
Functional descended	$\mathrm{KL}(\cdot \,\|\, \pi)$	$\mathrm{KL}(\cdot \,\|\, \pi)$ — the same one
Velocity field	$-f_\rho = -\nabla \log(\rho/\pi)$	$-P_\rho f_\rho$, the kernel smoothing of it
Dissipation rate	$\mathcal{I}(\rho \,\|\, \pi) = \|f_\rho\|^2_{L^2(\rho)}$	$\mathrm{KSD}^2(\rho \,\|\, \pi) = \|P_\rho f_\rho\|^2_{\mathcal{H}^d}$
Rate requires	$\mathrm{LSI}(\alpha)$	Stein-LSI(α_S); implies $\mathrm{LSI}(\alpha_S/\kappa)$, hence never weaker (Corollary 5.19)
Algorithm	Langevin, ULA, MALA	SVGD

5.2.3 Which geometry, exactly

Theorem 2.10 characterised the Wasserstein metric by an infimum of $L^2(\rho)$ velocity norms over solutions of the continuity equation. The Stein geometry is defined by replacing that norm, and nothing else.

Definition 5.9 (Stein metric tensor). For $\rho \in \mathcal{P}_2(\mathbb{R}^d)$ and a signed measure σ with $\sigma(\mathbb{R}^d) = 0$, set

$$g^S_\rho(\sigma,\sigma) := \inf\left\{ \|\phi\|^2_{\mathcal{H}^d} \;:\; \phi \in \mathcal{H}^d,\ \sigma + \nabla\cdot(\rho\phi) = 0 \right\}, \tag{5.17}$$

with the convention $\inf \varnothing = +\infty$.

Theorem 5.10 (Stein-norm steepest-descent structure of $\mathrm{KL}(\cdot \,\|\, \pi)$) ESTABLISHED. *Let $\rho \in \mathcal{P}_2(\mathbb{R}^d)$ satisfy the hypotheses of Lemma 5.6 and suppose $\mathrm{KSD}(\rho\|\pi) < \infty$. Then:*

(i) ϕ_ρ^* is the direction of steepest descent of $\mathrm{KL}(\cdot \,\|\, \pi)$ among perturbations of unit $\mathcal{H}^d$ norm: for $T_\varepsilon = \mathrm{id} + \varepsilon\phi$,

$$\frac{\mathrm{d}}{\mathrm{d}\varepsilon}\bigg|_{\varepsilon=0} \mathrm{KL}((T_\varepsilon)_\# \rho \,\|\, \pi) = -\mathbb{E}_\rho[\mathcal{A}_\pi \phi] = \left(f_\rho, \phi\right)_{L^2(\rho)}, \tag{5.18}$$

whose minimum over $\|\phi\|_{\mathcal{H}^d} \le 1$ is $-\mathrm{KSD}(\rho\|\pi)$. When $\mathrm{KSD}(\rho\|\pi) > 0$ it is attained at $\phi_\rho^*/\mathrm{KSD}(\rho\|\pi)$; when $\mathrm{KSD}(\rho\|\pi) = 0$ the first variation vanishes in every admissible direction and the minimum is 0, with no distinguished minimiser;

(ii) ϕ_ρ^* attains the infimum in (5.17) for the tangent vector $\sigma = -\nabla\cdot(\rho\phi_\rho^*)$ it generates, so that $g_\rho^S(\sigma, \sigma) = \mathrm{KSD}^2(\rho\|\pi)$.

Both parts are statements about a variational problem in $\mathcal{H}^d$ at a single ρ, and both are proved immediately below, before the metric-space reading is stated.

Proof of Theorem 5.10. (i) is a computation. Under T_ε the pushforward density is $\rho_\varepsilon = \rho \circ T_\varepsilon^{-1} \left|\det \nabla T_\varepsilon\right|^{-1}$; differentiating $\mathrm{KL}(\rho_\varepsilon \,\|\, \pi) = \int \rho \log \rho - \int \rho \log \pi \circ T_\varepsilon - \int \rho \log \left|\det \nabla T_\varepsilon\right|$ at $\varepsilon = 0$ gives $-\int \rho(\langle s_\pi, \phi\rangle + \nabla\cdot\phi) = -\mathbb{E}_\rho[\mathcal{A}_\pi\phi]$, and Lemma 5.6 rewrites it. Minimisation of a linear functional over a ball is Lemma 5.6 again.

(ii) Suppose $\psi \in \mathcal{H}^d$ also satisfies $\nabla\cdot(\rho\psi) = \nabla\cdot(\rho\phi_\rho^*)$. Then $\left(f_\rho, \psi - \phi_\rho^*\right)_{L^2(\rho)} = 0$ by (5.12), since $\mathbb{E}_\rho[\mathcal{A}_\pi(\psi - \phi_\rho^*)]$ depends on $\psi - \phi_\rho^*$ only through $\nabla\cdot(\rho(\cdot))$. Hence by (5.14) and the representer property, $\left(\phi_\rho^*, \psi - \phi_\rho^*\right)_{\mathcal{H}^d} = 0$, so $\|\psi\|_{\mathcal{H}^d}^2 = \left\|\phi_\rho^*\right\|_{\mathcal{H}^d}^2 + \left\|\psi - \phi_\rho^*\right\|_{\mathcal{H}^d}^2 \ge \left\|\phi_\rho^*\right\|^2$.

Both parts are complete as stated. ■

The sentence usually attached to these two — that SVGD "is the gradient flow of $\mathrm{KL}(\cdot \,\|\, \pi)$ in the Stein geometry" — is a statement of a different kind, about a curve in a metric space rather than about a variational problem at a point. It is separated here rather than appended, because the two have different proof obligations and only the first is discharged in this volume.

Interpretive Reading 5.11 (SVGD as a formal gradient flow in the Stein geometry) INTERPRETIVE. Theorem 5.10 proves the fixed-law variational statements that motivate the phrase "gradient flow of $\mathrm{KL}(\cdot \,\|\, \pi)$ in the Stein geometry". Turning those statements into a metric-space gradient-flow theorem would additionally require a distance induced by g^S, a compatible notion of absolutely continuous curve and metric derivative, and a tangent-space identification matching (5.17). Those structures are not constructed in this volume. The phrase is therefore used here as a formal geometric reading: at the level at which g^S has been introduced, the flow (5.15) has metric speed, metric slope and dissipation all equal to $\mathrm{KSD}(\rho_t\|\pi)$ and its square, in the same pattern as the energy dissipation equality of Section 2.2. Nothing in this volume is derived from the reading.

Remark 5.12 (Why the split, and what depends on which half). Theorem 5.10 is a pair of finite computations: a first variation and an attained infimum, both at a fixed ρ, both proved. Interpretive Reading 5.11 quantifies over curves and presupposes a metric space this volume does not construct, which is why it is tagged INTERPRETIVE rather than carried as a theorem. Merging the two would advertise the proof strength of the first for the second, which is precisely the substitution Table 11.2 calls a *quantifier error*. Nothing downstream uses Interpretive Reading 5.11: the dissipation identity (5.16), Corollary 5.19 and Proposition 5.15 all rest on Theorem 5.10 and on computations in $\mathcal{H}^d$. Table 11.3 carries the variational statement as row A6; the metric-space reading is held outside the canonical table as unresolved relation U1 of Section 11.8. See also Remark 5.13.

Remark 5.13 (What "gradient flow in the Stein geometry" is asserting). Interpretive Reading 5.11 follows from Theorem 5.10 and Proposition 5.8 *given* its hypothesis — that $(\mathcal{P}_2(\mathbb{R}^d), g^S)$ is a metric space in which the metric derivative is computed by (5.17) — and that hypothesis is the whole difficulty. The Stein construction does not in general induce a complete metric on $\mathcal{P}_2(\mathbb{R}^d)$, the tangent-space identification is delicate, and the standard phrase does not by itself say which of the two possible readings is meant: a rigorous metric-space gradient-flow theorem, or a formal Riemannian description in which "gradient flow" abbreviates the pair of variational facts already proved. This volume does not construct that metric space, and so uses the phrase in the second sense only: read Interpretive Reading 5.11 as the organising picture and Theorem 5.10 as the mathematics.

5.2.4 What the squared discrepancy is not

Proposition 5.14 (SVGD dissipates KL(‖) no faster than Langevin at the same law) DERIVED. *For every ρ satisfying the hypotheses of Lemma 5.6,*

$$\mathrm{KSD}^2(\rho \,\|\, \pi) \;\le\; \kappa\, \mathcal{I}(\rho \,\|\, \pi), \tag{5.19}$$

with κ as in (5.8). Consequently, if $\kappa \le 1$ — as for the Gaussian kernel — the flow (5.15) is not the W_2 gradient flow of $\mathrm{KL}(\cdot \,\|\, \pi)$ *at any ρ where the inequality is strict.*

Proof. By Proposition 5.8, $\mathrm{KSD}^2(\rho\|\pi) = \langle f_\rho, P_\rho f_\rho \rangle_{L^2(\rho)^d}$, and P_ρ is a positive self-adjoint operator on $L^2(\rho;\mathbb{R}^d)$ with trace $\int k(x,x)\rho(\mathrm{d}x) \le \kappa$ per component, hence with operator norm at most κ. Therefore $\langle f_\rho, P_\rho f_\rho \rangle \le \kappa \|f_\rho\|^2_{L^2(\rho)} = \kappa\, \mathcal{I}(\rho \,\|\, \pi)$. The W_2 gradient flow of $\mathrm{KL}(\cdot \,\|\, \pi)$ dissipates at rate exactly $\mathcal{I}(\rho \,\|\, \pi)$ (Section 2.2); a flow dissipating at a strictly smaller rate at a given ρ is a different flow. ∎

This settles one half of the misstatement. The other half is the claim that the driving functional is KSD^2 rather than KL(‖), and it fails for a more basic reason.

Proposition 5.15 (The SVGD velocity need not be a gradient field) DERIVED. *Let $d \ge 2$, let k be the Gaussian kernel with bandwidth $h > 0$, and let ρ be absolutely continuous with finite first moment.*

(i) For any positive C^1 target π with score s_π, the antisymmetric part of $\nabla\phi^*_\rho$ is

$$\partial_j \phi^*_{\rho,i}(y) - \partial_i \phi^*_{\rho,j}(y) = -\frac{1}{h}\int_{\mathbb{R}^d} \big[s_{\pi,i}(x)(y_j - x_j) - s_{\pi,j}(x)(y_i - x_i)\big] k(x,y)\,\rho(\mathrm{d}x). \tag{5.20}$$

(ii) For the Gaussian target $\pi = \mathcal{N}(0, I)$, where $s_\pi(x) = -x$, this specialises to

$$\partial_j \phi^*_{\rho,i}(y) - \partial_i \phi^*_{\rho,j}(y) = +\frac{1}{h} \int_{\mathbb{R}^d} (x_i y_j - x_j y_i)\, k(x,y)\, \rho(\mathrm{d}x). \tag{5.21}$$

(iii) There exist π, k and ρ for which (5.21) is non-zero on an open set — Example 5.16 is one, with $\pi = \mathcal{N}(0, I)$. For such a ρ, ϕ^*_ρ is not of the form $-\nabla \Psi$ for any Ψ, and in particular it is not $-\nabla \frac{\delta \mathcal{G}}{\delta \rho}$ for any functional $\mathcal{G}$ — $\mathrm{KSD}^2(\cdot \| \pi)$ included.

The quantifier in (iii) is existential and cannot be made universal: at $\rho = \pi = \mathcal{N}(0, I)$ the right-hand side of (5.21) vanishes identically, as it must, since there $\phi^*_\rho \equiv 0$. What fails is therefore the claim that the SVGD field is always a gradient, which is what a variational description of the flow would require.

Proof. Write $\phi^*_\rho(y) = \int \Phi(x, y)\rho(\mathrm{d}x)$ with, by (5.14) and the integration by parts of Lemma 5.6, $\Phi(x, y) = \left[s_\pi(x) + \frac{y-x}{h}\right]k(x, y)$ for the Gaussian kernel. Fix x and put $w(y) = s_\pi(x) + (y - x)/h$, so $\nabla_y w = h^{-1}I$ is symmetric and contributes nothing to the antisymmetric part; the remaining contribution is $w_i \partial_j k - w_j \partial_i k$ with $\partial_j k = -h^{-1}(y_j - x_j)k$. With $u := y - x$,

$$w_i u_j - w_j u_i = \left(s_{\pi,i}(x) + \tfrac{u_i}{h}\right)u_j - \left(s_{\pi,j}(x) + \tfrac{u_j}{h}\right)u_i = s_{\pi,i}(x)u_j - s_{\pi,j}(x)u_i,$$

the $u_i u_j$ terms cancelling. Hence the antisymmetric part of $\nabla_y \Phi(x, y)$ equals

$$-\frac{k(x,y)}{h}\left[s_{\pi,i}(x)(y_j - x_j) - s_{\pi,j}(x)(y_i - x_i)\right].$$

Integrating in x against ρ gives (5.20), which is (i) and holds for every target.

For (ii) put $s_\pi(x) = -x$. The bracket becomes $-x_i(y_j - x_j) + x_j(y_i - x_i) = x_j y_i - x_i y_j = -(x_i y_j - x_j y_i)$, and the leading $-1/h$ turns this into $+1/h$, which is (5.21); the two sign reversals are the step at which it is easiest to lose one.

For (iii), a C^1 field whose antisymmetric derivative part is non-zero on an open set is not a gradient, by equality of mixed partials, and Example 5.16 exhibits such a ρ. Conversely, taking $\rho = \pi = \mathcal{N}(0, I)$ in (5.21) and using $\int x\, k(x, y)\pi(\mathrm{d}x) = g(y)\, y/(1 + h)$ makes the integrand antisymmetric in (i, j) under a symmetric measure, so the curl vanishes: no universal statement is available. ■

Example 5.16 (The curl in closed form). Take $d = 2$, $\pi = \mathcal{N}(0, I)$ so $s_\pi(x) = -x$, the Gaussian kernel with bandwidth h, and $\rho = \mathcal{N}(m, I)$ with $m = (\mu, 0)$. The Gaussian integral in (5.21) is elementary: $\int k(x, y)\rho(\mathrm{d}x) = \left(\frac{h}{h+1}\right)^{d/2} e^{-|y-m|^2/2(h+1)} =: g(y) > 0$ and $\int x\, k(x, y)\rho(\mathrm{d}x) = g(y)\, \frac{hm+y}{h+1}$, whence

$$\partial_1 \phi^*_{\rho,2}(y) - \partial_2 \phi^*_{\rho,1}(y) = -\frac{\mu\, y_2}{h+1}\, g(y), \tag{5.22}$$

which is non-zero for every y with $y_2 \neq 0$ as soon as $\mu \neq 0$, and which vanishes identically at $\mu = 0$, that is at $\rho = \pi$ — the boundary case that forbids the universal quantifier in Proposition 5.15(iii). Note that the sign here is the one produced by (5.21) with $i = 2$, $j = 1$:

the two negative contributions $\int x_1 k\rho \ni h\rho$ and the leading index order cancel to leave a single minus. The rotational component is largest when the particle cloud is far from the target, decays as $\rho \to \pi$, and vanishes identically in $d = 1$, where every field is a gradient. The last remark explains why the error corrected here survives one-dimensional intuition and one-dimensional experiments intact.

Open Problem 5.17 (Is the SVGD curve a Wasserstein gradient flow of something?) OPEN PROBLEM. Proposition 5.15 rules out that the SVGD *velocity field* is, for every ρ, minus the gradient of a first variation. It does not rule out that the *curve* (ρ_t) solving (5.15) coincides with the W_2 gradient flow of some functional $\mathcal{G} \neq \mathrm{KL}(\cdot \| \pi)$, because only the $L^2(\rho_t)$-tangential part of the velocity moves the measure, and that part *is* a gradient. Proposition 5.14 settles the case $\mathcal{G} = \mathrm{KL}(\cdot \| \pi)$ and nothing more. Is there a kernel, a target, and a functional $\mathcal{G}$ for which the two curves agree; and if not, what is the obstruction? A negative answer would presumably proceed by showing that the energy dissipation equality fails for every candidate $\mathcal{G}$, which requires controlling $\Pi_\rho \phi^*_\rho$ rather than ϕ^*_ρ.

What does not follow

Nothing here identifies SVGD as the Wasserstein gradient flow of $\mathrm{KSD}^2(\cdot \| \pi)$. What is exact is the pair of statements proved above: the pointwise SVGD direction is the steepest descent of $\mathrm{KL}(\cdot \| \pi)$ in the Stein tangent norm (Theorem 5.10), and KSD^2 is the corresponding dissipation rate (Proposition 5.8) — occupying exactly the slot the relative Fisher information occupies in Section 2.2. Reading a dissipation rate as the functional being descended would equally license calling the Fokker–Planck equation the gradient flow of the Fisher information, and the vocabulary should not permit that.

Two arguments that do not close the question, and it is worth saying why. That KSD^2 is the dissipation of $\mathrm{KL}(\|)$ does not logically prevent it from also being the driving functional of some *other* gradient flow. And Proposition 5.15, which shows the raw field ϕ^*_ρ need not be a gradient, compares the wrong object: only the $L^2(\rho)$-tangential part of a velocity moves the measure, so a curve can be a W_2 gradient flow while a non-tangential representative of its velocity is not a gradient. Ruling out KSD^2 as a W_2 driving functional therefore requires comparing $\Pi_\rho \phi^*_\rho$ with the Wasserstein velocity of $\delta\,\mathrm{KSD}^2/\delta\rho$ at some ρ, which is Open Problem 5.17 and is open. Until that is done the honest status of the claim is *not established, not refuted*.

There is no general exponential decay of KSD^2 along the flow. Proposition 5.8 gives monotonicity of $\mathrm{KL}(\|)$ and says nothing about the monotonicity, let alone the rate, of KSD^2 itself. An exponential rate for $\mathrm{KL}(\|)$ requires the Stein log-Sobolev inequality of Definition 5.18, which is an assumption on the pair (π, k) and is never weaker than the ordinary log-Sobolev inequality (Corollary 5.19). Formulas asserting $\mathrm{KSD}^2(\rho_t \| \pi) \le e^{-ct}\mathrm{KSD}^2(\rho_0 \| \pi)$ with a universal c do not follow from anything in this section.

KSD does not always control weak convergence. For a light-tailed kernel and a heavy-tailed target there exist sequences ρ_n with $\mathrm{KSD}(\rho_n \| \pi) \to 0$ that do not converge weakly to π: mass can escape to infinity in a region where the kernel no longer sees it. A small kernel Stein discrepancy is therefore evidence of convergence only relative to a kernel–target pair for which this failure has been excluded, and reporting KSD as a convergence diagnostic without naming the pair is not meaningful.

The mean-field flow is not the algorithm. Everything in this section concerns (5.15), which is a $n \to \infty$, $\eta \to 0$ idealisation. Finite n introduces an interaction error, finite η a discretisation error, and interchanging the two limits is not justified by anything proved here: the two errors have different sources and must be controlled separately before an exchange is legitimate (Proposition 5.22). Section 5.3 treats them separately, and no bound proved here applies to a run with 50 particles.

5.2.5 Rates, and why there is no free one

Where they fail. Proposition 5.14 is vacuous where $\kappa > 1$, since then the inequality permits a larger dissipation than Langevin's; for the Gaussian kernel $\kappa = 1$ and the comparison is informative. Proposition 5.15 fails in $d = 1$, where every field is a gradient — which is why the error it corrects survives one-dimensional intuition and one-dimensional experiments intact — and its closed form is specific to the Gaussian kernel, though the mechanism is not.

What this does not say. Together they rule out that SVGD is the W_2 gradient flow of $\mathrm{KL}(\cdot \,\|\, \pi)$, and that its velocity field is minus the gradient of any first variation. They do not rule out that the *curve* coincides with the W_2 gradient flow of some other functional; that is Open Problem 5.17.

Definition 5.18 (Stein log-Sobolev inequality). The pair (π, k) satisfies a *Stein log-Sobolev inequality* with constant $a_S > 0$ if

$$\mathrm{KL}(\rho \,\|\, \pi) \;\le\; \frac{1}{2a_S}\,\mathrm{KSD}^2(\rho \,\|\, \pi) \qquad \text{for all admissible } \rho. \tag{5.23}$$

Corollary 5.19 (Rate, and the price of the hypothesis) DERIVED. *If (5.23) holds then, along (5.15), $\mathrm{KL}(\rho_t \,\|\, \pi) \le e^{-2a_S t}\mathrm{KL}(\rho_0 \,\|\, \pi)$. Moreover (5.23) implies the ordinary log-Sobolev inequality (2.14) for π with constant a_S/κ: as a hypothesis on (π, k), the Stein condition is therefore never weaker than the Wasserstein one. Whether it is strictly stronger — that is, whether some π satisfying (2.14) admits no kernel with a positive Stein constant — is not settled by strictness of (5.19) at a particular ρ, and is not claimed here.*

Proof. The rate is Grönwall applied to Proposition 5.8, exactly as in Section 2.2. For the second claim, chain (5.23) with (5.19): $\mathrm{KL}(\rho \,\|\, \pi) \le \frac{1}{2a_S}\mathrm{KSD}^2 \le \frac{\kappa}{2a_S}\mathcal{I}(\rho \,\|\, \pi)$, which is (2.14) with $a = a_S/\kappa$. ■

Notes on Lemma 5.6, Proposition 5.8, Theorem 5.10 and Corollary 5.19

Assumed. For all four: a bounded $C^{1,1}$ kernel with $\kappa < \infty$, a target π positive and C^1, and enough decay for the integration by parts of Lemma 5.6. For the dynamical statements in addition: the mean-field equation (5.15), not the n-particle algorithm; a positive classical solution with enough decay for two integrations by parts; a bounded $C^{1,1}$ kernel with $\kappa < \infty$; for the rate, (5.23), which is a joint hypothesis on the target and the kernel.

Where it fails. (5.23) is restrictive. By Corollary 5.19 it forces an ordinary log-Sobolev inequality, so it inherits every obstruction to that one — multimodality above all — and adds its own: the kernel must be able to see the region where ρ and π differ, so a bandwidth much smaller than the separation between mass of ρ and mass of π destroys it even for a strongly log-concave target. This is not a defect of the analysis; it is the bandwidth-selection problem appearing in the constant, which is where it belongs.

> **Dimension.** Nothing in (5.16) depends explicitly on d, and this must not be read as dimension-free behaviour. The operator norm bound κ is dimension-free, but the *gap* in (5.19) is not: it is governed by how much of f_ρ survives smoothing by P_ρ, and for a fixed bandwidth that fraction degrades as d grows. Principle 5.20 and Proposition 5.29 take this up.

Principle 5.20 (Smoothing degrades with dimension at fixed bandwidth) INTERPRETIVE. At fixed bandwidth h and fixed particle number n, the repulsive component of the SVGD field weakens as d grows: for particles drawn from a product-like law, $|x_i - x_j|^2$ concentrates around a value proportional to d, so $k(x_i, x_j)$ and its gradient are near-constant across pairs and the interaction that separates particles is lost. The observable consequence is particle collapse — a set of particles that tracks the mode of π and underestimates its spread.

This is an interpretive claim, not a theorem: it organises a phenomenon and identifies its mechanism, and nothing later in this volume depends on it. The median heuristic for h is the standard response and is a rescaling that keeps the typical kernel argument of order one; whether it is sufficient is a quantitative question that Open Problem 5.30 states and does not answer.

Operational reading

The dissipation identity says what to monitor, and the two objects in it must be kept apart. Along the *mean-field* flow, $\mathrm{KSD}^2(\rho_t \| \pi)$ is exactly $-\frac{d}{dt}\mathrm{KL}(\rho_t \| \pi)$. What a run computes is $\mathrm{KSD}^2(\hat{\rho}_n \| \pi)$, evaluated at the atomic empirical measure of n particles — a different law, at which $\mathrm{KL}(\hat{\rho}_n \| \pi)$ is typically infinite for a continuous target, so the identity cannot be read off it literally.

Read the computable quantity as a descent diagnostic: an estimate of the Stein signal in the current configuration. A plateau says that this estimated signal is small. Turning that into a statement about $-\frac{d}{dt}\mathrm{KL}(\rho_t \| \pi)$ requires controlling the finite-particle and discretisation errors, which Section 5.2 does not do. Even then, a small dissipation is not a converged objective.

That is one step beyond the identity, and this box stops there. Whether SVGD or a Langevin scheme is preferable for a given problem depends on the particle budget, the smoothness of V, the kernel, and the accuracy metric; Proposition 5.14 compares only the instantaneous dissipation of the two mean-field flows at the same law, which is one input among several and is settled against Table 5.2 and Table 4.1.

Atlas entries generated by this section

R9. SVGD velocity = steepest-descent direction of $\mathrm{KL}(\cdot \| \pi)$ in the Stein tangent norm. VARIATIONAL

Exact conditions. The mean-field equation (5.15); bounded $C^{1,1}$ kernel; regularity sufficient for Lemma 5.6; the metric (5.17) and no other.

What goes wrong outside them. Replacing the Stein norm by $L^2(\rho)$ gives the Langevin flow, not this one; at finite n or finite η the identification is an approximation whose error is not controlled by anything in this section.

R10. SVGD = Wasserstein gradient flow of $\mathrm{KSD}^2(\cdot \| \pi)$. UNRESOLVED

Status. Neither asserted nor refuted here, and recorded so that a reader arriving with the claim finds it addressed. Two facts are established and neither settles it. KSD^2 is the *dissipation rate* of $\mathrm{KL}(\cdot\,\|\,\cdot)$ along the flow (Proposition 5.8), which is a different role from that of a driving functional. And a raw representative of the SVGD velocity need not be a gradient field (Proposition 5.15) — but only the $L^2(\rho)$-tangential part of a velocity moves the measure, so that proposition does not exclude a curve-level W_2 gradient-flow representation. Settling the question means comparing the tangential velocity $\Pi_\rho \phi^*_\rho$ with the Wasserstein gradient of the first variation of KSD^2; that comparison is not carried out in this volume, and is Open Problem 5.17. The relation is carried as U3 in Section 11.8.

R11. KSD small = close to the target. ANALOGY

Exact conditions. Kernel–target pairs for which KSD is known to control weak convergence.

What goes wrong outside them. Light-tailed kernel with heavy-tailed target: $\mathrm{KSD} \to 0$ along sequences that do not converge weakly.

5.3 From Particles to the Mean Field

Section 5.2 studied the equation (5.15), which is what the n-particle algorithm becomes as $n \to \infty$ and $\eta \downarrow 0$. This section is about the gap.

5.3.1 The finite system and its equilibria

Definition 5.21 (n-particle SVGD). For particles $x_1, \dots, x_n \in \mathbb{R}^d$ write $\hat\rho_n = \frac{1}{n}\sum_j \delta_{x_j}$. The continuous-time n-particle SVGD flow is

$$\dot x_i = \phi^*_{\hat\rho_n}(x_i) = \frac{1}{n}\sum_{j=1}^{n}\Big[s_\pi(x_j)k(x_j, x_i) + \nabla_{x_j}k(x_j, x_i)\Big], \qquad i = 1, \dots, n. \tag{5.24}$$

Proposition 5.22 (Fixed points of the finite system, and the two limits) DERIVED. *A configuration $(x_1, \dots, x_n)$ is a fixed point of (5.24) if and only if*

$$\sum_{j=1}^{n}\Big[s_\pi(x_j)k(x_j, x_i) + \nabla_{x_j}k(x_j, x_i)\Big] = 0 \qquad \text{for every } i. \tag{5.25}$$

For $n = 1$ and any kernel with $\nabla_x k(x, y)\big|_{y=x} = 0$ — in particular the Gaussian kernel — this reduces to $s_\pi(x_1) = 0$: every fixed point of the one-particle flow is a critical point of $\log \pi$, and is not a sample. Consequently, if the one-particle trajectory converges, its limit lies in that critical set; and more generally, for finite n, any limit of $\hat\rho_n(t)$ as $t \to \infty$ must be the empirical measure of a configuration solving (5.25), which is a quadrature object and not a sample from π. Taking the two limits in the other order gives a different kind of object, and what that requires is Corollary 5.27 below.

Proof. Equation (5.25) is (5.24) set to zero, multiplied by n. For $n = 1$ the sum has one term, $s_\pi(x_1)k(x_1, x_1) + \nabla_x k(x, x_1)|_{x=x_1}$, and the second summand vanishes by hypothesis while

$k(x_1, x_1) > 0$, leaving $s_\pi(x_1) = 0$. If a trajectory of (5.24) converges then, the right-hand side being continuous, its limit is a zero of the field, hence solves (5.25); this is the standard argument and it is conditional on convergence, which nothing here supplies. ■

Remark 5.23 (What is deterministic, and what is not). Given the initial configuration, (5.24) is a deterministic ordinary differential equation, so the whole trajectory — and any limit it has — is a deterministic function of $(x_1(0), \dots, x_n(0))$. When those points are drawn at random, as they are in practice, the limiting configuration is a random variable: what (5.25) characterises is a set in which that random variable must lie, conditional on the initial draw. "The finite system converges to a deterministic configuration" is therefore a statement conditional on the initialisation, and it is not a statement that the algorithm's output has no randomness left in it.

The fixed configurations of (5.25) are best thought of as quadrature rules: point sets, determined by the initialisation, whose empirical measure integrates a class of test functions well. That is a different object from a sample, and it is the correct mental model for what a finite SVGD run produces when it settles. It is also why the variance of an SVGD ensemble is not an estimate of the variance of π without further argument.

What connects the finite system to the mean-field flow is a stability estimate for the flow itself, together with the accuracy of the empirical measure as an approximation of the initial law. Those are two separate quantities with two separate rates, and keeping them separate is the whole content of this subsection. The bridge between them is an exact statement, and it is worth isolating before the estimate that uses it.

Lemma 5.24 (The particle system is the mean-field equation started from an empirical measure) DERIVED. *Let $x_1, \dots, x_n$ solve (5.24) on $[0, T]$ and let $\hat\rho_n(t) = \frac{1}{n}\sum_j \delta_{x_j(t)}$. Then $\hat\rho_n$ is a measure-valued solution of (5.15) on $[0, T]$ with initial datum $\hat\rho_n(0)$: for every $\varphi \in C_c^\infty(\mathbb{R}^d)$,*

$$\frac{\mathrm{d}}{\mathrm{d}t}\langle \hat\rho_n(t), \varphi\rangle = \left\langle \hat\rho_n(t), \left\langle \nabla\varphi, \phi^*_{\hat\rho_n(t)} \right\rangle \right\rangle. \tag{5.26}$$

Proof. Differentiating $\langle \hat\rho_n(t), \varphi\rangle = \frac{1}{n}\sum_j \varphi(x_j(t))$ and substituting (5.24) gives $\frac{1}{n}\sum_j \left\langle \nabla\varphi(x_j), \phi^*_{\hat\rho_n}(x_j)\right\rangle$, which is the right-hand side of (5.26); and (5.26) is the weak form of $\partial_t\rho = -\nabla\cdot(\rho\,\phi^*_\rho)$. ■

Remark 5.25 (Why this is exact here and not in Chapter 8). The velocity in (5.24) is evaluated at the empirical measure of the system itself and each particle carries no independent noise, so no averaging is involved and Lemma 5.24 is an identity rather than an approximation. The interacting diffusions of Definition 8.1 have an independent Brownian motion per particle; their empirical measure does *not* solve the McKean–Vlasov equation, and the gap is exactly what Proposition 8.2 estimates. The two chapters therefore need different arguments, and the finite-n error here is entirely an initial-data error.

Proposition 5.26 (Stability of the mean-field flow) DERIVED. *Suppose the velocity map $\Phi : (y, \rho) \mapsto \phi^*_\rho(y)$ satisfies, for some $L < \infty$ and some $p \geq 1$,*

$$\left|\Phi(y, \rho) - \Phi(\tilde y, \tilde\rho)\right| \;\leq\; L\big(|y - \tilde y| + W_p(\rho, \tilde\rho)\big). \tag{5.27}$$

Then any two measure-valued solutions of (5.15) with values in $\mathcal{P}_p(\mathbb{R}^d)$ satisfy

$$W_p(\rho_t, \tilde{\rho}_t) \;\le\; e^{2Lt}\, W_p(\rho_0, \tilde{\rho}_0), \qquad t \ge 0. \tag{5.28}$$

By Lemma 5.24 the empirical measure of the n-particle system is such a solution, so taking $\tilde{\rho}_t = \hat{\rho}_n(t)$ gives, for $t \le T$,

$$W_p(\hat{\rho}_n(t), \rho_t) \;\le\; e^{2LT}\, W_p(\hat{\rho}_n(0), \rho_0) : \tag{5.29}$$

the finite-particle error is the initial empirical-measure error, propagated. Its rate is therefore exactly the rate at which the empirical measure of the initial sample approximates ρ_0 in W_p — the rate of Table 8.1, not $n^{-1/2}$.

Proof. Let $(X_0, \tilde{X}_0)$ be an optimal coupling of ρ_0 and $\tilde{\rho}_0$ for W_p, and let $X_t, \tilde{X}_t$ solve the characteristic equations $\dot{X} = \Phi(X, \rho_t)$ and $\dot{\tilde{X}} = \Phi(\tilde{X}, \tilde{\rho}_t)$ from them. Once $\rho_\cdot$ is regarded as given, $t \mapsto \rho_t$ solves the *linear* continuity equation with the Lipschitz velocity field $v_t = \Phi(\cdot, \rho_t)$, whose measure-valued solution with a given initial datum is unique and is the pushforward of that datum along the flow of v; hence $\mathrm{Law}(X_t) = \rho_t$, and likewise $\mathrm{Law}(\tilde{X}_t) = \tilde{\rho}_t$. By (5.27), $\frac{d}{dt}\left|X_t - \tilde{X}_t\right| \le L\left(\left|X_t - \tilde{X}_t\right| + W_p(\rho_t, \tilde{\rho}_t)\right)$. Taking L^p norms, using that the pair is a coupling so that $W_p(\rho_t, \tilde{\rho}_t) \le \left\|X_t - \tilde{X}_t\right\|_{L^p}$, gives $\frac{d}{dt}\left\|X_t - \tilde{X}_t\right\|_{L^p} \le 2L\left\|X_t - \tilde{X}_t\right\|_{L^p}$, and Proposition 1.4 together with optimality of the initial coupling gives (5.28). ■

Corollary 5.27 (The iterated limit, and what it needs) DERIVED. *Assume the Lipschitz condition (5.27) of Proposition 5.26, and assume the initial configuration is chosen so that*

$$W_p(\hat{\rho}_n(0), \rho_0) \longrightarrow 0 \qquad (n \to \infty), \tag{5.30}$$

which holds almost surely for independent samples from ρ_0 with a finite p-th moment. Then for every fixed t,

$$\lim_{n \to \infty} \hat{\rho}_n(t) = \rho_t \quad \text{in } W_p,$$

where (ρ_t) is the mean-field flow (5.15) from ρ_0. If in addition the Stein-LSI hypothesis of Corollary 5.19 holds, then $\mathrm{KL}(\rho_t \,\|\, \pi) \to 0$ as $t \to \infty$, and hence

$$\lim_{t \to \infty} \lim_{n \to \infty} \hat{\rho}_n(t) = \pi, \tag{5.31}$$

the inner limit in W_p and the outer limit in relative entropy — hence also in total variation, by Pinsker. Upgrading the outer limit to W_p needs a further hypothesis, a uniform p-th moment bound along the flow being the usual one.

Proof. The first claim is (5.29) with $\tilde{\rho}_0 = \hat{\rho}_n(0)$: the right-hand side is $e^{2Lt} W_p(\hat{\rho}_n(0), \rho_0)$, which tends to zero for fixed t by (5.30). The second is Corollary 5.19. The two together give (5.31), in the topologies named. ■

> Caution
>
> Neither limit may be taken for granted from the other, and the order is not a formality. Proposition 5.22 shows that the *inner* limit $t \to \infty$ at fixed n lands on a solution of the fixed-point equation (5.25) — a quadrature configuration, not a sample. The corollary above takes $n \to \infty$ first, and it

needs two things that the fixed-point analysis does not supply: convergence of the initial empirical measure, (5.30), without which even the $t = 0$ inner limit is the wrong law; and the stability estimate (5.28), which is what carries that initial closeness forward. Both hypotheses are on the finite-n side; the mean-field convergence rate of Corollary 5.19 supplies neither.

Remark 5.28 (When the hypothesis holds). For the SVGD field (5.14), (5.27) holds with $p = 1$ when $k \in C_b^{1,1}$ and s_π is *bounded and* Lipschitz, since then $x \mapsto s_\pi(x)k(x,y) + \nabla_x k(x,y)$ is bounded Lipschitz uniformly in y and (5.24) is an integral against ρ. Boundedness of the score is a real restriction: it excludes the Gaussian target. For an unbounded score the same argument runs on a set carrying most of the mass, at the cost of a constant depending on the moments of ρ_0 and $\tilde{\rho}_0$, and the conclusion is unchanged in form.

5.3.2 Dimension and the bandwidth

Principle 5.20 asserted that the repulsive part of the field weakens as d grows at fixed bandwidth. In the Gaussian case that is a computation.

Proposition 5.29 (Kernel collapse at fixed bandwidth) DERIVED. *Let x_i, x_j be independent $\mathcal{N}(0, I_d)$ and let k be the Gaussian kernel with fixed bandwidth h. Then $|x_i - x_j|^2/d \to 2$ almost surely as $d \to \infty$, and consequently*

$$\frac{1}{d}\log k(x_i, x_j) \longrightarrow -\frac{1}{h} \qquad \textit{almost surely,} \tag{5.32}$$

equivalently $k(x_i, x_j) = e^{-d/h + o(d)}$, while $k(x_i, x_i) = 1$. The same exponential rate governs the repulsion term $\nabla_{x_j} k(x_j, x_i) = -h^{-1}(x_j - x_i)k(x_j, x_i)$:

$$\frac{1}{d}\log\left\|\nabla_{x_j} k(x_j, x_i)\right\| \longrightarrow -\frac{1}{h} \qquad \textit{almost surely,} \tag{5.33}$$

the polynomial factor $h^{-1}|x_j - x_i| \sim h^{-1}\sqrt{2d}$ being invisible on this scale. So at fixed bandwidth the interaction between distinct particles is exponentially small in d, while the self-term $k(x_i, x_i)$ is 1.

> Caution
>
> The conclusion is about an *exponential rate*, and it cannot be sharpened to a multiplicative one. The law of large numbers gives $|x_i - x_j|^2 = 2d + o(d)$, so the exponent is determined only up to $o(d)$; the typical fluctuation is in fact of order $\sqrt{d}$, which contributes a factor $e^{O(\sqrt{d})}$. Writing $k(x_i, x_j) = e^{-d/h}(1 + o(1))$ would assert that this factor tends to one, and it does not. Statements of the form "$\frac{1}{d}\log(\cdot)$ converges" are the right currency here, and the comparison the section needs — interaction against self-term — is already decided at that resolution.

Proof. $|x_i - x_j|^2 = \sum_{r=1}^d (x_{ir} - x_{jr})^2$ is a sum of d independent copies of a χ_1^2 variable scaled by 2, so the strong law gives $|x_i - x_j|^2/d \to 2$ almost surely. Hence $\frac{1}{d}\log k(x_i, x_j) = -|x_i - x_j|^2/(2hd) \to -1/h$, which is (5.32). The gradient formula is a direct computation, and its norm is $h^{-1}|x_j - x_i|\, k(x_j, x_i)$; taking $\frac{1}{d}\log$ of it adds $\frac{1}{d}\log(h^{-1}|x_j - x_i|)$, which tends to zero because $|x_j - x_i|$ grows like $\sqrt{2d}$, giving (5.33). ∎

The median heuristic sets h to the median of $|x_i - x_j|^2$, which by Proposition 5.29 is of order d; that restores kernel values of order one and, with them, the repulsion. What it does not restore is discrimination: with $h \asymp d$ the kernel is nearly constant across the typical range of pairwise distances, so the field (5.24) becomes close to the same average drift applied to every particle. The quantitative version of that statement is not proved here and is the content of Open Problem 5.30.

Notes on Proposition 5.22, Proposition 5.26 and Proposition 5.29

Assumed. Proposition 5.22: only that $\nabla_x k(x,y)|_{y=x} = 0$, which holds for any kernel of the form $k(x,y) = \kappa(|x-y|^2)$. Lemma 5.24: nothing beyond existence of the trajectories on $[0,T]$; it is an identity, and it is false for a system with independent per-particle noise (Remark 5.25). Proposition 5.26: the Lipschitz property (5.27) of the velocity map, which by Remark 5.28 needs a bounded Lipschitz score and excludes the Gaussian target without a moment-dependent constant. Proposition 5.29: independent Gaussian particles, which is the initialisation and not the state of the system after any positive time.

What the collapse statement covers. It is a statement about independent Gaussian points at fixed bandwidth. It makes Principle 5.20 quantitative in that case and does not establish it in general: after time zero the particles are dependent and not Gaussian, and no result here controls $|x_i - x_j|$ then. Principle 5.20 remains interpretive for that reason.

Two errors, not one. Proposition 5.26 is a statement about the flow and carries no n at all; n enters only through Lemma 5.24, which identifies the particle system as the same flow started from $\hat{\rho}_n(0)$. The finite-particle error is therefore $W_p(\hat{\rho}_n(0), \rho_0)$ propagated, and for an initial sample drawn independently from ρ_0 that quantity is the empirical-measure error of Table 8.1: dimension-cursed in W_2 for $d > 4$, and equal to $n^{-1/2}$ only for the tagged-particle object, which is not what (5.29) bounds. A dimension-free $n^{-1/2}$ rate for $W_2(\hat{\rho}_n(t), \rho_T)$ cannot hold for such an initialisation, because setting $t = 0$ would contradict the empirical rate.

What the two limits statement does not say. It does not say the finite system is bad, it gives no rate for the discrepancy, and — most importantly — it does not say that the finite system converges at all. What it says is that any $t \to \infty$ limit at fixed n is constrained by (5.25) and is therefore not a sample from π, so every guarantee in Section 5.2 is a statement about a horizon on which (5.29) is still informative.

Fixed point is not convergence. Equation (5.25) characterises the zeros of a vector field. Existence of a $t \to \infty$ limit is a separate question requiring a Lyapunov function, compactness, or a Łojasiewicz-type inequality for the finite system, none of which is established here; the proposition is therefore phrased conditionally throughout. Nor is the limit configuration deterministic as a random variable when the particles are initialised at random, which is Remark 5.23.

What is not proved about the two limits. That the configurations solving (5.25) fail to converge to π as $n \to \infty$ is not established here, and it may be false for good kernels: an atomic measure can converge weakly to π, and a quadrature rule is designed to. What is established is that at each finite n any limit solves (5.25) rather than being a sample, and that at $n = 1$ it is a Dirac at a critical point of $\log \pi$. Constructing a target and a kernel for which the configurations have a non-π limit is open.

Open Problem 5.30 (Bandwidth and dimension) OPEN PROBLEM. Under the median heuristic, h scales with d and the kernel values are of order one. Does the resulting mean-field flow retain a dimension-independent Stein logarithmic Sobolev constant for a fixed target family — say Gaussians with bounded condition number — or does the constant degrade with d despite the rescaling? Equivalently, is the loss identified in Proposition 5.29 removed by the heuristic or merely relocated? Proposition 5.14 bounds KSD^2 by $\kappa\, \mathcal{I}(\,\|\,)$ with $\kappa = 1$

for the Gaussian kernel and says nothing about the gap, which is where the answer lives.

5.4 Natural Gradient and the Fisher Metric

Section 2.2.1 promised that the Fisher information *matrix* would appear here under the symbol $G(\theta)$ and would not be confused with the relative Fisher information $\mathcal{I}(\cdot \,\|\, \pi)$ of Chapter 2. This section keeps that promise and completes the list of geometries.

Definition 5.31 (Fisher information matrix and the Fisher–Rao metric). For a family $\{q_\varphi\}$ with $q_\varphi > 0$ and smooth in φ, the *Fisher information matrix* is

$$G(\varphi)_{ij} := \int q_\varphi \, \partial_{\varphi_i} \log q_\varphi \, \partial_{\varphi_j} \log q_\varphi \, dx, \tag{5.34}$$

a $p \times p$ positive semidefinite matrix on the parameter space. The associated Riemannian metric on Φ is the *Fisher–Rao metric*.

Proposition 5.32 (Natural gradient flow) DERIVED. *Assume $G(\varphi) \succ 0$. The steepest-descent flow of $\varphi \mapsto \mathrm{KL}(q_\varphi \,\|\, \pi)$ with respect to the Fisher–Rao metric is*

$$\dot{\varphi} = -\, G(\varphi)^{-1} \, \nabla_\varphi \mathrm{KL}(q_\varphi \,\|\, \pi)\,, \tag{5.35}$$

and it is invariant under smooth reparameterisation of the family. Moreover G is the Hessian of the relative entropy along the family in the sense that $G(\varphi) = \nabla^2_{\varphi'} \mathrm{KL}(q_\varphi \,\|\, q_{\varphi'}) \big|_{\varphi' = \varphi}$.

Proof. The first claim is the computation of Proposition 5.3 with G_W replaced by G, since the constraint defining steepest descent is now $\langle \dot{\varphi}, G \dot{\varphi} \rangle \le 1$; invariance follows because G transforms as a metric tensor. For the second, expand $\mathrm{KL}(q_\varphi \,\|\, q_{\varphi'})$ to second order in $\varphi' - \varphi$: the zeroth and first order terms vanish, and the second-order term is $\frac{1}{2} \langle \delta, G \delta \rangle$ by differentiating $\int q_\varphi \log(q_\varphi / q_{\varphi'})$ twice and using $\int \partial_i q_\varphi = 0$. ∎

What the Hessian identity does and does not say. G is the Hessian of $\mathrm{KL}(q_\varphi \,\|\, q_{\varphi'})$ in the *second* argument at coincidence. It is not the Hessian of the objective $\mathrm{KL}(q_\varphi \,\|\, \pi)$, so (5.35) is not Newton's method, and the two agree only when the target lies in the family and the residual vanishes.

What it does not say. Nothing about convergence: invariance is a statement about coordinates and carries no rate. Nothing about cost: forming and inverting G is $O(p^3)$ in general, and every practical variant approximates it, at which point the invariance of Proposition 5.32 no longer holds for the implemented flow.

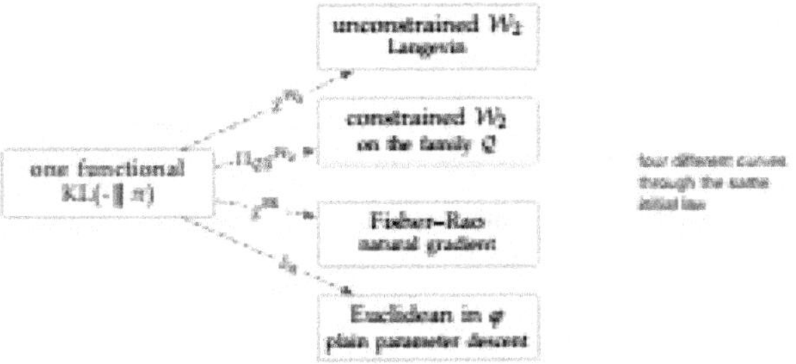

Figure 5.1: The same functional, descended in four metrics. Only the norm used to measure a velocity changes; the resulting trajectories differ, and two of the four are not invariant under a smooth reparameterisation of the family. What this figure does not imply: that the four are approximations of one another, or that any is preferable. Choosing a metric is a modelling decision, and reporting a trajectory without naming the metric is the geometry substitution catalogued in Table 11.2.

Table 5.2: Four geometries, one functional, one family. Every row descends $\mathrm{KL}(\cdot \,\|\, \pi)$; they differ only in the norm used to measure a velocity, and they trace four different curves. The last column is the property that decides whether a reported trajectory is a property of the method or of the coordinates.

Norm on velocities	Flow	Where it lives	Invariant under reparameterisation?
$L^2(\rho)$, unrestricted	Fokker–Planck (2.11), i.e. Langevin (Theorem 2.13)	All of $\mathcal{P}_2(\mathbb{R}^d)$; leaves any family at once	Not applicable: no parameters
$L^2(q_\varphi)$, restricted to $T_\varphi Q$	Constrained flow (5.3) (Proposition 5.3)	The family Q	Yes
Fisher–Rao on Φ	Natural gradient (5.35) (Proposition 5.32)	The family Q	Yes, and differs from the row above whenever $G \neq G_W$
Euclidean on Φ	$\dot{\varphi} = -\nabla_\varphi \mathrm{KL}(\,\|\,)$	The family Q	No (Example 5.4)
$\mathcal{H}^d$, unrestricted	SVGD mean-field flow (5.15) (Theorem 5.10)	All of $\mathcal{P}_2(\mathbb{R}^d)$; no family	Not applicable; depends on the kernel instead

The Fisher information matrix is not the relative Fisher information. $G(\varphi)$ of (5.34) is a $p \times p$ matrix on a parameter space, defined for a family and depending on its parameterisation through the chain rule. $\mathcal{I}(\nu \,\|\, \pi)$ of (2.14) is a scalar functional of a pair of measures, defined without any family. The two are related only through a specific parameterisation and are not interchangeable in any formula. The convention box of Section 2.2.1 fixed this and it is repeated here because the notation is the trap.

Natural gradient is not "the" gradient. By Table 5.2 it is the gradient in one of at least four geometries, and it is not the one that makes the flow a restriction of the Langevin dynamics — that is the second row, with G_W and not G. Both are invariant and they are different flows. Which is appropriate is a modelling choice about what a unit of movement should cost, and this volume does not make it.

Invariance is not optimality. Rows two and three are invariant and row four is not, which settles that row four's trajectory is partly an artefact of coordinates. It does not follow that rows two or three converge faster, reach a better stationary point, or are cheaper per step; G and G_W must be formed and inverted, and for a family with many parameters that cost dominates.

5.5 Chapter Summary

Chapter 5 treats variational inference and particle methods as descent of one functional in different geometries. Proposition 5.3 identifies the constrained Wasserstein flow as a preconditioned parameter flow and shows it is invariant under smooth reparameterisation while the plain parameter flow is not; Example 5.4 exhibits two parameterisations whose vector fields are not collinear, so the orbits differ rather than the clocks.

For SVGD the chapter proves what is provable at a point and holds the rest apart. Theorem 5.10 is a pair of variational statements in the reproducing-kernel Hilbert space; the metric-space reading is Interpretive Reading 5.11, tagged INTERPRETIVE, since this volume does not construct $(\mathcal{P}_2(\mathbb{R}^d), g^S)$. Proposition 5.15 computes the curl of the SVGD velocity and shows it need not be a gradient, with the quantifier existential rather than universal — at $\rho = \pi$ the curl vanishes, as it must. Whether KSD^2 drives some other flow is open (Open Problem 5.17).

Exercises

§5.2

Tier I Direct application.

5.1 *One particle*

Show that for $\rho = \delta_a$ and the Gaussian kernel the SVGD field satisfies $\phi^*_{\delta_a}(a) = s_\pi(a)$, so that a single particle performs gradient ascent on $\log \pi$ and converges to a mode rather than to a sample. Compute $\mathrm{KSD}^2(\delta_a \| \pi)$ for $\pi = \mathcal{N}(0,1)$ in $d = 1$ and observe that it does not vanish as $a \to 0$. Explain what the residual measures.

Hint. $\mathrm{KSD}^2(\delta_a \| \pi) = k_\pi(a,a)$ with k_π the Stein kernel; the residual is the $\partial_x \partial_y k$ term.

5.2 *The two dissipation rates on a Gaussian*

Let $\pi = \mathcal{N}(0,1)$ and $\rho = \mathcal{N}(m,1)$ in $d = 1$. Compute $\mathcal{I}(\rho \| \pi)$ exactly, compute $\mathrm{KSD}^2(\rho \| \pi)$ for the Gaussian kernel with bandwidth h, and verify (5.19) with $\kappa = 1$. Determine the limit of the ratio as $h \to \infty$ and as $h \to 0$, and say which regime recovers Langevin.

Tier II Completing the derivation.

5.3 *The steepest-descent computation*

Carry out the differentiation in Theorem 5.10(i) in full, including the term $\frac{d}{d\varepsilon}\big|_0 \log|\det(I + \varepsilon\nabla\phi)| = \nabla\cdot\phi$. Identify where the boundedness of $\nabla\phi$ is used, and state what happens for a kernel whose RKHS contains functions with unbounded gradient.

5.4 *Minimal-norm representation*

Prove Theorem 5.10(ii) without the Pythagoras shortcut, by writing the constraint $\nabla\cdot(\rho(\phi - \phi^*_\rho)) = 0$ explicitly and identifying the orthogonal complement in $\mathcal{H}^d$ of the set of fields satisfying it. Compare with the analogous statement for $\mathrm{Tan}_\rho\mathcal{P}_2(\mathbb{R}^d)$ in Theorem 2.10 and say precisely which object plays the role of "gradient fields" in the Stein geometry.

Tier III Probing the boundary. Each of these asks where a statement stops being true.

5.5 *Verifying the curl, and locating the rotation*

Verify (5.22) by direct computation. Then plot the rotational component for $\mu \in \{0.5, 2, 5\}$ and $h \in \{0.1, 1, 10\}$ and describe how it scales. Explain why the field becomes a gradient in the two limits $\mu \to 0$ and $h \to \infty$, and why neither limit is available during a run that has not yet converged.

5.6 *A dissipation rate is not a functional*

Suppose someone asserts that the Fokker–Planck equation of Section 2.2 is the W_2 gradient flow of $\mathcal{I}(\cdot\,\|\,\pi)$. Refute it by computing the first variation of $\mathcal{I}(\cdot\,\|\,\pi)$ for a Gaussian family and comparing $-\nabla\delta\mathcal{I}(\,\|\,)/\delta\rho$ with $-f_\rho$. Then state, in one sentence, why the SVGD claim examined in Section 5.2.4 has exactly the same shape — and why, unlike this one, it is left open rather than refuted.

5.7 *Where KSD stops being a diagnostic*

Let π be a Student t distribution with two degrees of freedom on $\mathbb{R}$ and k the Gaussian kernel. Construct a sequence ρ_n with $\mathrm{KSD}(\rho_n\|\pi) \to 0$ and $\rho_n \not\to \pi$. Then state what property of the kernel would have prevented it, and check whether the inverse multiquadric kernel has that property.

Hint. Move a vanishing fraction of mass to a location $t_n \to \infty$ chosen so that both $k(t_n,\cdot)$ and $s_\pi(t_n)k(t_n,\cdot)$ are negligible.

5.8 *Why the two limits may not be interchanged freely*

Consider n particles with step size η. Give a heuristic scaling argument for the interaction error at finite n and for the discretisation error at finite η, and exhibit two joint limits $(n,\eta) \to (\infty, 0)$ along which the resulting laws differ. State which of the results of this section survive each limit.

Tier IV Computational.

5.9 *Measuring the dissipation identity, and the collapse*

Take π a mixture of two Gaussians in $\mathbb{R}^d$, separated by 4 in the first coordinate and standard in the rest.

(a) For $d = 2$ and $n = 200$, run SVGD with the median-heuristic bandwidth and estimate both sides of (5.16) numerically: $\mathrm{KSD}^2(\hat{\rho}_n \| \pi)$ from its closed form, and $\frac{d}{dt}\mathrm{KL}(\,\|\,)$ by finite differences against a reference $\mathrm{KL}(\,\|\,)$ computed on a grid. Report the agreement and the step size at which it breaks down.

(b) Run Langevin from the same initialisation and plot both dissipation rates against $\mathrm{KL}(\,\|\,)$. Confirm (5.19) empirically and report the size of the gap.

(c) Fix $n = 50$ and vary $d \in \{2, 10, 50, 200\}$. Measure the standard deviation of pairwise particle distances and the fraction of the target's variance recovered, with and without the median heuristic.

(d) Which of the three effects — finite n, finite η, and the dimensional degradation of Principle 5.20 — does each figure isolate? Name, for each, the experiment you would have to run to separate it from the other two.

Acceptance criteria: (a) reports the breakdown step size, not only the agreement; (b) reports the gap as a number; (d) is answered with the design of the separating experiment.

Tier I Direct application.

5.10 *The two Gaussian trajectories*

Integrate (5.5) and (5.6) from $\varsigma(0) = 3$ and plot both against t. Report the times at which each first reaches $\varsigma = 1.1$ and the ratio of the two. Then verify that the two curves in the (m, ς) plane coincide as sets and differ as parameterised curves.

5.11 *Fisher matrix for the Gaussian family*

Compute $G(\varphi)$ from (5.34) for $q_{m,\varsigma} = \mathcal{N}(m, \varsigma^2)$ in the coordinates (m, ς) and in (m, v) with $v = \varsigma^2$. Verify that (5.35) gives the same curve in both, and compare it with $G_W = I$ from Example 5.4. State in one sentence what the comparison shows about rows two and three of Table 5.2.

Tier II Completing the derivation.

5.12 *Steepest descent on a submanifold*

Fill in the variational argument of Proposition 5.3: maximise a linear functional over the ellipsoid $\langle \dot{\varphi}, G_W \dot{\varphi} \rangle \leq 1$ and verify the normalisation. Then show that the maximal rate of decrease is $(\langle \nabla_\varphi \mathrm{KL}(\,\|\,), G_W^{-1} \nabla_\varphi \mathrm{KL}(\,\|\,) \rangle)^{1/2}$, and identify this quantity as the metric slope of Theorem 2.13 restricted to $\mathcal{Q}$.

5.13 *The Hessian identity*

Prove the second claim of Proposition 5.32 by expanding $\mathrm{KL}(q_\varphi \| q_{\varphi'})$ to second order, and explain why the same expansion of $\mathrm{KL}(q_{\varphi'} \| q_\varphi)$ gives the same matrix. Say what this implies about which of the two arguments the Fisher–Rao metric is insensitive to, and to what order.

Tier III Probing the boundary.

5.14 *A family where the two invariant flows differ*

Take the family of Gaussians on $\mathbb{R}^2$ with covariance $\varsigma^2 R_\theta \operatorname{diag}(1, \varrho) R_\theta^\top$, parameterised by $(\varsigma, \varrho, \theta)$. Compute G and G_W and show they are not proportional. Then exhibit an initial condition from which the natural-gradient and constrained-Wasserstein flows reach different stationary points of $\mathrm{KL}(\cdot \,\|\, \pi)$ for a bimodal π.

5.15 *Two particles, one fixed point*

Solve (5.25) for $n = 2$, $d = 1$, $\pi = \mathcal{N}(0, 1)$ and the Gaussian kernel with bandwidth h. Show that the fixed configuration is symmetric, $x_2 = -x_1$, and compute x_1 as a function of h. Compare the empirical variance x_1^2 with 1, and describe the limits $h \to 0$ and $h \to \infty$. State which of the two limits recovers a sample and which recovers a mode.

5.16 *Where propagation of chaos stops being useful*

Using (5.29) together with the empirical rate (8.6), and taking T to be the time at which Corollary 5.19 says the mean-field flow has reached accuracy ε, solve for the n needed to keep the particle error below ε as well. Report the dependence on ε, on a_ς / L and on d, and say what happens when $L > a_\varsigma$. The dimension enters through the empirical rate and not through the stability estimate; say which of the two dominates for $d > 4$.

Tier IV Computational.

5.17 *Four geometries on one problem*

Take π a two-dimensional Gaussian with condition number $\kappa = 50$ and $\mathcal{Q}$ the family of diagonal Gaussians, which does not contain π.

(a) Integrate all four parametric flows of Table 5.2 from a common start and plot the trajectories in the $(\varsigma_1, \varsigma_2)$ plane. Report whether they reach the same stationary point.

(b) Repeat row four in three parameterisations — ς, ς^2 and $\log \varsigma$ — and report the three trajectories and the three times to a fixed accuracy.

(c) Run n-particle SVGD for $n \in \{2, 10, 100\}$ and compare its empirical covariance with that of π and with the variational optima from (a). Report the deficit at each n.

(d) State which of your measurements are properties of the method and which are properties of the coordinates, and which of (a)–(c) tests Proposition 5.22.

Acceptance criteria: (b) reports three distinct trajectories with their times; (d) classifies every measurement.

CHAPTER 6

Stochastic Optimization Dynamics

Training is a discrete stochastic iteration, and this chapter asks what continuous-time objects describe it. The answer is: several, in different regimes, and none of them in the regime where training actually runs.

That sounds like a negative result and is not meant as one. The Euclidean gradient flow of Section 1.2 is exact as $\eta \to 0$; the stochastic modified equation of Section 6.2 is a weak approximation of a stated order at small η; the edge-of-stability phenomenon of Section 6.4 is what is measured at the step sizes people use. Each is a correct statement about its own regime. What this chapter is careful about is the boundaries between them, because almost every claim about "the SDE view of SGD" that turns out to be wrong is a correct claim from one regime asserted in another. Section 6.5 is a list of the specific crossings, and Table 6.1 is the map.

6.1 Stochastic Gradient Descent and the Structure of Its Noise

Definition 6.1 (Stochastic gradient descent). Let $L(\theta) = \frac{1}{n}\sum_{i=1}^{n} \ell_i(\theta)$ with $\ell_i \in C^2(\mathbb{R}^p)$. Given a batch B_k of size B drawn uniformly with replacement and independently across k, stochastic gradient descent is

$$\theta_{k+1} = \theta_k - \eta\, \hat{g}(\theta_k, B_k), \qquad \hat{g}(\theta, B) = \frac{1}{B}\sum_{i \in B} \nabla \ell_i(\theta). \tag{6.1}$$

Proposition 6.2 (Mean, covariance, and rank of the gradient noise) DERIVED. *With sampling with replacement,* $\mathbb{E}[\hat{g}(\theta, B)] = \nabla L(\theta)$ *and*

$$\Sigma(\theta) := \mathrm{Cov}(\hat{g}(\theta, B)) = \frac{1}{B} C(\theta), \qquad C(\theta) = \frac{1}{n}\sum_{i=1}^{n} (\nabla \ell_i - \nabla L)(\nabla \ell_i - \nabla L)^{\top}. \tag{6.2}$$

Moreover $\operatorname{rank} C(\theta) \le n - 1$. *Each individual realisation of the noise* $\hat{g} - \nabla L$ *is a linear combination of at most* B *centred per-example gradients, but the span of all possible realisations is contained in* $\operatorname{span}\{\nabla \ell_i - \nabla L\}_{i=1}^{n}$, *of dimension at most* $\min(p, n-1)$, *with no bound in terms of* B*: even at* $B = 1$, *different draws produce noise vectors spanning up to* $n - 1$ *dimensions. Consequently for* $p > n$ *the noise is supported on a proper subspace of* $\mathbb{R}^p$ *at every* θ, *whatever the batch size. For a least-squares loss* $\ell_i(\theta) = \frac{1}{2} r_i(\theta)^2$,

$$C(\theta) = \frac{1}{n}\sum_i r_i^2 \nabla r_i \nabla r_i^{\top} - \nabla L \nabla L^{\top}, \tag{6.3}$$

which vanishes at any interpolating parameter, where every residual r_i *is zero.*

Proof. Unbiasedness is linearity of the expectation over a uniform draw. For the covariance, the B summands are independent and identically distributed with covariance $C(\theta)$, so their average has covariance C/B. The rank bound holds because C is a sample covariance of n vectors, hence supported on the span of $\{\nabla \ell_i - \nabla L\}$, of dimension at most $n-1$; a single draw of $\hat{g}$ lies in the span of B of them. Equation (6.3) is the definition of C with $\nabla \ell_i = r_i \nabla r_i$. ∎

Notes on Proposition 6.2

Where the rank bound bites, and where it does not. $\operatorname{rank} C \le n-1$ is an identity about a sample covariance of n vectors, so for an over-parameterised model the instantaneous noise lives on a subspace of dimension at most $n-1$ out of p. That rules out the isotropic full-rank Langevin approximation pointwise, which is what Proposition 6.5 needs. It does *not* rule out that the process has a full-support law: a degenerate diffusion whose noise directions rotate with the state can reach every direction through the drift, as Hörmander's condition formalises, and the kinetic dynamics of Section 4.2 is the standard example. Whether the law has full support is a controllability question, and $\operatorname{rank} C(\theta) < p$ neither settles it nor is settled by it.

What the vanishing of the noise does and does not say. Equation (6.3) says C vanishes at an interpolating parameter. Near one, with residual scale $r \to 0$ and bounded Jacobians, the orders are

$$\|C\| = O(r^2), \qquad \|\nabla L\| = O(r), \qquad \text{so} \qquad \frac{\|C\|}{\|\nabla L\|} = O(r) \to 0,$$

but the comparison that governs one step of (6.1) is between the gradient and the *standard deviation* of the increment, and $\|C\|^{1/2} / \|\nabla L\| = O(1)$: the noise amplitude is generically of the same order as the gradient, not smaller. Covariance and amplitude are different quantities with different orders, and Section 6.3 depends on which one is meant. All of this is a statement about C near an interpolating parameter, not about whether training reaches one.

The gradient noise is not isotropic, not state independent, and not of full rank. By (6.2) its covariance is a state-dependent matrix; by the rank bound it is degenerate whenever the parameter count exceeds the sample count, which is the regime of interest; and by (6.3) it collapses at an interpolating solution rather than persisting. Every one of the three is assumed away by the reading of (6.1) as a discretisation of an isotropic Langevin dynamics, and Section 6.3 shows what each assumption is doing.

Anisotropy is not a nuisance parameter. $C(\theta)$ is, for least-squares losses, closely related to the Gauss–Newton matrix, hence aligned with the curvature rather than independent of it. A model of the noise as independent of $\nabla^2 L$ is not a simplification of the truth in a controlled direction; by Proposition 6.6 it changes whether a Gibbs stationary law exists at all.

Sampling with replacement is a modelling choice. Real training samples without replacement within an epoch. Within a single batch this replaces $\Sigma = C/B$ by the exact finite-population expression

$$\Sigma_{\text{wor}} = \frac{n-B}{B(n-1)} C = \frac{n-B}{n-1} \cdot \frac{C}{B},$$

so the correction factor is $(n-B)/(n-1)$ and not $1 - B/n$; the two agree only to leading order in B/n, and they differ decisively at $B = n$, where the exact factor is zero — a full batch has no sampling noise at all — while $1 - B/n$ also vanishes but for the wrong reason and with the wrong neighbourhood. More importantly, no scalar correction captures what actually changes: without replacement the increments are *dependent across steps* within an epoch, which is a property of the process and not of a single covariance. Every statement below is for the independent case of Definition 6.1.

6.2 Stochastic Modified Equations

6.2.1 The approximation and its order

Proposition 6.3 (Stochastic modified equations) ESTABLISHED. *Assume $L \in C^4$ with bounded derivatives of order 1 to 4, and Σ bounded, Lipschitz and C^2. Fix $T > 0$ and set $t = k\eta$. Then the solution Θ of*

$$d\Theta_t = -\nabla L(\Theta_t)\,dt + \sqrt{\eta\,\Sigma(\Theta_t)}\,dW_t, \qquad \Theta_0 = \theta_0, \tag{6.4}$$

is a weak approximation of order 1 to (6.1) in the sense of Definition 1.47: for every f in a class of smooth test functions with polynomial growth there is $C_f(T)$ with

$$\left|\mathbb{E}f(\theta_{\lfloor T/\eta \rfloor}) - \mathbb{E}f(\Theta_T)\right| \;\le\; C_f(T)\,\eta. \tag{6.5}$$

Replacing the drift by $-\nabla\left(L + \frac{\eta}{4}\left|\nabla L\right|^2\right)$ raises the order to 2 when Σ is constant. For a state-dependent Σ the second-order matching involves derivatives of the covariance as well, and the corrected drift above is not by itself sufficient; the statement is therefore made at first order in that generality.

The factor η inside the square root in (6.4) is the whole content of the approximation: one step of (6.1) moves the parameter by $-\eta\nabla L$ plus a fluctuation of size $\eta\sqrt{\Sigma}$, and over $1/\eta$ steps those fluctuations accumulate to $\sqrt{\eta\Sigma}$ in the rescaled time. Everything the chapter says about temperature comes from that η, and Section 6.3 says under what conditions it can be called a temperature.

6.2.2 What the approximation misses, computed

Equation (6.5) is a statement at a fixed horizon T. The quantity most often wanted is the stationary law, which is a $T \to \infty$ statement; the fixed-horizon result does not licence that exchange. In the quadratic case the discrepancy can be computed exactly, and it is not zero.

Proposition 6.4 (Stationary covariance: iteration against modified equation) DERIVED. *Let $L(\theta) = \frac{1}{2}\langle\theta, H\theta\rangle$ with $H \succ 0$ and let Σ be constant and commuting with H. In the eigendirection of H with eigenvalue λ and noise variance ς^2, and assuming $\eta\lambda < 2$:*

$$s_\infty^{\mathrm{SGD}} = \frac{\eta\,\varsigma^2}{\lambda\,(2-\eta\lambda)}, \qquad s_\infty^{\mathrm{SME}} = \frac{\eta\,\varsigma^2}{2\lambda}, \qquad \frac{s_\infty^{\mathrm{SGD}}}{s_\infty^{\mathrm{SME}}} = \frac{2}{2-\eta\lambda} = 1 + \frac{\eta\lambda}{2} + O(\eta^2\lambda^2). \tag{6.6}$$

The relative error is $\eta\lambda/(2-\eta\lambda)$ — the same expression as the unadjusted Langevin bias (4.3) — so it is $O(\eta)$ for $\eta\lambda$ small and diverges as $\eta\lambda \uparrow 2$, which is exactly the edge-of-stability threshold of Proposition 6.9(ii).

Proof. For (6.1) with a quadratic loss the iteration is $\theta_{k+1} = (I - \eta H)\theta_k + \eta\zeta_k$ with ζ_k centred of covariance Σ. In an eigendirection the stationary variance satisfies $s = (1-\eta\lambda)^2 s + \eta^2\varsigma^2$, so $s(2\eta\lambda - \eta^2\lambda^2) = \eta^2\varsigma^2$, giving the first expression; the geometric factor $|1-\eta\lambda| < 1$ is what $\eta\lambda < 2$ supplies. For (6.4) the equation in that direction is $d\Theta = -\lambda\Theta dt + \sqrt{\eta}\varsigma\, dW$, an Ornstein–Uhlenbeck process whose stationary variance is $\eta\varsigma^2/(2\lambda)$ by Proposition 1.43. The ratio and its expansion follow. ■

> Interpretation
>
> The modified equation is a device for reading discrete dynamics with continuous tools, and it is licensed over a *fixed horizon* and to a *stated order*. Two things follow. The step size appears twice — it sets the discretisation error and it scales the injected noise — so it cannot be tuned as if it were only a time step. And the approximation is asked to hold as $\eta \to 0$ over bounded time, which is precisely not the regime in which a stationary distribution or an edge-of-stability threshold is observed; those require the limits to be taken in the other order.

What weak order does and does not control. It controls $\mathbb{E}f(\theta_k)$ against $\mathbb{E}f(\Theta_t)$ for smooth f at a fixed horizon. It does not control the trajectory — there is no coupling between the two processes and no pathwise statement — and it does not control the stationary law, which lives at $T = \infty$. Proposition 6.4 is the exact measure of the second failure in the one case where it can be computed, and the answer is the same $\eta\lambda/(2-\eta\lambda)$ that appeared in Proposition 4.2: the modified equation and the algorithm agree to first order in η and disagree about where they end up.

Where the approximation has no valid order estimate. The constant in (6.5) degrades as $\eta\lambda \to 2$, and (6.6) shows the stationary discrepancy diverging there. The edge-of-stability regime of Section 6.4 is defined by $\eta S \approx 2$; the modified equation therefore has nothing to say about it, and Table 6.1 records this as the rule that rows two and three do not connect.

6.3 Stationary Behaviour and Effective Temperature

If (6.4) had a Gibbs stationary law $\propto e^{-\beta L}$, much would follow: a variational characterisation, a free energy, the whole of Chapter 2. It has one under a condition, and the condition is restrictive.

Proposition 6.5 (Isotropic noise gives a temperature) [DERIVED]. *Suppose $\Sigma(\theta) = \varsigma^2 I$ is constant and isotropic. Then (6.4) is the Langevin dynamics (2.10) for the potential $V = 2L/(\eta\varsigma^2)$, and its stationary law is*

$$\pi_\eta(\theta) \propto \exp\left(-\frac{2}{\eta\varsigma^2}L(\theta)\right), \qquad \text{that is } \beta_{\text{eff}} = \frac{2}{\eta\varsigma^2} = \frac{2B}{\eta\varsigma_1^2}, \tag{6.7}$$

where ς_1^2 is the per-sample gradient variance, using $\Sigma = C/B$ from (6.2). In particular the pair (η, B) enters only through η/B.

Proof. With $a = \eta\varsigma^2 I$ constant and drift $-\nabla L$, the criterion of Proposition 2.5 for $\pi \propto e^{-U}$ requires $-\nabla L = -\frac{1}{2}a\nabla U$, that is $\nabla U = 2\nabla L/(\eta\varsigma^2)$, which is solved by $U = 2L/(\eta\varsigma^2)$. Since the drift is then a gradient with respect to the rescaled potential, Proposition 2.5(iv) applies and the dynamics is reversible with stationary law (6.7). ■

Proposition 6.6 (Anisotropic noise generally destroys detailed balance) DERIVED. *Let Σ be constant, symmetric and positive definite. A positive density $\pi \propto e^{-U}$ with $U \in C^2$ is a reversible stationary density for (6.4) if and only if*

$$\nabla U = \frac{2}{\eta}\Sigma^{-1}\nabla L, \tag{6.8}$$

and for $L \in C^3$ on a simply connected domain such a U exists if and only if $\Sigma^{-1}\nabla^2 L(\theta)$ is symmetric for every θ, equivalently if and only if Σ commutes with $\nabla^2 L(\theta)$ pointwise. If the commutation fails, a stationary density may still exist — and for a positive definite quadratic L, for which the drift $-\nabla^2 L$ is stable, one always does — but no stationary state is reversible, and every one of them carries a non-zero stationary probability current in the sense of Proposition 2.5(ii).

What does not follow from the failure of (6.8) is that no stationary law is of the form $e^{-\beta L}$. Stationarity of $\pi \propto e^{-\beta L}$ requires only that the flux $e^{-\beta L}(\frac{\eta\beta}{2}\Sigma - I)\nabla L$ be divergence free, which is a weaker condition than its vanishing; ruling it out is a separate computation, done for the Gaussian case in Example 6.7. Stationary, reversible *and* Gibbs with a scalar temperature *are three properties, and this proposition settles the second.*

Proof. By Proposition 2.5 with $a = \eta\Sigma$ constant, a density e^{-U} is stationary and the dynamics reversible with respect to it exactly when the drift equals $-\frac{1}{2}a\nabla U$, that is when $\nabla U = \frac{2}{\eta}\Sigma^{-1}\nabla L$. Such a U exists if and only if the right-hand side is a gradient, and a C^1 field F on a simply connected domain is a gradient if and only if $\partial_j F_i = \partial_i F_j$; here $\partial_j F_i = \frac{2}{\eta}(\Sigma^{-1}\nabla^2 L)_{ij}$, so the condition is the symmetry of $\Sigma^{-1}\nabla^2 L$. Since $\nabla^2 L$ is symmetric, $\Sigma^{-1}\nabla^2 L$ is symmetric if and only if $\Sigma^{-1}\nabla^2 L = \nabla^2 L\,\Sigma^{-1}$, that is Σ commutes with $\nabla^2 L$. If no such U exists then no stationary density is reversible — the last claim of the proposition — and by Proposition 2.6 the stationary flux $j[\pi]$ is divergence free and non-zero. ∎

Example 6.7 (Two dimensions, no temperature). Take $p = 2$, $\Sigma = \mathrm{diag}(1, \epsilon)$ with $\epsilon \neq 1$ and $\epsilon > 0$, and $L(\theta) = \frac{1}{2}\langle\theta, H\theta\rangle$ with $H = \begin{pmatrix} 1 & \gamma \\ \gamma & 1 \end{pmatrix}$ and $0 < |\gamma| < 1$, so that $H \succ 0$ and the loss is confining — without that restriction H has a non-positive eigenvalue, the linear dynamics does not have a stationary law at all, and the example would be about the wrong failure. Then $\Sigma^{-1}H = \begin{pmatrix} 1 & \gamma \\ \gamma/\epsilon & 1/\epsilon \end{pmatrix}$ is not symmetric, so by Proposition 6.6 no stationary state of the modified equation is reversible. A stationary law nevertheless exists and is Gaussian — the equation is linear — with covariance solving the Lyapunov equation $H\Sigma_\infty + \Sigma_\infty H = \eta\Sigma$, which has a unique solution because $H \succ 0$. That covariance is not proportional to H^{-1} — a direct computation, since $\Sigma_\infty \propto H^{-1}$ would force $\Sigma \propto I$ — so here the stationary law is genuinely not $\propto e^{-\beta L}$ for any β, which is the separate claim that Proposition 6.6 does not supply. By Proposition 2.6 it carries a non-zero divergence-free current.

The example is worth reading twice. It exhibits a stationary density of the form e^{-U} — every non-degenerate Gaussian is one — for a dynamics with no reversible stationary state at all, and it separately checks that this U is not proportional to L. *Stationary*, *reversible* and *Gibbs with a scalar temperature* are three different properties; the first implies neither of the others, and each exclusion needs its own computation.

Scope of Proposition 6.5, Proposition 6.6 and Example 6.7

Assumed. A *constant* noise covariance throughout. For state-dependent $\Sigma(\theta)$ the reversibility condition acquires the divergence term $\frac{1}{2}\nabla\cdot s$ of Proposition 3.2 and becomes $\nabla U = \frac{2}{\eta}\Sigma^{-1}(\nabla L + \frac{\eta}{2}\nabla\cdot\Sigma)$: a *different* condition, not a uniformly stronger one. Neither direction of implication is asserted — the extra term can in principle cancel against the commutator obstruction for particular Σ, and constructing such a case is not attempted here — so Proposition 6.6 should be read as the constant case and not as a favourable extreme of a monotone family.

What the effective temperature is worth. Under isotropy, (6.7) is exact for the modified equation, and with it come the free energy of Chapter 2, the rates of Section 2.4, and the linear scaling rule η/B — all of them statements about (6.4), not about (6.1), and inheriting the horizon restriction of Proposition 6.3 together with the stationary discrepancy of Proposition 6.4. Two approximations are stacked and each has its own regime.

Where isotropy fails. By Proposition 6.2 the covariance is a sample covariance aligned with the per-example gradients, hence in general neither isotropic nor commuting with $\nabla^2 L$. Proposition 6.6 is therefore the typical case and Proposition 6.5 the exceptional one. The correct reading is that an effective temperature is an approximation whose validity is the commutation condition, and that condition is checkable. Note what the typical case does *not* say: a stationary law usually exists, and Example 6.7 exhibits one. What fails is that it is reversible and that it is a function of L alone.

The scalar-temperature claim is the example's, not the proposition's. Proposition 6.6 rules out a reversible e^{-U}; that a stationary law fails to be $\propto e^{-\beta L}$ is checked in Example 6.7 by solving the Lyapunov equation, and a reader wanting the statement for another L must redo that step.

What would falsify them. For Proposition 6.6, a constant Σ not commuting with $\nabla^2 L$ for which a C^2 potential U with $\nabla U = \frac{2}{\eta}\Sigma^{-1}\nabla L$ nevertheless exists.

What does not follow

The learning rate is not the temperature. By Proposition 6.5 the identification $\beta_{\text{eff}} = 2B/(\eta\sigma_1^2)$ holds for the modified equation with isotropic constant noise, and by Proposition 6.6 no reversible such identification exists otherwise, with Example 6.7 ruling out the non-reversible possibility in the Gaussian case. What survives in general is a modified equation with a stationary law that exists, is not reversible, is not a function of L alone, and carries a circulating flux — so the free energy of Chapter 2 is not available, the variational picture is not available, and the phrase "SGD samples from a tempered posterior" has no referent.

The linear scaling rule inherits every hypothesis above it. That η and B enter through η/B follows from (6.7), hence from isotropy, from constancy of Σ, from the weak-order approximation of Proposition 6.3, and from a fixed horizon. It is a statement about a stationary law of an approximating equation, and it says nothing about the transient behaviour that training actually exhibits.

A stationary law is not where training stops. Even where (6.7) holds, training is run for a finite number of steps from a specific initialisation and is stopped by a schedule, not by equilibration. By Example 2.29 the time to equilibrate can be exponential in a barrier height, so the stationary law may be irrelevant to what a run produces.

6.4 Edge of Stability: Phenomenon, Theory, and Limits

The preceding sections approximated discrete optimisation by continuous dynamics: the Euclidean gradient flow as the $\eta \to 0$ limit, and the stochastic modified equation as a weak approximation valid to a stated order in η. This section is about the regime in which both approximations are known to fail, and in which trained networks are usually run.

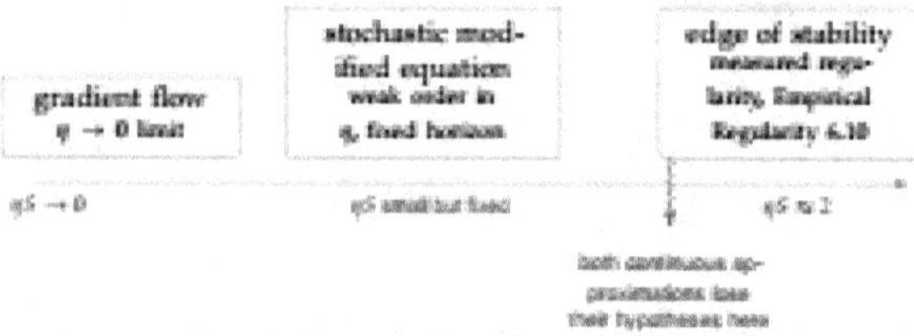

Figure 6.1: Three regimes, distinguished by ηS with S the sharpness. Each has its own object and its own class of statement; the jagged line marks where the constants in the continuous approximations degrade. What this figure does not imply: that the regimes are connected by a limit one may take inside a proof. A statement established in one regime is not evidence for the others, which is the substitution catalogued in Table 6.2. Adaptive optimisers and normalisation layers move the threshold and are not on this axis.

The organising fact is a threshold. For a β-smooth loss, every classical guarantee for gradient descent assumes $\eta\beta < 2$; the descent lemma of Proposition 6.9 is the whole of that theory in one line. What is measured in practice is that training does not remain in that regime: the sharpness rises until η times it reaches roughly 2, and then stays there. The classical theory does not describe a defect of training; training leaves the domain on which the classical theory was proved.

Because the material spans a measured regularity, a theorem about an idealisation, and an unresolved causal question, the section keeps the three apart and marks each. That separation is the point of the section as much as the content is.

6.4.1 The classical threshold, and the whole of what it gives

Definition 6.8 (Sharpness). For $L \in C^2(\mathbb{R}^p)$ the sharpness at θ is $S(\theta) := \lambda_{\max}(\nabla^2 L(\theta))$, the largest eigenvalue of the Hessian. Gradient descent with step $\eta > 0$ is the iteration $\theta_{k+1} = \theta_k - \eta \nabla L(\theta_k)$.

Proposition 6.9 (Descent lemma, and its exact converse on a quadratic) CLASSICAL.

(i) Let ∇L be β-Lipschitz. If $0 < \eta < 2/\beta$ then for every k

$$L(\theta_{k+1}) \leq L(\theta_k) - \eta\left(1 - \tfrac{\eta\beta}{2}\right)\|\nabla L(\theta_k)\|^2, \tag{6.9}$$

and the coefficient is strictly positive, so the loss decreases monotonically.

(ii) Let $L(\theta) = \frac{1}{2}\langle\theta, H\theta\rangle$ *with* $H \succeq 0$ *symmetric and* $\lambda := \lambda_{\max}(H)$*. Gradient descent converges from every initialisation if and only if* $\eta\lambda < 2$*; if* $\eta\lambda > 2$ *the component of* θ_k *along a top eigenvector grows geometrically at rate* $\eta\lambda - 1 > 1$*.*

Proof. (i) β-Lipschitz gradient gives $L(\theta') \leq L(\theta) + \langle\nabla L(\theta), \theta' - \theta\rangle + \frac{\beta}{2}\|\theta' - \theta\|^2$; substitute $\theta' = \theta - \eta\nabla L(\theta)$ and collect. (ii) In the eigenbasis the iteration decouples into $\theta^{(i)}_{k+1} = (1 - \eta\lambda_i)\theta^{(i)}_k$, which converges for every initial value iff $|1 - \eta\lambda_i| \leq 1$ for all i with equality only where $\lambda_i = 0$, i.e. iff $\eta\lambda < 2$; for $\eta\lambda > 2$ the top coordinate is multiplied by $1 - \eta\lambda$, of modulus $\eta\lambda - 1 > 1$. ∎

The two halves of Proposition 6.9 meet at $\eta S = 2$, and it is worth saying plainly what that means for the rest of the chapter: *the classical theory of gradient descent is a theory of the regime* $\eta S < 2$. Everything proved in Section 1.2 and Section 6.2 — the gradient-flow limit, the modified equation, the convergence rates — carries a hypothesis that places it inside that regime. None of it is wrong. It simply does not reach the regime described next.

6.4.2 The phenomenon

Empirical Regularity 6.10 (Progressive sharpening and the stability ceiling) ESTABLISHED · EMPIRICAL REGULARITY *In full-batch gradient descent at a fixed step size* η*, on a range of network architectures, data sets and twice-differentiable losses, the following is measured: (a) the sharpness* $S(\theta_k)$ *increases over much of training, from an initialisation at which* $\eta S \ll 2$*; (b) once* $\eta S(\theta_k)$ *reaches approximately 2 it ceases to increase and thereafter fluctuates just above that value for the remainder of training; and (c) throughout that period the training loss decreases over long windows while failing to decrease monotonically over short ones, in contradiction to (6.9). The regularity is measured by Cohen et al. [8], whose architectures, losses and sharpness definition fix its scope.*

Scope of Empirical Regularity 6.10

What kind of claim this is. A measured regularity across systems, carrying the evidence-modality badge and no proof. It is reported here because it is stable across many reported settings and because the rest of the section is organised around it; it is not a theorem, nothing later in this volume depends on it, and it is not stated for "all networks". Its scope is the set of settings in which it has been measured.

Where it is reported not to hold, or to change form. With adaptive optimisers the operative threshold is not $2/\eta$ and is set by the preconditioner; with normalisation layers the sharpness is not scale-invariant and the reported curves depend on the parameterisation (Proposition 6.13); with mini-batch gradients the ceiling sits elsewhere (Section 6.4.5); with a decaying schedule the phenomenon tracks the schedule rather than a fixed value. A reader should treat "$2/\eta$" as the full-batch constant-step value and nothing more general.

What would falsify it. As a regularity, a well-specified full-batch constant-step setting in the reported class in which sharpness stabilises away from $2/\eta$ by a factor bounded away from one, measured with a sharpness estimator whose bias is reported. Note that (a) and (b) can fail independently: they are two claims, and only (b) has a mechanism in Section 6.4.3.

Two things in Empirical Regularity 6.10 are usually run together and should not be. Part (b), the ceiling, has a mechanism that can be proved inside an idealisation — this is the content of Section 6.4.3. Part (a), the rise, does not. Why sharpness should increase at all is not explained by any argument in this volume.

Open Problem 6.11 (Progressive sharpening) OPEN PROBLEM. Identify a class of losses and initialisations, containing a recognisable model of a trained network, in which $S(\theta_k)$ provably increases along the gradient-descent trajectory while $\eta S < 2$. The mechanism of Theorem 6.12 operates only above the threshold and is silent here; a satisfactory answer would explain why the trajectory is driven towards the boundary of the regime in which its own analysis is valid.

6.4.3 A model in which the ceiling is a theorem

The following is the smallest model in which the stabilising feedback can be exhibited completely. It is a two-parameter loss with a manifold of global minima and a sharpness that varies along it — the two features the mechanism needs, and the only two.

Theorem 6.12 (Self-stabilisation in the two-parameter model) MODEL-DEPENDENT. *Let $L(x, y) = \frac{1}{2}x^2y^2$ on $\mathbb{R}^2$ and run gradient descent with step $\eta > 0$:*

$$x_{k+1} = x_k(1 - \eta y_k^2), \qquad y_{k+1} = y_k(1 - \eta x_k^2). \tag{6.10}$$

The set of global minima is $\{xy = 0\}$; along $\{x = 0\}$ the sharpness is $S = y^2$. Write $\sigma_k := \eta y_k^2$ for the normalised sharpness, so that the linear-stability threshold of Proposition 6.9(ii) is $\sigma = 2$. Assume $x_0 \neq 0$, $y_0 > 0$ and $\eta x_k^2 < 1$ for all $k \leq K$. Then:

(i) (Sharpness never increases.) *$\sigma_{k+1} = \sigma_k(1 - \eta x_k^2)^2 \leq \sigma_k$, with equality only if $x_k = 0$; the decrement is governed by x_k^2, so sharpness is reduced exactly in proportion to the instability.*

(ii) (Instability is self-limiting.) *If $\sigma_k \geq 2 + \delta$ for every $k \leq K$ with $\delta > 0$, then $|x_{k+1}| \geq (1 + \delta)\,|x_k|$ for those k, whence*

$$K \leq \frac{\log(1/(\eta x_0^2))}{2\log(1 + \delta)}. \tag{6.11}$$

Consequently no trajectory of (6.10) remaining in the region $\eta x^2 < 1$ can stay above the threshold by a fixed margin for more than (6.11) steps: the sharpness is driven back to the threshold by the very oscillation that exceeding it creates.

Proof. (i) Squaring the second equation of (6.10) and multiplying by η gives $\sigma_{k+1} = \sigma_k(1 - \eta x_k^2)^2$; under $0 < \eta x_k^2 < 1$ the factor lies in $(0, 1)$. (ii) The first equation gives $|x_{k+1}| = |x_k|\,|1 - \sigma_k| = |x_k|\,(\sigma_k - 1)$ when $\sigma_k \geq 2$, and $\sigma_k - 1 \geq 1 + \delta$. Iterating, $|x_K| \geq |x_0|\,(1 + \delta)^K$; combining with the standing assumption $\eta x_K^2 < 1$, that is $|x_K| < \eta^{-1/2}$, gives $(1 + \delta)^K < \eta^{-1/2}/|x_0|$, which is (6.11). ∎

The model behind Theorem 6.12

The model. $L(x,y) = \frac{1}{2}x^2y^2$ on $\mathbb{R}^2$: full-batch, deterministic, constant step, exact arithmetic, and a standing restriction to the region $\eta x^2 < 1$ in which (6.10) preserves the sign of y. Two parameters, a one-dimensional manifold of minima, and a sharpness that varies along it.

Exact inside it. Both displays, with no error term: the normalised sharpness is non-increasing at every step, and an excursion above $2+\delta$ has the explicitly bounded length (6.11). The proof is four lines and contains no approximation.

What does not transfer. (1) *The rise.* Nothing here says σ_t increases when $\sigma_t < 2$; part (i) says it never increases at all. The model exhibits the ceiling and not the approach to it, and Open Problem 6.11 is what is missing. (2) *The alignment.* The mechanism works because the unstable coordinate x is exactly the coordinate whose growth reduces the sharpness y^2. In a general C^3 loss the gradient of the sharpness need not be aligned with the top eigenvector, and where it is not, growth of the unstable component does not reduce sharpness. That alignment is a property of this model, not a consequence of smoothness. (3) *Fluctuation about the threshold.* The model gives monotone non-increase, not the hovering just above $2/\eta$ that Empirical Regularity 6.10(b) reports. The two are compatible but they are different statements, and the second is not proved here or anywhere in this volume. (4) *Stochasticity.* Adding gradient noise changes the operative threshold, not merely the constants; see Section 6.4.5.

6.4.4 Sharpness and generalisation

It is frequently asserted that operating at the edge of stability biases training towards flatter minima, and that flatter minima generalise better, so that a large step size improves test error through this chain. The first link is the subject of Theorem 6.12 within its model. The second link, in its usual form, is not merely unproved: it is false, for a reason that costs three lines.

Proposition 6.13 (Sharpness is not a function of the learned function). *Let $f_{w_1,w_2}(x) = w_2 w_1 x$ and $L(w_1,w_2) = \frac{1}{2}(w_2w_1 - y)^2$ for a fixed $y \neq 0$. At any global minimum, $w_1w_2 = y$ and*

$$\nabla^2 L = \begin{pmatrix} w_2^2 & w_1w_2 \\ w_1w_2 & w_1^2 \end{pmatrix}, \qquad S = w_1^2 + w_2^2. \tag{6.12}$$

The reparameterisation $(w_1, w_2) \mapsto (cw_1, w_2/c)$, $c > 0$, leaves f_{w_1,w_2} and L unchanged and maps S to $c^2w_1^2 + w_2^2/c^2$, which ranges over $[2|y|, \infty)$ as c ranges over $(0,\infty)$. Hence the set of parameters representing one and the same function contains points of arbitrarily large sharpness, and sharpness determines no property of the learned function.

Proof. At a minimum the residual vanishes, so the terms of $\nabla^2 L$ involving it drop and (6.12) follows; its eigenvalues are 0 and $w_1^2 + w_2^2$. The map leaves w_1w_2 invariant, hence leaves f and L invariant, and $c^2w_1^2 + w_2^2/c^2 \geq 2|w_1w_2| = 2|y|$ by the arithmetic–geometric mean inequality, with equality at $c^2 = |w_2/w_1|$ and divergence as $c \to 0$ or $c \to \infty$. The same construction applies verbatim to any positively homogeneous activation. ∎

What survives this is a narrower statement, and it is worth writing as a conjecture with the test that would refute it, rather than as a principle that cannot be checked.

Conjecture 6.14 (Trajectory-restricted flatness) CONJECTURE. **Among the minima that full-batch gradient descent at step η actually reaches from a standard initialisation, those reached at larger η have lower test error, and the effect is mediated by a reparameterisation-invariant measure of flatness — one computed on a normalised parameterisation, or on the function rather than the parameters.**

Refuting test. Fix the architecture, data, and final training loss. Exhibit two runs differing only in η whose reached minima have equal values of the proposed invariant measure and materially different test error, or equal test error and materially different values of the measure. Either outcome refutes the mediation claim while leaving the raw correlation between η and test error untouched.

Where it fails, and where it does not. At a non-minimum the residual terms survive and the closed form changes; the conclusion does not, because the reparameterisation still preserves the function while changing the Hessian. Architectures whose parameterisation is already scale-fixed — by weight normalisation, or by a constraint — are outside the construction, and for them sharpness is a function of the function; that is the class in which Conjecture 6.14 could be tested most cleanly.

What this does not say. It does not say that flatness is irrelevant, nor that the correlation between step size and test error is spurious. It says that the raw largest Hessian eigenvalue cannot be the mediating quantity, because it is not a function of what is being generalised.

Edge of stability is not a theorem about neural networks. Empirical Regularity 6.10 is a measured regularity in a reported class of settings, and Theorem 6.12 is a theorem about a two-parameter loss. Neither is a statement about arbitrary twice-differentiable losses or about "two-layer networks" in general, and this volume does not contain one.

Sharpness is not a property of the function. By Proposition 6.13, any claim of the form "flatter minima generalise better", stated with the raw Hessian eigenvalue, is false as written: the quantity on the left can be changed by a factor of a million without changing the network's output on any input. Claims of this kind require a scale-invariant measure, and which measure is not settled; Conjecture 6.14 states the surviving version together with what would kill it.

Neither continuous-time approximation reaches this regime. The gradient flow is the $\eta \to 0$ limit; the stochastic modified equation of Proposition 6.3 is a weak approximation of a stated order in η, which is a statement about expectations of test functions over a fixed time horizon and carries a smoothness hypothesis with a constant that degrades as $\eta S \to 2$. Using either to interpret behaviour at $\eta S \approx 2$ is an extrapolation outside the hypotheses of the theorem that produced them. Table 6.1 sorts the four regimes, and no result in this volume crosses a row of it.

The threshold $2/\eta$ is specific to full-batch constant-step descent. It comes from $|1 - \eta\lambda| \leq 1$ and nothing else. Adaptive preconditioning replaces λ by a preconditioned eigenvalue; momentum replaces the condition by one on a 2×2 companion matrix; mini-batching changes the notion of stability altogether (Section 6.4.5). Reporting "$2/\eta$" in any of those settings is a category error, not a constant that needs adjusting.

6.4.5 Mini-batch gradients: a different threshold

Empirical Regularity 6.15 (Stochastic edge of stability) ESTABLISHED. EMPIRICAL REGULARITY *For mini-batch stochastic gradient descent at fixed step size and batch size, the sharpness at which training stabilises is measured to sit below the full-batch value $2/\eta$, by a gap that widens as the batch size decreases. The operative stability condition involves the covariance of the gradient noise in addition to the Hessian, and is therefore not obtained from Proposition 6.9(ii) by adjusting a constant. See Cohen et al. [8]; the mini-batch threshold is measured separately in the literature and its scope is that of the experiments reporting it.*

Scope of Empirical Regularity 6.15

What kind of claim this is. A measured regularity with the evidence-modality badge, on the same footing as Empirical Regularity 6.10 and with the same scope restriction: the settings in which it has been measured, at fixed step size and fixed batch size.

Where it is reported to change form. The gap is reported to widen as the batch size falls, which means the claim is not a single number but a family indexed by batch size; quoting it without the batch size is quoting nothing. Under gradient accumulation, or with a batch size comparable to the data set, it degenerates to Empirical Regularity 6.10.

What this does not say. It gives no stability threshold. The second statement — that the operative condition involves the noise covariance — is a statement about the *form* of the condition, not a derivation of it, and this volume proves no such condition.

What would falsify it. A well-specified mini-batch setting in the reported class in which the stabilised sharpness sits at or above the full-batch value, measured with an estimator whose bias is reported.

The distinction matters more than it may appear. The full-batch analysis identifies a deterministic linear instability: a direction along which the iteration map has a multiplier of modulus greater than one. The stochastic picture replaces this by a condition on the growth of a second moment, in which the noise injected along a direction interacts with the curvature in that direction; a mode can be second-moment unstable while the deterministic multiplier is inside the unit disc. Carrying the sentence "sharpness stabilises at $2/\eta$" from one setting to the other is exactly the regime substitution that Table 6.2 catalogues.

Table 6.1: The four regimes of Chapter 6, and what is proved in each. The rightmost column is the operative one: a result proved in one row may not be quoted in another, and every claim in this chapter belongs to exactly one row. Rows three and four are where trained networks are usually run and where the least is proved.

Regime	Object	What is proved	What may not be carried out of it
$\eta \to 0$	Euclidean gradient flow $\dot{\theta} = -\nabla L$	Exact continuous-time convergence statements; the loss is monotone	Any statement about a fixed η; the flow has no stability threshold at all

continued overleaf

Table 6.1: **(continued)**

Regime	Object	What is proved	What may not be carried out of it
Small η, $\eta S \ll 2$	Stochastic modified equation	Weak approximation of order 1, or 2 with the drift correction (Proposition 6.3): expectations of smooth test functions over a bounded horizon	Pathwise statements; behaviour as $\eta S \to 2$, where the constant in the order estimate degrades; anything at the threshold
Full batch, $\eta S \approx 2$	Discrete map, deterministic	Proposition 6.9(ii) on quadratics; Theorem 6.12 in a two-parameter model; the rest is Empirical Regularity 6.10, measured	Nothing in this row is a general theorem about networks; the ceiling has a mechanism only inside a model, and the rise has none
Mini-batch, fixed η	Discrete map, stochastic	Empirical Regularity 6.15, measured; the stability condition is on a second moment and involves the noise covariance	The full-batch threshold $2/\eta$; any diffusion-limit reasoning, which belongs to row two

Operational reading

Read the ceiling as a constraint the optimiser imposes on itself, and read η as selecting where that constraint binds: at a fixed step size the trajectory cannot settle at a minimum sharper than roughly $2/\eta$, because by Proposition 6.9(ii) such a point is linearly unstable and the iteration is pushed away from it. That is a statement about which minima are reachable, and it is as far as this box goes.

The step after it — that the reachable minima are therefore better ones — is Conjecture 6.14, is not established, and by Proposition 6.13 is false in its usual formulation. Tuning η against validation error remains sound practice; explaining the result by a flatness mechanism does not follow from anything in this section.

Atlas entries generated by this section

R12. Discrete gradient descent = its continuous-time limit. LIMIT

Exact conditions. $\eta \to 0$ for the gradient flow; for a fixed small η, weak approximation of a stated order over a bounded horizon, under the smoothness hypotheses of Proposition 6.3, and with ηS bounded away from 2.

What goes wrong outside them. At $\eta S \approx 2$ the approximation has no valid order estimate; the flow has no instability, so it cannot represent a phenomenon defined by one. Rows one and two of Table 6.1 do not reach rows three and four.

R13. Large step size = flat minima = better generalisation. ANALOGY

Exact conditions. None. The middle term is not reparameterisation invariant (Proposition 6.13), so the chain is not well formed as stated; Conjecture 6.14 is the surviving version and carries a refuting test.

What goes wrong outside them. Sharpness at a fixed function can be made arbitrarily large;

any conclusion drawn from its value alone is drawn from a choice of coordinates.

6.5 A Catalogue of Regime Substitutions

Table 6.1 sorted the four regimes and stated the rule: a result proved in one row may not be quoted in another. This section names the specific crossings that this chapter's material invites, because a rule stated abstractly is obeyed and a rule stated with its instances is enforced.

Each row of Table 6.2 is a sentence that is true in one place and false where it is usually said. None of them is a straw man; each corresponds to a claim that the material of Section 6.2 and Section 6.3 makes tempting, and each is refuted by a result in this chapter rather than by an objection.

Table 6.2: Six substitutions, with the regime each statement is true in, the regime it is usually asserted in, and the result that separates them. The second and third columns are rows of Table 6.1.

Statement	True in	Usually asserted in	What separates them
"SGD follows a stochastic differential equation."	Small η, fixed horizon, weak sense	Fixed η, until convergence	Proposition 6.3 is a fixed-T statement with $C_f(T)$ exponential in T
"The stationary law of SGD is $e^{-2L/\eta\sigma^2}$."	The modified equation, isotropic constant noise	The iteration, arbitrary noise	Proposition 6.4 for the first gap, Proposition 6.6 for the second
"The learning rate is a temperature."	$\Sigma \propto I$ constant	Any training run	Proposition 6.6: a reversible Gibbs law exists iff Σ commutes with $\nabla^2 L$; otherwise the stationary state circulates
"Noise drives the iterate towards flatter minima."	Statements about a stationary law, where one exists	The transient of a finite run	The notes on Proposition 1.8: even the exact flow's energy argument is about a functional, not a trajectory
"The same SDE explains the edge of stability."	Nowhere	The regime $\eta S \approx 2$	Proposition 6.4: the discrepancy diverges as $\eta\lambda \uparrow 2$
"Flatness explains generalisation."	No regime; the quantity is not a function of the model	All of them	Proposition 6.13: sharpness at a fixed function is unbounded

Operational reading

The practical form of the rule is a question to ask of any statement about training dynamics: *at what step size, over what horizon, and about which object?* The three answers place a claim in a row of Table 6.1, and two claims in different rows are not evidence for or against one another.

That is one step beyond the table, and this box stops there. What actually happens during training at the step sizes in use is an empirical question on which this volume reports the regularities of Section 6.4 and nothing more.

6.6 Chapter Summary

Chapter 6 separates the regimes in which a continuous-time model of stochastic gradient descent is licensed. The gradient flow is the $\eta \to 0$ limit; the stochastic modified equation is a weak approximation of stated order over a fixed horizon; the stationary analysis is at fixed step; and the edge of stability is a measured regularity at $\eta S \approx 2$ where both continuous approximations lose their hypotheses. Figure 6.1 places the three along one axis, and Table 6.2 catalogues the substitutions between them.

The chapter's sharpest statement is negative and exact: Proposition 6.6 gives the condition under which stochastic gradient descent has a temperature at all — the noise covariance must commute with the Hessian — and Example 6.7 shows by a Lyapunov computation that the stationary law of a Gaussian counterexample is not proportional to $e^{-\beta L}$ for any β. *Stationary, reversible* and *Gibbs with a scalar temperature* are three properties, and each exclusion needs its own computation.

Exercises

§6.4

Tier I Direct application.

6.1 *The threshold with momentum*

For heavy-ball descent $\theta_{k+1} = \theta_k - \eta \nabla L(\theta_k) + \mu(\theta_k - \theta_{k-1})$ on the quadratic of Proposition 6.9(iii), write the iteration as a 2×2 companion matrix in each eigendirection and derive the stability condition. Show that it reduces to $\eta\lambda < 2$ at $\mu = 0$ and to $\eta\lambda < 2(1+\mu)$ in general. State what this does to the number "$2/\eta$" as a diagnostic.

6.2 *The toy model by hand*

Take $\eta = 1$, $x_0 = 10^{-3}$, $y_0 = \sqrt{3}$ in (6.10), so $c_0 = 3$. Iterate until $c_k < 2$ and compare the number of steps with the bound (6.11) at $\delta = 1$. Explain why the bound is loose and what it is loose by.

Tier II Completing the derivation.

6.3 *The descent lemma is tight*

Show that (6.9) is attained with equality on a quadratic in the top eigendirection, and deduce that no bound of the same form can hold for $\eta\beta > 2$. Then exhibit a β-smooth L

and a step size $\eta \in (2/\beta, 3/\beta)$ for which gradient descent converges anyway, and explain why this does not contradict Proposition 6.9(i).

6.4 *Where the alignment enters*

In Theorem 6.12, replace $L(x, y) = \frac{1}{2}x^2y^2$ by $L(x, y) = \frac{1}{2}x^2\psi(y)$ with $\psi > 0$ smooth. Compute the analogue of part (i) and identify the condition on ψ under which the growth of $|x|$ still reduces the sharpness. Then construct a ψ for which it increases it, and describe what the trajectory does.

Hint. The relevant object is the sign of $\psi'(y)$ relative to the direction in which y is pushed by $-\eta x^2\psi'(y)/2$.

Tier III Probing the boundary. Each of these asks where a statement stops being true.

6.5 *Rescaling a network's sharpness*

Take a two-layer ReLU network with a scalar output, trained to a global minimum on a small data set. Apply the reparameterisation $(W_1, W_2) \mapsto (cW_1, W_2/c)$ for $c \in \{10^{-2}, 1, 10^2\}$, verify numerically that the function is unchanged on the data and on fresh inputs, and report the sharpness at each. Then report the sharpness of a normalised version — for instance with the layer norms equalised — and say which of the two quantities Conjecture 6.14 requires.

6.6 *Two regimes, one formula*

Fit the stochastic modified equation (6.4) to a small network at η chosen so that $\eta S \approx 0.2$, and verify its weak approximation order numerically on a smooth test function. Then increase η until $\eta S \approx 2$ and repeat the measurement. Report at what value of ηS the fitted order departs from its theoretical value, and state which hypothesis of the approximation theorem your measurement is detecting the failure of.

6.7 *The gap that batch size opens*

Design an experiment that distinguishes the two candidate explanations for Empirical Regularity 6.15: that the mini-batch ceiling is the full-batch ceiling displaced by a noise-dependent constant, and that it is set by a different stability condition altogether. State what each explanation predicts as the batch size is varied at fixed η, and identify the measurement that separates them. You are not asked to run it.

Tier IV Computational.

6.8 *Measuring all four rows of Table 6.1*

Use a small fully connected network on a subset of a standard data set, with a smooth loss, and a power-iteration estimator for $S(\theta_k)$ whose bias you report.

(a) Full batch, η small. Confirm monotone decrease of the loss and $\eta S < 2$ throughout; this is row one to two.

(b) Full batch, η ten times larger. Plot $\eta S(\theta_k)$ and the loss on the same axis. Report where ηS first reaches 2, whether it then exceeds it, and by how much; report the fraction of steps on which the loss increases.

(c) Mini-batch at the same η, for batch sizes $\{8, 64, 512, \text{full}\}$. Report the stabilised value of ηS against batch size.

(d) Repeat (b) after the reparameterisation of Exercise 6.5 with $c = 10$. The loss curve should be identical and the sharpness curve should not. Say what that shows about which of the two curves is a property of training and which is a property of coordinates.

Acceptance criteria: the estimator bias in the preamble is reported as a number; (b) reports the overshoot as a fraction, not only as a picture; (d) is answered in two sentences.

Tier I Direct application.

6.9 *The noise vanishes at interpolation*

For a linear model $f_\theta(x) = \langle\theta, x\rangle$ and squared loss, compute $C(\theta)$ from (6.3) explicitly and verify that it vanishes at any θ fitting all n points exactly. Then show that for $p > n$ such a θ exists, and describe the set of them.

6.10 *Both stationary variances*

Verify (6.6) numerically in $d = 1$ with $\lambda = 1$, $\varsigma = 1$ and $\eta \in \{0.05, 0.5, 1.5, 1.95\}$, by simulating both (6.1) with Gaussian noise and (6.4). Report the measured ratio against $2/(2 - \eta\lambda)$ at each η.

Tier II Completing the derivation.

6.11 *Sampling without replacement*

Redo Proposition 6.2 for sampling without replacement within a batch and show that $\Sigma = \frac{1}{B}\left(1 - \frac{B-1}{n-1}\right)C$. Identify the value of B at which the noise vanishes and explain why, and say which hypothesis of Definition 6.1 the without-replacement scheme violates across steps.

6.12 *The order-two drift correction*

Show that the modified drift $-\nabla\left(L + \frac{\eta}{4}|\nabla L|^2\right)$ of Proposition 6.3 arises from expanding one step of (6.1) to second order in η and matching the generator. Identify the term it cancels and verify that it is a gradient, so that the corrected equation is still of the form (6.4) with a modified potential.

Tier III Probing the boundary.

6.13 *A stationary law that circulates*

In the setting of Example 6.7 with $\varepsilon = \frac{1}{4}$ and $\gamma = \frac{1}{2}$, solve the Lyapunov equation for Σ_∞ and verify that it is not proportional to H^{-1}. Then compute the stationary flux $J[\pi]$ from (2.6) and confirm that it is divergence free and non-zero. State which part of Proposition 2.6 your computation instantiates.

6.14 *Where the two limits cross*

Using (6.5) with $C_\eta(T) = C_0 e^{LT}$ and (6.6), determine for which (η, T) the fixed-horizon error is smaller than the stationary discrepancy, and sketch the region in the (η, T) plane. State which of the two errors a long training run is dominated by.

6.15 *Isotropy by preconditioning*

Suppose Σ is constant and known. Show that the preconditioned iteration $\theta_{k+1} = \theta_k - \eta\Sigma^{-1}\hat{g}(\theta_k)$ has a modified equation whose stationary law *is* Gibbs, and identify the potential. Then say what this costs — which quantity must be estimated, and what happens when Σ is singular, as Proposition 6.2 says it is whenever $p > n$.

Tier IV Computational.

6.16 *Three regimes on one problem*

Take a small network on a subset of a standard data set with a smooth loss.

(a) Estimate $C(\theta)$ at initialisation and at the end of training, by sampling per-example gradients. Report its effective rank and the alignment of its top eigenvectors with those of $\nabla^2 L$. Report whether the commutation condition of Proposition 6.6 is close to holding.

(b) At a small η, compare $\mathbb{E}f(\theta_k)$ with $\mathbb{E}f(\Theta_t)$ for f a smooth test function, using many runs, and fit the order in η. Confirm the exponent of Proposition 6.3.

(c) Increase η until $\eta S \approx 2$ and repeat (b). Report the ηS at which the fitted order departs from 1.

(d) State which row of Table 6.1 each of (a)–(c) belongs to, and which row of Table 6.2 your measurement in (c) is testing.

Acceptance criteria: (a) reports the alignment as a number; (c) reports the departure point; (d) names a row for each part.

CHAPTER 7

Stochastic Control and Reinforcement Learning

The three preceding chapters shared a shape: a dynamics is given, and one studies where its law goes. Control reverses the question. The dynamics is partly ours to choose, a cost says what we want, and the object of study is the choice.

That reversal changes the mathematics more than it changes the notation. Nothing descends: the optimum is characterised by a *backward* equation running from the terminal condition, and there is in general no functional of the current state whose decrease certifies progress. That is the third row of Table 11.1, and Section 7.1 is where it is proved rather than asserted.

Two connections to the rest of the volume make this chapter more than an excursion. When the control cost is quadratic and enters through the same channel as the noise, the problem becomes a relative entropy on path space (Section 7.2) — the same object as Theorem 3.14 — and its solution is an exponential tilt of a reference process, which is the construction Chapter 10 builds on. And under a matrix condition identified in Section 7.4, the backward equation linearises, which is the only exact bridge in this volume between control and the generative constructions of Chapter 3.

7.1 Control and the Hamilton–Jacobi–Bellman Equation

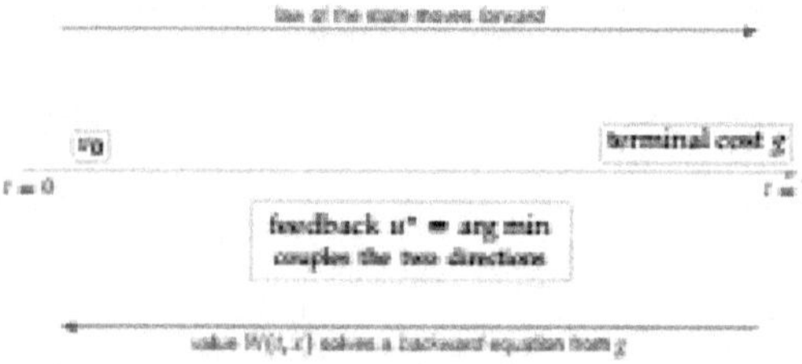

Figure 7.1: A controlled system runs in two directions at once: the law of the state forward from ν_0, the value backward from the terminal cost, coupled through the optimal feedback. What this figure does not imply: that the pair descends a functional. There is no forward steepest descent here — the optimum is characterised by the backward equation (Proposition 7.2) — and a forward descent argument answers a different question, which is the entry for the controlled row of Table 11.1.

Definition 7.1 (Controlled diffusion and value function). Let $U \subseteq \mathbb{R}^m$ be closed, and for a measurable feedback control $u : [0, T] \times \mathbb{R}^d \to U$ let X^u solve

$$dX_t = b(t, X_t, u(t, X_t))\,dt + \sigma(t, X_t)\,dW_t, \qquad X_s = x. \tag{7.1}$$

The cost and the value function are

$$J(s,x;u) = \mathbb{E}\Big[\int_s^T f(t,X_t^u,u(t,X_t^u))\,dt + g(X_T^u)\Big], \qquad V(s,x) = \inf_u J(s,x;u), \tag{7.2}$$

the infimum being over admissible feedback controls, that is those for which (7.1) has a unique weak solution and the cost is finite.

Proposition 7.2 (Verification theorem) DERIVED. *Suppose $W \in C^{1,2}([0,T)\times\mathbb{R}^d)\cap C([0,T]\times\mathbb{R}^d)$ solves the Hamilton–Jacobi–Bellman equation*

$$\begin{aligned} \partial_t W + \tfrac{1}{2}\operatorname{tr}\big(a\,\nabla^2 W\big) + \inf_{u\in U}\Big\{\langle b(t,x,u),\nabla W\rangle + f(t,x,u)\Big\} &= 0, \\ W(T,\cdot) &= g, \end{aligned} \tag{7.3}$$

with $a = \sigma\sigma^T$, and suppose W has polynomial growth of order q. Restrict attention to admissible controls satisfying, in addition to Definition 7.1,

$$\mathbb{E}\sup_{t\le T}|X_t^u|^q < \infty \qquad \text{and} \qquad \mathbb{E}\int_0^T \Big|\sigma^T\nabla W(t,X_t^u)\Big|^2\,dt < \infty, \tag{7.4}$$

which hold in particular under the coefficient hypotheses of Theorem 1.41. Then $W \le J(\cdot,\cdot;u)$ for every such u, so $W \le V$. If moreover the infimum in (7.3) is attained at a measurable $u^*(t,x)$ for which (7.1) is well posed, then $W = V$ and u^* is optimal.

Proof. Fix an admissible u and apply Theorem 1.33 to W along X^u on $[s, T\wedge\tau_n]$, with τ_n the exit time from a ball of radius n:

$$\begin{aligned} W(T\wedge\tau_n, X^u_{T\wedge\tau_n}) = W(s,x) &+ \int_s^{T\wedge\tau_n}\Big(\partial_t W + \langle b^u,\nabla W\rangle + \tfrac{1}{2}\operatorname{tr}(a\nabla^2 W)\Big)dt \\ &+ \int_s^{T\wedge\tau_n}\Big\langle \sigma^T\nabla W, dW_t\Big\rangle. \end{aligned}$$

By (7.3) the integrand of the first integral is at least $-f(t,X_t^u,u_t)$, because the infimum over U is no larger than the value at u_t. The stochastic integral is a true martingale up to τ_n — the integrand is bounded on the ball — hence has zero mean. Taking expectations, $\mathbb{E}W(T\wedge\tau_n, X^u_{T\wedge\tau_n}) \ge W(s,x) - \mathbb{E}\int_s^{T\wedge\tau_n} f$. Now let $n\to\infty$. This is the step that consumes (7.4): the first bound gives, with the polynomial growth of W, uniform integrability of $W(T\wedge\tau_n, X^u_{T\wedge\tau_n})$, so the left-hand side converges to $\mathbb{E}W(T,X_T^u) = \mathbb{E}g(X_T^u)$; the second makes the stochastic integral a true martingale on all of $[s,T]$, so no mass is lost in the limit. Neither follows from Definition 7.1 alone, which asks only for a unique weak solution and a finite cost, and this is why they are hypotheses of the proposition rather than remarks in the proof. With $W(T,\cdot) = g$ one obtains $J(s,x;u) \ge W(s,x)$. Taking the infimum over u gives $V \ge W$.

If the infimum is attained at u^* and the corresponding equation is well posed, the same computation with $u = u^*$ has equality at every step, so $J(s,x;u^*) = W(s,x)$, whence $V \le W$ and the two agree. ∎

Interpretation

The Hamilton–Jacobi–Bellman equation is dynamic programming written infinitesimally. The value function propagates *backward* from the terminal cost, because what an optimal decision needs to know is what happens after it; the state law propagates *forward*, because that is the direction in which the system runs. The control couples the two, and this is why a controlled system is not a forward descent of anything (Figure 7.1) — there is no functional of the current law whose steepest descent produces the optimal drift.

Notes on Proposition 7.2

Assumed. That a $C^{1,2}$ solution of (7.3) exists; that the pointwise infimum is attained at a measurable feedback for which the closed-loop equation is well posed; and the moment and square-integrability conditions (7.4), without which the localising sequence τ_n cannot be removed. The first two are genuine and neither is generic; the third is usually invisible because it is inherited from the coefficient hypotheses of Theorem 1.41, but the admissible class of Definition 7.1 does not supply it, and the comparison $W \leq V$ is over whichever class does.

Which direction this is. The verification theorem is the easy half of dynamic programming: from a smooth solution to optimality. The other half — that V itself solves (7.3) — requires the dynamic programming principle and, in general, V is not $C^{1,2}$ and (7.3) holds only in the viscosity sense. This volume uses only the direction proved here, because every construction it makes exhibits a candidate W and verifies it.

Where smoothness fails. With a degenerate a, with a state constraint, or with a control set on whose boundary the infimum is attained, V typically has kinks and Proposition 7.2 does not apply. The entropy-regularised problems of Section 7.2 are better behaved precisely because the regulariser makes the inner infimum smooth and interior.

What does not follow

Optimal control is not descent. Equation (7.3) runs backwards from $W(T,\cdot) = g$, and its solution at time t depends on the whole future. No functional of the current state decreases along the optimal trajectory in the manner of Proposition 2.12, and there is no gradient flow of which the optimal control is the velocity. This is the content of the third row of Table 11.1, and it is the reason the language of Chapter 2 does not transfer here: a forward descent argument answers a different question.

The value function is not a free energy. It has a variational characterisation — it is an infimum — but over *controls*, not over *measures*, and the infimum is attained at a policy rather than at a distribution. In Section 7.2 a genuine variational characterisation over measures does appear, and it appears only after the cost is restricted to a quadratic form entering through the noise channel.

7.2 Entropy-Regularised Control and KL-Regularised Policies

7.2.1 The Gibbs variational principle

One identity does all the work in this section, and it is elementary.

Lemma 7.3 (Gibbs variational principle) DERIVED. *Let* $\mathbb{R}$ *be a probability measure on a* *measurable space and let* Φ *be measurable with* $\mathbb{E}_{\mathbb{R}} e^{-\Phi} < \infty$. *Then*

$$\inf_{\mathrm{Q} \ll \mathrm{R}} \Big\{ \mathrm{KL}(\mathrm{Q} \,\|\, \mathrm{R}) + \mathbb{E}_{\mathrm{Q}}[\Phi] \Big\} = -\log \mathbb{E}_{\mathrm{R}}\big[e^{-\Phi}\big], \tag{7.5}$$

and the infimum is attained at the unique Q^* *with* $d\mathrm{Q}^*/d\mathrm{R} = e^{-\Phi}/\mathbb{E}_{\mathrm{R}}[e^{-\Phi}]$.

Proof. Let $Z = \mathbb{E}_{\mathrm{R}} e^{-\Phi}$ and $d\mathrm{Q}^* = Z^{-1} e^{-\Phi} d\mathrm{R}$. For any $\mathrm{Q} \ll \mathrm{R}$ with finite relative entropy,

$$\mathrm{KL}(\mathrm{Q} \,\|\, \mathrm{R}) + \mathbb{E}_{\mathrm{Q}}\Phi = \mathbb{E}_{\mathrm{Q}} \log \frac{d\mathrm{Q}}{d\mathrm{R}} + \mathbb{E}_{\mathrm{Q}}\Phi = \mathbb{E}_{\mathrm{Q}} \log \frac{d\mathrm{Q}}{d\mathrm{Q}^*} - \log Z = \mathrm{KL}(\mathrm{Q} \,\|\, \mathrm{Q}^*) - \log Z,$$

using $d\mathrm{Q}^*/d\mathrm{R} = Z^{-1} e^{-\Phi}$. The first term is non-negative and vanishes exactly at $\mathrm{Q} = \mathrm{Q}^*$. ■

7.2.2 The discrete statement, which is exact

Proposition 7.4 (KL-regularised policy optimisation) [illegible]. *Let* π_{ref} *be a probability distribution on a measurable space* $\mathcal{Y}$, *let* $r : \mathcal{Y} \to \mathbb{R}$ *satisfy* $\mathbb{E}_{\pi_{\mathrm{ref}}} e^{r/\beta} < \infty$ *for some* $\beta > 0$, *and consider*

$$\max_{\pi \ll \pi_{\mathrm{ref}}} \Big\{ \mathbb{E}_{\pi}[r] - \beta\, \mathrm{KL}(\pi \,\|\, \pi_{\mathrm{ref}}) \Big\}. \tag{7.6}$$

The maximum is $\beta \log \mathbb{E}_{\pi_{\mathrm{ref}}} e^{r/\beta}$ *and is attained uniquely at*

$$\pi^*(dy) = \frac{1}{Z_\beta} e^{r(y)/\beta}\, \pi_{\mathrm{ref}}(dy), \qquad Z_\beta = \mathbb{E}_{\pi_{\mathrm{ref}}} e^{r/\beta}. \tag{7.7}$$

Equivalently, the reward is recovered from the optimal policy by

$$r(y) = \beta \log \frac{\pi^*(y)}{\pi_{\mathrm{ref}}(y)} + \beta \log Z_\beta, \tag{7.8}$$

so that r *is determined by* π^* *up to an additive constant.*

Proof. Lemma 7.3 with $\Phi = -r/\beta$, after dividing (7.6) by β and changing sign. Equation (7.8) is (7.7) solved for r. ■

Proposition 7.4 is exact, has a two-line proof, and involves no dynamics, no continuous time, and no stochastic calculus. It is a canonical static variational identity underlying KL-regularised policy optimisation: the optimum is a tilt of the reference policy, and (7.8) says that fitting a reward and fitting a policy are the same problem up to a constant. What it is not is an account of any particular training pipeline. A method that optimises a sampled surrogate, over a parametric policy class, with a moving reference and an estimated reward, is related to this identity in the way a restricted-class objective is related to a population one — which is the subject of Theorem 3.20 in another chapter, and the same caution applies.

7.2.3 The path-space statement

Proposition 7.5 (Entropy-regularised control is a tilt of a reference process) [illegible]. *Let $\mathbb{R}$ be the law on $C([0,T];\mathbb{R}^d)$ of the uncontrolled reference $\mathrm{d}X_t = b(t,X_t)\mathrm{d}t + \sigma(t,X_t)\mathrm{d}W_t$ started from a fixed $X_0 = x_0$, and, for a control u, let $\mathbb{Q}^u$ be the law of $\mathrm{d}X_t = (b+\sigma u)(t,X_t)\mathrm{d}t + \sigma(t,X_t)\mathrm{d}W_t$ from the same x_0. Write σ^+ for the Moore–Penrose pseudoinverse and $\Pi := \sigma^+\sigma$ for the orthogonal projection onto $(\ker\sigma)^\perp$. Then, by Proposition 10.3 applied with $v = \sigma u$ — not by Theorem 3.14, whose statement fixes an invertible σ —*

$$\mathrm{KL}(\mathbb{Q}^u \,\|\, \mathbb{R}) = \tfrac{1}{2}\,\mathbb{E}_{\mathbb{Q}^u}\int_0^T |\Pi\, u(t,X_t)|^2\,\mathrm{d}t \;\le\; \tfrac{1}{2}\,\mathbb{E}_{\mathbb{Q}^u}\int_0^T |u(t,X_t)|^2\,\mathrm{d}t, \tag{7.9}$$

with equality in the second place if and only if u takes values in $(\ker\sigma)^\perp$ almost everywhere — in particular whenever σ is injective, where $\Pi = I$ and (7.9) is the familiar identity. Consequently the control problem satisfies the lower bound

$$\inf_u \mathbb{E}\Big[\tfrac{1}{2}\int_0^T |u|^2\,\mathrm{d}t + g(X_T)\Big] \;\ge\; \inf_{\mathbb{Q}\ll\mathbb{R}}\{\mathrm{KL}(\mathbb{Q}\,\|\,\mathbb{R}) + \mathbb{E}_{\mathbb{Q}}g(X_T)\} \;=\; -\log\mathbb{E}_{\mathbb{R}}e^{-g(X_T)}, \tag{7.10}$$

the equality being Lemma 7.3 with $\Phi = g(X_T)$, whose minimiser is

$$\frac{\mathrm{d}\mathbb{Q}^*}{\mathrm{d}\mathbb{R}} = \frac{e^{-g(X_T)}}{\mathbb{E}_{\mathbb{R}}e^{-g(X_T)}}. \tag{7.11}$$

The bound is attained, and no invertibility is needed for this. If $h(t,x) := \mathbb{E}_{\mathbb{R}}[e^{-g(X_T)} \mid X_t = x]$ is positive and $C^{1,2}$, then $\mathbb{Q}^*$ is the Doob h-transform of $\mathbb{R}$ and is generated by the admissible feedback

$$u^*(t,x) = \sigma(t,x)^\top \nabla \log h(t,x), \tag{7.12}$$

so the infimum over feedback controls equals $-\log\mathbb{E}_{\mathbb{R}}e^{-g(X_T)}$.

Proof. Equation (7.9) is Proposition 10.3 with $v = \sigma u$, for which the range condition (10.7) is automatic; the integrand is $|\sigma^+\sigma u|^2 = |\Pi u|^2$, reducing to $|\sigma^{-1}\sigma u|^2 = |u|^2$ and hence to (3.17) when σ is invertible. That the *minimal* norm is the right one is forced by the identity itself — u and $u+\zeta$ with $\sigma\zeta = 0$ generate the same drift, hence the same path law, hence the same left-hand side, while $\mathbb{E}\int|u+\zeta|^2$ can be made arbitrarily large. The map $u \mapsto \mathbb{Q}^u$ sends admissible controls *into* the set of path measures absolutely continuous with respect to $\mathbb{R}$ with finite relative entropy, so (7.10) follows from Lemma 7.3 by taking the infimum over the smaller set on the left; the passage from $\frac{1}{2}\mathbb{E}\int|u|^2$ to $\mathrm{KL}(\mathbb{Q}^u\,\|\,\mathbb{R})$ uses the inequality in (7.9) and so is valid for a degenerate σ as well. For attainability, $h(t,X_t)$ is a $\mathbb{R}$-martingale by the tower property, so $\mathrm{d}\mathbb{Q}^*/\mathrm{d}\mathbb{R}|_{\mathcal{F}_t} = h(t,X_t)/h(0,X_0)$, and Theorem 10.1 identifies the drift under $\mathbb{Q}^*$ as $b + \sigma\sigma^\top\nabla\log h$ — the drift generated by (7.12). That feedback takes values in $\mathrm{range}(\sigma^\top) = (\ker\sigma)^\perp$, so $\Pi u^* = u^*$ and its cost is $\frac{1}{2}\mathbb{E}\int_0^T|\sigma^\top\nabla\log h|^2\mathrm{d}t = \mathrm{KL}(\mathbb{Q}^*\,\|\,\mathbb{R})$ by (7.9): the optimal control is automatically the minimal-norm representative, which is why attainability survives a degenerate σ. Equality therefore holds in (7.10). ■

Remark 7.6 (Two statements about a degenerate σ, and only one of them is a loss). Two clauses must be kept apart when $\ker\sigma \neq \{0\}$, because one of them is false and the other is not.

False: the energy of an arbitrary control is the path entropy. The path law depends on u only through $v = \sigma u$, so u and $u + \zeta$ with $\sigma\zeta = 0$ give the same Q^{u} and the same relative entropy while $\mathbb{E}\int |u + \zeta|^2$ is unbounded above. No identity of the form $\mathrm{KL}(Q^{u} \,\|\, \mathbb{R}) = \frac{1}{2}\mathbb{E}\int |u|^2$ can hold for every admissible u; what holds is (7.9), with the projection, and the entropy is the *minimal* Cameron–Martin energy $\frac{1}{2}\mathbb{E}\int |\sigma^{+}v|^2$ among representations of the same drift increment.

True: attainability survives. The optimal drift increment supplied by the h-transform is $a\nabla \log h = \sigma\sigma^{\top}\nabla \log h$, which lies in range$(\sigma)$ by construction and is realised by

$$u^{*} = \sigma^{\top}\nabla \log h \in (\ker \sigma)^{\perp}, \tag{7.13}$$

whether or not σ is invertible; and since u^{*} is already the minimal-norm representative, its quadratic cost coincides with the path entropy. So the bound (7.10) is still attained, by (7.13), and the degeneracy costs a *representation* — uniqueness of u — rather than the optimum. The invertibility hypothesis belongs to the state-space formula $\theta = \sigma^{-1}(b^{1} - b^{2})$ of Proposition 10.2, where an arbitrary drift difference must be pulled back through σ; here no such pullback is needed, because the drift difference is manufactured from σ in the first place, and the statement that covers this case is Proposition 10.3. Keeping the two formulations apart is the difference between a control problem that is well posed under actuation constraints and one that appears not to be.

Remark 7.7 (Why the initial state is fixed). The deterministic $X_0 = x_0$ is not a simplification. A control cannot change where the process starts, so the admissible path measures all share the initial law, whereas the infimum on the right of (7.10) ranges over all $Q \ll \mathbb{R}$ — and by Proposition 10.13(i) its minimiser, the global tilt (7.11), moves the initial law by the factor $h_0(X_0) = \mathbb{E}_{\mathbb{R}}[e^{-g(X_T)} \mid X_0]$. With X_0 deterministic h_0 is trivially constant and the minimiser is admissible, which is what makes the bound tight. For a genuinely random $X_0 \sim \nu_0$ the two differ: the attainable value is $\mathbb{E}_{\nu_0}[-\log h_0(X_0)]$, the unrestricted Gibbs bound is $-\log \mathbb{E}_{\nu_0} h_0(X_0)$, and Jensen's inequality separates them unless h_0 is ν_0-a.s. constant. The admissible optimum is then the conditionally normalised transform (10.20) rather than the global tilt, and $-\log \mathbb{E}_{\mathbb{R}} e^{-g(X_T)}$ is a strict lower bound.

Remark 7.8 (The terminal law of the optimal process). By (7.11) the time-T marginal of Q^{*} has density proportional to $q_T^{\mathrm{ref}}(x)e^{-g(x)}$ with respect to Lebesgue measure, where q_T^{ref} is the terminal law of the reference. It is *not* proportional to e^{-g}. Choosing $g = -\log p_{\mathrm{data}}$ therefore does not produce p_{data} at time T; it produces something proportional to $q_T^{\mathrm{ref}}\, p_{\mathrm{data}}$. Correcting for this requires either a density ratio in the terminal cost or a hard marginal constraint, and those are two different problems; Proposition 10.13 separates them.

Notes on Proposition 7.4, Proposition 7.5 and Lemma 7.3

Assumed. Lemma 7.3: only integrability of $e^{-\Phi}$. Proposition 7.4: nothing further — it is a statement about measures on an arbitrary space and holds verbatim for a distribution over token sequences. Proposition 7.5: identical diffusion coefficients, a control entering through the noise channel, a quadratic cost with the identity as its weight, finiteness of the path-space relative entropy, and — for the attainability clause only — a positive $C^{1,2}$ function h, and enough regularity to apply Theorem 10.1. An invertible σ is *not* assumed: (7.9) is stated with the projection Π precisely so that the degenerate case is covered, and it reduces to the familiar identity when σ is injective (Remark 7.6). Also a deterministic initial state: with a random X_0 the equality in (7.10) fails, by Remark 7.7.

Two directions, two arguments. Equation (7.10) is an inequality because the controls form a *subset* of the admissible path measures: no claim is made that every path measure of finite relative entropy is generated by a Markov feedback, and in general it is not — a path-dependent change of measure requires a path-dependent drift. Equality holds because the particular minimiser is a Doob h-transform, generated by (7.12), and that feedback is minimal-norm, so equality holds for a singular σ too. What a singular σ does change is the *cost formula for an arbitrary control*: σu determines only Πu, several controls give the same law, and the entropy is the minimal Cameron–Martin norm rather than $|u|^2$ — which is why (7.9) carries the projection.

Where the path-space version stops. All four hypotheses are used. If the drift increment has a component outside $\operatorname{Ran}\sigma$ on a set of positive $dt \otimes Q^u$ occupation measure — a control acting through a channel the noise does not reach — then Q^u is not absolutely continuous with respect to $\mathbb{R}$ and $\mathrm{KL}(Q^u \,\|\, \mathbb{R}) = \infty$; (7.9) is then not merely inexact but has an infinite left-hand side. That is an *oriented* statement, in the direction the cost is written; mutual singularity would need a separating condition visible from both laws and is not claimed (Remark 10.4). A non-identity weight R in the cost changes the constant, and Section 7.4 says exactly which combinations survive.

What Proposition 7.4 does not say. It is a statement about a single decision with a fixed reference and a fixed reward. It says nothing about how π^* is computed, nothing about estimating r from comparisons, and nothing about sequential decisions: for a multi-step problem the same identity applies at each step to the *soft* value function, not to the immediate reward, and the two coincide only in the single-step case.

The continuous-time version is a modelling layer, not a limit. Proposition 7.4 is exact for a distribution over sequences of tokens. Proposition 7.5 is exact for a diffusion. Nothing in this volume derives the second from the first. Writing an autoregressive process as a diffusion requires a state space, a scaling of the step, and a noise structure, *none* of which the discrete statement supplies and each of which is a hypothesis about the model rather than a consequence of it. Table 11.3 records the relation as exact in discrete time and analogical in continuous time, and this section is the reason.

The tilt is not the target. By Remark 7.8 the optimal process reaches $q_T^{\mathrm{ref}} e^{-g}$ and not e^{-g}. This is the most frequently repeated error in the area, it survives because the two agree when the reference terminal law is uniform, and no reference process on $\mathbb{R}^d$ has a uniform terminal law.

A soft optimum is not a sample. Equation (7.7) is the maximiser of a regularised objective. It is a distribution, and drawing from it is a separate problem — Chapter 4 — whose difficulty is governed by the properties of r and β and is not addressed by the variational identity.

7.3 Continuous-Time Reinforcement Learning

Reinforcement learning replaces the known model of Definition 7.1 by samples. Two constructions carry over exactly, and one warning is needed before the section can be read as practical advice.

Proposition 7.9 (Exploratory Hamilton–Jacobi–Bellman equation) DERIVED. *Replace the deterministic feedback of Definition 7.1 by a randomised policy* $\varpi(da \mid t, x)$ *and add an entropy reg-*

ulariser relative to a reference policy ϖ_0, so that the running cost becomes $\int f \, d\varpi + \beta \mathrm{KL}(\varpi \,\|\, \varpi_0)$. Then the inner minimisation in (7.3) is solved in closed form: with $h(t,x,u) = \langle b(t,x,u), \nabla W \rangle + f(t,x,u)$,

$$\varpi^*(\mathrm{d}u \mid t,x) \;\propto\; \varpi_0(\mathrm{d}u \mid t,x)\, e^{-h(t,x,u)/\beta}, \tag{7.14}$$

and the equation becomes

$$\partial_t W - \beta \log \mathbb{E}_{\varpi_0}\left[e^{-h(t,x,u)/\beta}\right] + \tfrac{1}{2}\,\mathrm{tr}(a\nabla^2 W) = 0, \qquad W(T,\cdot) = g. \tag{7.15}$$

Proof. For fixed (t,x) the inner problem is $\inf_{\varpi}\{\mathbb{E}_{\varpi} h + \beta \mathrm{KL}(\varpi \,\|\, \varpi_0)\}$, which is Lemma 7.3 with $\mathbb{R} = \varpi_0$ and $\Phi = h/\beta$ after dividing by β; the minimiser is (7.14) and the value is $-\beta \log \mathbb{E}_{\varpi_0} e^{-h/\beta}$. Substituting into (7.3) gives (7.15). ■

Proposition 7.10 (Policy evaluation is a martingale condition) DERIVED. *Fix a randomised policy ϖ and let X be the corresponding closed-loop diffusion, with $X_0 \sim \nu_0$. Let $W \in C^{1,2}$ satisfy $W(T,\cdot) = g$ and suppose $\int_0^T \mathbb{E}\left|\sigma^\top \nabla W(t,X_t)\right|^2 \mathrm{d}t < \infty$, so that the stochastic integral below is a true martingale rather than a local one. Then*

$$M_t := W(t,X_t) + \int_0^t \bar{f}(s,X_s)\,\mathrm{d}s \qquad \text{is a martingale} \tag{7.16}$$

if and only if the evaluation equation $\partial_t W + \mathcal{L}^{\varpi} W + \bar{f} = 0$ holds $\mathrm{d}t \otimes \mathrm{Law}(X_t)$-almost everywhere, where $\bar{f}(s,x) = \int f(s,x,u)\varpi(\mathrm{d}u \mid s,x) + \beta \mathrm{KL}(\varpi \,\|\, \varpi_0)(s,x)$. If in addition the residual is continuous and $\mathrm{Law}(X_t)$ has full support for every $t > 0$ — which holds under uniform ellipticity — the equation holds pointwise on $[0,T] \times \mathbb{R}^d$ and W is the value function of ϖ.

Proof. By Theorem 1.33, $\mathrm{d}M_t = (\partial_t W + \mathcal{L}^{\varpi} W + \bar{f})\mathrm{d}t + \langle \sigma^\top \nabla W, \mathrm{d}W_t \rangle$ with $\mathcal{L}^{\varpi}$ the generator of the closed-loop diffusion; under the square-integrability hypothesis the second term is a true martingale. Hence M is a martingale precisely when the finite-variation part vanishes, that is when $\mathbb{E}\int_0^T \left|\partial_t W + \mathcal{L}^{\varpi} W + \bar{f}\right|(t,X_t)\mathrm{d}t = 0$, which says the residual is zero for $\mathrm{d}t \otimes \mathrm{Law}(X_t)$-almost every (t,x) — *along the occupation measure of X*, and not a priori anywhere else. This is the only conclusion the martingale property can carry: a residual supported off the reachable set is invisible to M. Adding continuity of the residual and full support of $\mathrm{Law}(X_t)$ upgrades "almost everywhere" to "everywhere", and the resulting equation with its terminal condition characterises the value function of ϖ by Proposition 7.2 applied with the single control ϖ. ■

Proposition 7.11 (Why continuous-time policy evaluation is not small-step policy evaluation) DERIVED. *Let W be a candidate with Bellman residual $\varepsilon(t,x) = \partial_t W + \mathcal{L}^{\varpi} W + \bar{f}$, and let*

$$\delta_\Delta = W(t+\Delta, X_{t+\Delta}) - W(t,X_t) + \int_t^{t+\Delta} \bar{f}(s,X_s)\,\mathrm{d}s \tag{7.17}$$

be the temporal-difference increment over a step Δ. Then

$$\mathbb{E}\left[\delta_\Delta \mid X_t = x\right] = \varepsilon(t,x)\,\Delta + o(\Delta), \qquad \mathrm{Var}(\delta_\Delta \mid X_t = x) = \left|\sigma^\top \nabla W\right|^2(t,x)\,\Delta + o(\Delta), \tag{7.18}$$

so the signal-to-noise ratio of a single increment is $O(\sqrt{\Delta})$ and vanishes as $\Delta \downarrow 0$. Aggregated over $n = T/\Delta$ increments of one trajectory the ratio is $O(\sqrt{T})$, independent of Δ.

Proof. Apply Theorem 1.33 to W over $[t, t+\Delta]$: the drift contributes $\int_t^{t+\Delta} \varepsilon(s, X_s)\mathrm{d}s = \varepsilon(t,x)\Delta + o(\Delta)$ by continuity, and the martingale part has zero mean and variance $\mathbb{E}\int_t^{t+\Delta} \left|\sigma^\top \nabla W\right|^2 \mathrm{d}s = \left|\sigma^\top \nabla W\right|^2 (t,x)\Delta + o(\Delta)$ by the Itô isometry (1.22). The ratio of mean to standard deviation is therefore $\varepsilon\sqrt{\Delta}/\left|\sigma^\top \nabla W\right|$. Summing $n = T/\Delta$ increments with the martingale parts orthogonal, as in Proposition 1.49, multiplies the mean by n and the standard deviation by $\sqrt{n}$, giving a ratio proportional to $\sqrt{n\Delta} = \sqrt{T}$. ∎

Notes on Proposition 7.9, Proposition 7.10 and Proposition 7.11

What the signal-to-noise statement does and does not say. It says the information in a *single* transition vanishes with the step, and that the information in a *fixed horizon* does not. The consequence is for estimators that treat transitions as independently informative — per-sample updates with function approximation — and not for estimators that integrate over a trajectory. Reading it as "continuous-time reinforcement learning is impossible" is wrong in the second direction; reading it as "the discrete algorithm converges to the continuous one as $\Delta \to 0$" is wrong in the first.

What is not addressed. How ε is estimated when W is a learned function; the bias introduced by bootstrapping; exploration, beyond the regulariser that makes (7.14) interior; and any convergence result for a learning algorithm, of which this volume proves none.

The martingale property sees only where the process goes. For Proposition 7.10 the equivalence is with the evaluation equation holding along the occupation measure $\mathrm{d}t \otimes \mathrm{Law}(X_t)$. Without ellipticity — a degenerate diffusion, or a policy confining X to a submanifold — W may be wrong off the reachable set and M will not detect it, which is exactly the situation of a learned critic evaluated outside the data distribution. The upgrade to a pointwise statement is a support hypothesis and is stated as one.

7.4 Linearly Solvable Control

Under a condition on the geometry of the actuation and the noise, the nonlinear equation (7.3) becomes linear. The condition is a matrix identity, and stating it as a scalar one is the error this section exists to prevent.

Proposition 7.12 (The logarithmic transform, and when it linearises) DERIVED. *Consider the controlled dynamics $\mathrm{d}X_t = \big(b(t,X_t) + B(t,X_t)u_t\big)\mathrm{d}t + \sigma(t,X_t)\mathrm{d}W_t$ with running cost $q(t,x) + \frac{1}{2}\langle u, Ru\rangle$ for a symmetric $R \succ 0$, and terminal cost g. Then:*

(i) the inner infimum in (7.3) is attained at

$$u^* = -R^{-1}B^\top \nabla W, \qquad \text{with value} \qquad -\tfrac{1}{2}\left\langle \nabla W, BR^{-1}B^\top \nabla W\right\rangle,$$

so that (7.3) reads

$$\partial_t W + \langle b, \nabla W\rangle - \tfrac{1}{2}\left\langle \nabla W, BR^{-1}B^\top \nabla W\right\rangle + \tfrac{1}{2}\operatorname{tr}\!\left(a\nabla^2 W\right) + q = 0; \tag{7.19}$$

(ii) the substitution $\Psi = e^{-W/\lambda}$ turns (7.19) into the linear equation

$$\partial_t \Psi + \langle b, \nabla \Psi\rangle + \tfrac{1}{2}\operatorname{tr}\!\left(a\nabla^2 \Psi\right) = \frac{q}{\lambda}\Psi, \qquad \Psi(T,\cdot) = e^{-g/\lambda}, \tag{7.20}$$

if and only if the matrix matching condition

$$\lambda B R^{-1} B^{\top} = a = \sigma\sigma^{\top}, \qquad \textit{equivalently} \qquad B R^{-1} B^{\top} = \frac{a}{\lambda}, \tag{7.21}$$

holds, for a scalar $\lambda > 0$, at every (t, x);

(iii) *under (7.21), Ψ has the Feynman–Kac representation $\Psi(t,x) = \mathbb{E}\big[\exp\big(-\lambda^{-1} g(X_T) - \lambda^{-1}\int_t^T q\,ds\big) \mid X_t = x\big]$ along the uncontrolled dynamics.*

Proof. (i) The map $u \mapsto \langle Bu, \nabla W\rangle + \frac{1}{2}\langle u, Ru\rangle$ is a strictly convex quadratic, minimised at $u^* = -R^{-1}B^{\top}\nabla W$ with the stated value.

(ii) With $W = -\lambda \log \Psi$ one has $\nabla W = -\lambda \nabla\Psi/\Psi$ and

$$\nabla^2 W = -\lambda \frac{\nabla^2\Psi}{\Psi} + \lambda \frac{\nabla\Psi\nabla\Psi^{\top}}{\Psi^2}, \qquad \text{so} \qquad \tfrac{1}{2}\operatorname{tr}(a\nabla^2 W) = -\frac{\lambda}{2}\frac{\operatorname{tr}(a\nabla^2\Psi)}{\Psi} + \frac{\lambda}{2}\frac{\langle\nabla\Psi, a\nabla\Psi\rangle}{\Psi^2}.$$

Since $\nabla W = -\lambda\nabla\Psi/\Psi$, the quadratic term of (7.19) is $-\frac{1}{2}\lambda^2\left\langle\nabla\Psi, BR^{-1}B^{\top}\nabla\Psi\right\rangle/\Psi^2$, so the two terms of (7.19) that are quadratic in $\nabla\Psi$ sum to

$$\frac{1}{2\Psi^2}\Big(\lambda\langle\nabla\Psi, a\nabla\Psi\rangle - \lambda^2\left\langle\nabla\Psi, BR^{-1}B^{\top}\nabla\Psi\right\rangle\Big) = \frac{\lambda}{2\Psi^2}\left\langle\nabla\Psi, \big(a - \lambda BR^{-1}B^{\top}\big)\nabla\Psi\right\rangle.$$

This vanishes for all $\nabla\Psi$ if and only if the symmetric matrices a and $\lambda BR^{-1}B^{\top}$ agree, which is (7.21); conversely if they differ at some (t,x) then choosing Ψ with a $\nabla\Psi$ in an eigendirection of the difference leaves a term quadratic in $\nabla\Psi$, so the equation is not linear. Multiplying what remains by $-\Psi/\lambda$ gives (7.20).

Note the direction of (7.21): a *larger* temperature λ requires a *smaller* actuation gain $BR^{-1}B^{\top} = a/\lambda$, because λ multiplies the control cost through $W = -\lambda\log\Psi$. The case $B = \sigma$, $R = I$ has $\lambda = 1$ and does not distinguish the two orientations, which is why a regression test for this identity must use $\lambda \neq 1$.

(iii) is the Feynman–Kac formula applied to (7.20), whose first-order and second-order parts are the generator of the uncontrolled dynamics. ■

How restrictive (7.21) is. It says the actuation directions, weighted by R^{-1}, must span the noise directions in the same proportions — with the actuation gain equal to a/λ, not to λa. Three consequences: the control cannot act in a direction the noise does not reach, so an under-actuated system is excluded; the relative cost of acting in two directions is fixed by the relative noise magnitudes, so R is not free; and λ is a single scalar, so no direction may be scaled independently. For the special case $B = \sigma$ and $R = I$ the condition holds with $\lambda = 1$, which is the setting of Proposition 7.5, and that is the only case this volume uses.

What replaces it when it fails. Nothing exact. Equation (7.19) remains a nonlinear equation, the Feynman–Kac representation is unavailable, and the connection to the path-space relative entropy of Proposition 7.5 is broken because the cost is no longer the relative entropy. Approximations exist and are not treated here.

What does not follow

Path-integral control is not isomorphic to diffusion modelling. The exact statement is Proposition 7.12 under (7.21), and under that condition the optimal process is the tilt (7.11) of a reference diffusion — which is a construction that also appears in Chapter 3. What is shared is that construction, not an isomorphism of problems: the generative model has no cost, no control channel, and no terminal objective, and the control problem has no data distribution. Table 11.3 records the relation as exact under (7.21), and the matrix form of that condition is what a scalar statement of it loses.

The linear equation is not a computational solution. Equation (7.20) is linear and lives on $\mathbb{R}^d$; solving it by a grid is exactly the exponential cost that Table 11.1 says continuous formulations do not remove. The Feynman–Kac representation replaces the grid by a Monte-Carlo estimate of $\mathbb{E}_{\mathbb{R}}[e^{-g(X_T)/\lambda}]$, and the practical obstruction is the variance of that estimator, which is not a corollary of linearity and is not bounded anywhere in this chapter. What can be said without a theorem is structural: the integrand is an exponential tilt, so as λ decreases the estimator concentrates its mass on rarer paths, and the same happens as the horizon lengthens and $g(X_T)$ spreads. Whether that degrades a particular estimator, and at what rate, is an importance-sampling variance question with its own hypotheses on g and $\mathbb{R}$; no quantitative variance statement is claimed here, and a reader who needs one should look for a bound on the second moment of the tilt rather than quote this paragraph.

7.5 Chapter Summary

Chapter 7 builds three levels and keeps them apart. Proposition 7.4 is a static variational identity: no dynamics, no time index, and it says nothing about a sequential decision process. Proposition 7.5 is its path-space counterpart, in which the entropy is paid in the control coordinates and the optimal feedback is the minimal-norm representative — so a degenerate diffusion coefficient costs uniqueness of the control, not attainability of the optimum. The two are related by analogy and not by a theorem, and Table 11.3 records that as three separate rows.

The verification theorem Proposition 7.2 carries the moment and square-integrability conditions its localisation argument actually consumes, and Proposition 7.12 states the matrix matching condition $\lambda B R^{-1} B^{\top} = a$ under which the Hamilton–Jacobi–Bellman equation linearises — a condition on the geometry of actuation, not on the cost.

Exercises

Tier I Direct application.

7.1 *Linear quadratic control*

For $b = Ax + Bu$, σ constant, $f = \frac{1}{2}(\langle x, Qx\rangle + \langle u, Ru\rangle)$ and $g = \frac{1}{2}\langle x, Q_T x\rangle$, verify by substitution into (7.19) that $W(t,x) = \frac{1}{2}\langle x, P_t x\rangle + c_t$ solves the equation, and derive the Riccati equation for P. Confirm Proposition 7.2 applies and identify u^*.

7.2 *The alignment formula*

Verify (7.8) directly, and show that adding a constant to r changes π^* not at all. State what this implies for the identifiability of a reward from demonstrations of an optimal policy.

Tier II Completing the derivation.

7.3 ***Sequential decisions***

Extend Proposition 7.4 to a finite-horizon Markov decision process with KL regularisation at each step, and show that the optimal policy is $\pi_k^*(\cdot \mid s) \propto \pi_{\mathrm{ref}}(\cdot \mid s)e^{Q_k(s,\cdot)/\beta}$ with Q the soft action-value function. Identify precisely where the single-step statement is applied and why r must be replaced by Q.

7.4 ***The matching condition in coordinates, with*** $\lambda \neq 1$

Write (7.21) out for $d = 2$, $m = 1$, $B = (1,0)^\top$, $R = \rho$ and $\sigma = \mathrm{diag}(\varsigma_1, \varsigma_2)$, and show it holds only if $\varsigma_2 = 0$, in which case it fixes $\lambda = \rho\varsigma_1^2$. Check that the opposite orientation $BR^{-1}B^\top = \lambda a$ would give $\lambda = 1/(\rho\varsigma_1^2)$ instead, so that the two disagree unless $\rho\varsigma_1^2 = 1$ — this is the regression test the case $B = \sigma$, $R = I$ cannot perform. Interpret: which physical situation does $\varsigma_2 = 0$ exclude, and what does it say about controlling a system with unactuated degrees of freedom?

Tier III Probing the boundary.

7.5 ***The tilt is not the target, quantitatively***

Let the reference be a standard Ornstein–Uhlenbeck process with $q_T^{\mathrm{ref}} = \mathcal{N}(0,1)$ and let $p_{\mathrm{data}} = \mathcal{N}(\mu,1)$. Take $g = -\log p_{\mathrm{data}}$ and compute the terminal marginal of (7.11) exactly. Report the discrepancy from p_{data} as a function of μ, and give the terminal cost that would produce p_{data} instead.

7.6 ***Where the value function is not smooth***

Take $d = 1$, $b = u$ with $U = [-1,1]$, $\sigma \equiv 0$, $f \equiv 0$ and $g(x) = |x|$. Compute V explicitly and show it is not differentiable on a set that the optimal trajectory reaches. State which hypothesis of Proposition 7.2 fails and what replaces the theorem.

7.7 ***Signal to noise at two resolutions***

For the Ornstein–Uhlenbeck reference with a quadratic W having a known Bellman residual ε, compute both quantities in (7.18) in closed form. Then compute the number of independent transitions needed to detect ε at a fixed confidence, as a function of Δ, and the number of *trajectories* of fixed length T needed. Explain why the two answers scale differently and which one an implementation is paying.

Tier IV Computational.

7.8 ***Solving one problem three ways***

Take $d = 2$, a double-well q, $b = 0$, $\sigma = I$, $R = I$, $B = I$, so that (7.21) holds with $\lambda = 1$.

(a) Solve (7.20) on a grid and read off $W = -\log \Psi$ and u^*.

(b) Estimate Ψ by the Feynman–Kac representation with 10^5 uncontrolled trajectories. Report the relative error against (a) as a function of the horizon, and the empirical variance of the estimator.

(c) Simulate the optimally controlled process from (a) and verify numerically that its terminal law is proportional to $q_T^{\mathrm{ref}} e^{-g}$ and not to e^{-g}.

(d) State which of (a)–(c) tests Proposition 7.12, which tests Remark 7.8, and which tests neither but measures the obstruction named in the last box of Section 7.4.

Acceptance criteria: (b) reports the variance growth with the horizon; (c) reports the discrepancy as a number; (d) classifies all three.

CHAPTER 8

Mean-Field Limits

A system of N exchangeable particles interacting through their empirical measure has, as $N \to \infty$, a limit that is a single nonlinear equation for one distribution. That statement is the subject of this chapter, and almost everything in it turns on one question: *in which metric?*

The question is not pedantic. The error between the N-particle system and its limit can be measured by coupling individual particles, by comparing empirical measures, by testing against smooth functions, or by relative entropy, and these four quantities have four different rates with four different dependencies on the dimension. One of them is $N^{-1/2}$ with a dimension-free constant. Another is $N^{-1/d}$ and cannot be improved. Reporting the first while measuring the second is the error Table 8.1 exists to prevent.

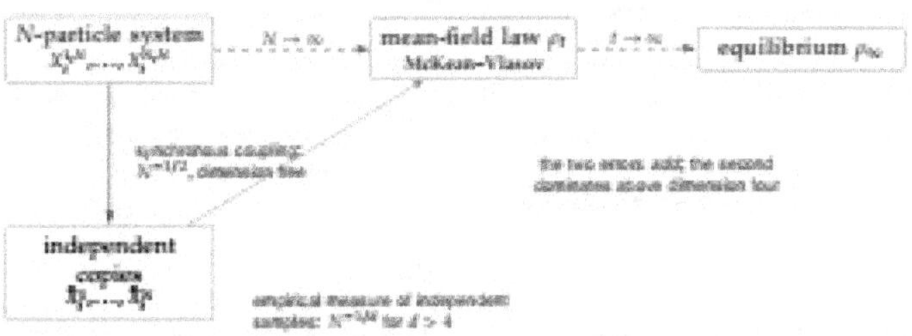

Figure 8.1: The route from a finite particle system to an equilibrium passes through two limits, and the finite-N error splits into two terms with different rates. The coupling term is dimension free; the sampling term is not, and it is present even with no dynamics at all. What this figure does not imply: that the two limits commute, or that a uniform-in-time statement follows from a finite-horizon one. Exchanging $N \to \infty$ and $t \to \infty$ requires a separate hypothesis (Section 8.3), and the horizontal arrows are limits, not identities.

The chapter's second theme is what mean-field convexity does and does not buy. The risk of a two-layer network is convex as a function of the measure of neurons; this is elementary and true, and it does not imply that the gradient flow converges, because the relevant convexity for a W_2 gradient flow is geodesic convexity and the two are different properties. Section 8.3 gives what is available instead: for the entropy-regularised problem there is a genuine convergence theorem, with a hypothesis that can be stated and is not automatic.

8.1 Interacting Particles and the McKean–Vlasov Limit

Definition 8.1 (Particle system and its mean-field limit). Let $b : \mathbb{R}^d \times \mathcal{P}_2(\mathbb{R}^d) \to \mathbb{R}^d$ be of *pairwise interaction form*

$$b(x,\mu) = b_0(x) + \int_{\mathbb{R}^d} K(x,y)\,\mu(dy), \tag{8.1}$$

with b_0 and K Lipschitz and K bounded. The N-particle system is

$$dX_t^{i,N} = b(X_t^{i,N}, \mu_t^N)dt + \sqrt{2}\,dW_t^i, \qquad \mu_t^N = \frac{1}{N}\sum_{j=1}^{N} \delta_{X_t^{j,N}}, \tag{8.2}$$

with independent Brownian motions W^i and *independent, identically distributed* initial data $X_0^{i,N} \sim \rho_0$ with finite second moment. The *McKean–Vlasov* equation is

$$d\bar{X}_t = b(\bar{X}_t, \rho_t)dt + \sqrt{2}\,dW_t, \qquad \rho_t = \mathrm{Law}(\bar{X}_t), \tag{8.3}$$

whose marginals solve the nonlinear Fokker–Planck equation $\partial_t\rho = -\nabla\cdot(b(\cdot,\rho)\rho) + \Delta\rho$.

Proposition 8.2 (Classical propagation of chaos) ESTABLISHED. *Assume the pairwise form (8.1) with b_0 Lipschitz and K bounded and Lipschitz, and iid initial data as in Definition 8.1. Couple (8.2) and (8.3) by using the same Brownian motions and the same initial values, $\bar{X}_0^i = X_0^{i,N}$, so that the $\bar{X}^i$ are independent copies of the McKean–Vlasov solution. Then for every T there is $C(T)$, growing exponentially in T and independent of d and N, with*

$$\max_{i\le N}\, \mathbb{E}\sup_{t\le T}\left|X_t^{i,N} - \bar{X}_t^i\right|^2 \le \frac{C(T)}{N}. \tag{8.4}$$

The synchronous-coupling argument in this form is due to Sznitman [26].

8.1.1 Four errors, four rates

Equation (8.4) is a statement about *tagged particles*. It is not a statement about the empirical measure, and the following makes the relation exact.

Proposition 8.3 (The two errors add) DERIVED. *Let $\bar{\mu}_t^N = \frac{1}{N}\sum_j \delta_{\bar{X}_t^j}$ be the empirical measure of the independent copies. Then*

$$\mathcal{W}_2(\mu_T^N, \rho_T) \le \underbrace{\left(\tfrac{1}{N}\textstyle\sum_j |X_T^{j,N} - \bar{X}_T^j|^2\right)^{1/2}}_{\text{coupling}} + \underbrace{\mathcal{W}_2(\bar{\mu}_T^N, \rho_T)}_{\text{sampling}}, \tag{8.5}$$

where the first term has expectation $O(N^{-1/2})$ by Proposition 8.2 with a dimension-free constant.

Proof. The triangle inequality for $\mathcal{W}_2$ gives $\mathcal{W}_2(\mu^N, \rho) \le \mathcal{W}_2(\mu^N, \bar{\mu}^N) + \mathcal{W}_2(\bar{\mu}^N, \rho)$. The pairing $X^{j,N} \leftrightarrow \bar{X}^j$ is a coupling of the two empirical measures, so $\mathcal{W}_2^2(\mu^N, \bar{\mu}^N) \le \frac{1}{N}\sum_j |X^{j,N} - \bar{X}^j|^2$, which is the first term. Its expectation is bounded using (8.4) and Jensen. ■

The second term of (8.5) involves no dynamics at all: it is the error of the empirical measure of N independent samples. Its rate is an external theorem, and it is stated separately so that the decomposition above — which is two lines of this volume's own — is not confused with a quoted result.

Proposition 8.4 (Empirical-measure rate in W_2) ESTABLISHED. *For $\bar{\mu}^N$ the empirical measure of N independent samples from ρ,*

$$\mathbb{E}\, W_2(\bar{\mu}^N, \rho) \lesssim \begin{cases} N^{-1/4}, & d < 4, \\ N^{-1/4} \log^{1/2} N, & d = 4, \\ N^{-1/d}, & d > 4, \end{cases} \tag{8.6}$$

uniformly *over the class*

$$\mathcal{P}_{q,M} = \left\{ \rho \in \mathcal{P}_2(\mathbb{R}^d) : \int |x|^q \rho(dx) \le M \right\}, \qquad q > 4, \tag{8.7}$$

the exponent q and the bound M entering the suppressed constant; and within $\mathcal{P}_{q,M}$ there are laws attaining each exponent, so no smaller exponent holds uniformly over the same class. The rate and the moment class are due to Fournier and Guillin [9].

The threshold on q is not decorative: for q close to 2 the heavy-tailed members of $\mathcal{P}_2(\mathbb{R}^d)$ converge strictly more slowly than (8.6), and the displayed exponents are then false uniformly. The cleanest version takes ρ compactly supported, where all three cases hold with a constant depending only on the diameter. For an individual ρ the rate can be strictly faster — it is zero for a Dirac, and it is governed by the intrinsic rather than the ambient dimension for a law on a low-dimensional set.

Corollary 8.5 (The sampling term dominates above dimension four) DERIVED. *Over $\mathcal{P}_{q,M}$ and for $d > 4$, combining (8.5) with (8.6) gives $\mathbb{E} W_2(\mu_t^N, \rho_t) \lesssim N^{-1/2} + N^{-1/d}$, in which the second term dominates; and no bound better than $N^{-1/d}$ holds uniformly, since the $t = 0$ case already reduces to Proposition 8.4. In particular $N^{-1/2}$ does not.*

Table 8.1: Four ways of measuring the same convergence, with four rates. Rows one and two are the ones routinely exchanged: the first is dimension free and the second is not, and by Corollary 8.5 the second dominates in dimension above four. A claim of "propagation of chaos at rate $N^{-1/2}$" is true of row one and false of row two.

Quantity	Rate	Dimension	What it needs
Tagged-particle coupling $\mathbb{E}\,\lvert X^{i,N} - \bar{X}^i \rvert^2$	$C(T)/N$ (Proposition 8.2)	Free of d; $C(T)$ exponential in T	A synchronous coupling: the same Brownian motions and initial values

continued overleaf

Table 8.1: (continued)

Quantity	Rate	Dimension	What it needs
Empirical measure $\mathbb{E}W_2(\mu^N, \rho)$	$N^{-1/d}$ for $d > 4$, uniformly over $\mathcal{P}_{q,M}$ of (8.7) with $q > 4$ (Proposition 8.4, Corollary 8.5)	Cursed; the exponent is sharp over that class, and can be better for an individual ρ	A q-th moment bound with $q > 4$, or compact support — not merely $\rho \in \mathcal{P}_2(\mathbb{R}^d)$, for which the exponents fail; no dynamics, since it already holds for independent samples
Weak error $\lvert \mathbb{E}\langle \mu^N, f\rangle - \langle \rho, f\rangle \rvert$	Typically $O(1/N)$ under stronger hypotheses than this chapter assumes; *not established here* (Remark 8.6)	Free of d in the results usually quoted; the constant depends on derivatives of f and on regularity in the measure variable	Smoothness of f is *not* sufficient: the rate needs differentiability of the coefficients in the measure argument, and does not follow from the synchronous coupling of Proposition 8.2
Relative entropy of the k-marginal	$O((k/N)^2)$ (Proposition 8.20)	Free of d under a uniform inequality	Stronger structural hypotheses; see Section 8.4

Remark 8.6 (The weak-error row is not a theorem of this chapter). The $O(1/N)$ weak rate is included because it is the quantity most often reported, and it must be read as a pointer rather than a claim. It does not follow from Proposition 8.2: the synchronous coupling controls $\mathbb{E}\left|X^{i,N} - \bar{X}^i\right|^2$, and feeding that into a test function gives $O(N^{-1/2})$, not $O(N^{-1})$. The improvement to $1/N$ comes from cancellation in the expectation, and establishing it requires differentiability of the coefficients with respect to the measure argument — a derivative on $\mathcal{P}_2(\mathbb{R}^d)$, with its own regularity theory — rather than smoothness of f alone. No such theorem is stated or proved in this volume, and none is quoted with a source; until one is, this row supplies vocabulary and a caution, not a rate that may be cited.

Interpretation

Two errors, and only one of them is about the dynamics. Coupling a particle to its own independent copy measures how far interaction has pushed it, and that error is $N^{-1/2}$ with a constant free of dimension. Comparing the empirical measure of N independent samples to their common law measures something else entirely — how well N points can represent a distribution — and that error is dimension-cursed, present even with no dynamics at all. The decomposition adds the two, and above dimension four the second dominates. The practical reading: a mean-field approximation can be excellent while the empirical measure it produces is still a poor description of the limit.

Notes on Proposition 8.2, Proposition 8.3, Proposition 8.4 and Corollary 8.5

Assumed, and by which statement. Proposition 8.2: the pairwise interaction form (8.1) with a bounded Lipschitz kernel; iid initial data; and a synchronous coupling, which requires the two systems to be built on the same probability space. Proposition 8.3: nothing further — it is the triangle inequality and one explicit coupling, and it is the only one of the three proved here. Proposition 8.4: a moment class, and an external source; it involves no dynamics and would hold verbatim for iid samples with no particle system in sight. Corollary 8.5: both of the preceding, over the same class — which is the step at which quoting two results for two different classes would produce a bound for neither.

Why the pairwise form, and not merely W_1-Lipschitz. A drift that is Lipschitz in the measure for W_1 is not enough for (8.4). Take $b(x,\mu) = W_1(\mu,\rho_0)e_1$, which is 1-Lipschitz in μ and constant in x: the mean-field drift is identically zero, while the N-particle drift at time zero has size $W_1(\mu_0^N,\rho_0)$, whose square is not $O(1/N)$ in high dimension by (8.6). The pairwise form avoids this because the discrepancy it produces is $\frac{1}{N}\sum_j K(x,\bar{X}^j) - \int K(x,y)\rho(dy)$, an average of N independent centred terms, of size $N^{-1/2}$ by the central limit theorem — and independence is exactly what the iid initial data buys.

Why exchangeable initial data is not enough. If the $X_0^{i,N}$ are merely exchangeable — in the extreme, all equal to one draw — then the comparators $\bar{X}^i$ are not independent, the average above is not a sum of independent terms, and Proposition 8.3 may not treat $\bar{\mu}^N$ as an empirical measure of an iid sample.

What the coupling bound is about. It compares one particle with one independent copy. It is not a statement about the empirical measure, and converting it into one costs (8.5), after which the sampling term takes over. The dimension-free constant belongs to the coupling and does not survive the conversion.

Where the exponential horizon bites. $C(T)$ in (8.4) grows exponentially in T under these hypotheses, so the bound is empty on time scales long enough for the limit equation to equilibrate. Section 8.4 gives the uniform-in-time statement and its stronger hypotheses; this section's result is a finite-horizon one.

8.2 Two-Layer Networks in the Mean-Field Scaling

Definition 8.7 (Mean-field parameterisation). Let $\phi(x;\theta)$ be a feature map, bounded and C^1 in θ. A two-layer network with m units is $f_\theta(x) = \frac{1}{m}\sum_{j=1}^{m}\phi(x;\theta_j) = \langle\mu^m,\phi(x;\cdot)\rangle$ with $\mu^m = \frac{1}{m}\sum_j \delta_{\theta_j}$. For $\mu \in \mathcal{P}(\mathbb{R}^d)$ set $f_\mu(x) = \int \phi(x;\theta)\,\mu(d\theta)$ and

$$R(\mu) = \mathbb{E}_{(x,y)}\big[\ell\big(f_\mu(x),y\big)\big]. \tag{8.8}$$

Proposition 8.8 (Flat convexity, and what it is not) DERIVED. *Suppose $\ell(\cdot,y)$ is convex for each y; for part (iii) suppose in addition that ℓ is differentiable in its first argument with $\mathbb{E}\big[\big|\partial_1\ell(f_\mu(x),y)\big|\sup_\theta|\phi(x;\theta)|\big] < \infty$, so that differentiation under the expectation is justified. Then:*

(i) R is convex along linear interpolations: $R\big((1-s)\mu_0+s\mu_1\big) \le (1-s)R(\mu_0)+sR(\mu_1)$;

(ii) R need not be convex along W_2 geodesics. Indeed already the linear functional $\mu \mapsto \int V\,d\mu$ is convex along W_2 geodesics if and only if V is convex, so any R of that form with V non-convex is flat-convex and not displacement convex;

(iii) the first variation is $\dfrac{\delta R}{\delta\mu}(\theta) = \mathbb{E}\big[\partial_1\ell\big(f_\mu(x),y\big)\,\phi(x;\theta)\big]$, defined up to an additive constant, and μ^* minimises R if and only if $\delta R/\delta\mu\,[\mu^*]$ attains its infimum over $\mathbb{R}^d$ on the support of μ^* — equivalently, is constant there and no smaller anywhere, a statement about the field and not about the constant.

Proof. (i) f_μ is linear in μ, so $s \mapsto f_{(1-s)\mu_0+s\mu_1}(x)$ is affine, and a convex function of an affine function is convex; taking expectations preserves convexity. (ii) Along the W_2

geodesic from μ_0 to μ_1 given by the optimal map T, $\mu_s = ((1-s)\mathrm{id} + sT)_{\#}\mu_0$, so $\int V \mathrm{d}\mu_s = \int V((1-s)\theta + sT\theta)\mu_0(\mathrm{d}\theta)$, which is convex in s for every μ_0 and every optimal T if and only if V is convex; taking μ_0, μ_1 Diracs gives the converse. (iii) Differentiating (8.8) in the direction χ with $\int \chi = 0$ gives $\int (\delta R/\delta\mu)\chi$ with the stated expression, and the optimality condition is the first-order condition for a minimum over probability measures. ■

Proposition 8.9 (Global optimality of a converged limit) ESTABLISHED. *Assume ℓ convex and differentiable in its first argument with the integrability of Proposition 8.8, and ϕ bounded, continuous, and C^1 in θ with bounded and continuous $\nabla_\theta\phi$. Let μ_t solve the W_2 gradient flow of R, that is the McKean–Vlasov equation with $b = -\nabla_\theta \delta R/\delta\mu$. Suppose that*

(a) $\mu_{t_k} \to \mu_\infty$ weakly along some $t_k \to \infty$, and the family $(\mu_t)_{t\geq 0}$ is tight;

(b) the dissipation vanishes along that sequence,

$$\int \left|\nabla_\theta \frac{\delta R}{\delta\mu}[\mu_{t_k}]\right|^2 \mathrm{d}\mu_{t_k} \longrightarrow 0, \tag{8.9}$$

which is available along some sequence whenever R is bounded below, since $\int_0^\infty$ of the left-hand side equals $R(\mu_0) - \lim_t R(\mu_t) < \infty$;

(c) μ_∞ has full support;

(d) the first variation is uniformly dominated: $\mathbb{E}\sup_{|z|\leq M} |\partial_1\ell(z, Y)| < \infty$ for every $M < \infty$, which together with boundedness of ϕ and $\nabla_\theta\phi$ gives a bound on $|\nabla_\theta \delta R/\delta\mu[\mu](\theta)|$ that is uniform in both θ and μ.

Then $\nabla_\theta \delta R/\delta\mu[\mu_\infty]$ vanishes identically, $\delta R/\delta\mu[\mu_\infty]$ is constant on $\mathbb{R}^d$, and μ_∞ is a global minimiser of R. See Nitanda, Wu and Suzuki [20].

Remark 8.10 (Which hypothesis does which work). Weak convergence alone does not give the conclusion, and it is worth seeing where each assumption enters. Under the stated boundedness and continuity, $\mu \mapsto \nabla_\theta \delta R/\delta\mu[\mu](\theta)$ is continuous for the weak topology, uniformly on compact sets in θ. That is a statement about compact sets only, and weak convergence controls nothing on the tails, so (d) is what closes the gap: it makes the integrands in (8.9) uniformly bounded, tightness from (a) makes the mass outside a large ball uniformly small, and the two together give uniform integrability. With (a), (b) and (d) the limit passes and $\nabla_\theta \delta R/\delta\mu[\mu_\infty] = 0$ μ_∞-almost everywhere. Without a uniform domination the compact-set convergence can be defeated by mass escaping to infinity carrying the dissipation with it, and the limit measure inherits nothing. Full support then upgrades "almost everywhere" to "everywhere", continuity and connectedness of $\mathbb{R}^d$ make the first variation constant, and Proposition 8.8(iii) — whose optimality condition is exactly *constant on the support and no smaller anywhere* — converts that into global optimality. Drop (b) and the limit need not be stationary; drop (c) and the first variation is controlled only where the limit already puts mass, which is what makes the statement useless for an atomic μ^*.

Interpretation

Convexity in the measure and convexity along transport are different properties, and the mean-field risk has the first but not the second. Mixing two networks and averaging their outputs is a convex operation — that is flat convexity, and it is elementary. Moving the neurons themselves along an optimal transport path is not, and it is the second notion that a W_2 gradient flow descends. So the familiar statement "the mean-field risk is convex" is true and buys much less than it appears to: it gives a convex *problem*, not a convergent *flow*.

Notes on Proposition 8.8 and Proposition 8.9

Assumed. Proposition 8.8: convexity of the loss in its first argument and linearity of f_μ in μ — that is, the *mean-field* parameterisation with the factor $1/m$, and no other. At fixed m the unscaled architecture is still an integral against the empirical measure, namely $m \int \phi \, d\mu^m$; what the $1/m$ buys is an *m-independent* functional of μ, so that one limit object serves every width. Without it the functional being minimised changes with m, there is no single R to be convex, and none of the section applies. Proposition 8.9: convergence of the flow along a sequence, tightness, a vanishing-dissipation sequence, and a full-support limit; boundedness and continuity of ϕ and $\nabla_\theta \phi$; and a W_2 gradient flow of R itself, without regularisation.

Which convexity is which. Part (i) is convexity along $s \mapsto (1-s)\mu_0 + s\mu_1$, a path that mixes the two measures. Part (ii) concerns convexity along W_2 geodesics, which transport mass rather than mixing it. The two paths join the same endpoints and are different curves in $\mathcal{P}_2(\mathbb{R}^d)$, and a functional convex along one may be concave along the other. Only the second is the hypothesis of Theorem 2.13 and of every rate in Table 2.2.

What the optimality statement is conditional on. Its antecedent — convergence to a full-support limit — is the difficult part, and the dynamics does not supply it: mass can escape to infinity and the support can collapse. The theorem is worth having because it says what remains to be shown, not because it shows it.

Convexity in the measure is not geodesic convexity. By Proposition 8.8 the risk is convex along linear interpolations for every architecture and every convex loss, and this is an elementary consequence of f_μ being linear in μ. It is the wrong convexity for the dynamics: the W_2 gradient flow converges exponentially under *displacement* convexity, by Theorem 2.13 and Table 2.2, and by part (ii) already the simplest possible risk fails it whenever the associated potential is non-convex. Every claim of global exponential convergence for mean-field training that cites flat convexity has substituted one property for the other.

The support condition is a hypothesis about the limit. Proposition 8.9 assumes that the flow converges along a sequence, that its dissipation vanishes along that sequence, and that the limit has full support. None of the three is established by the dynamics: mass can escape to infinity, and the support can collapse. What the result says is that *if* the flow converges to a stationary, spread-out limit *then* the limit is global — a conditional statement whose antecedent is the difficult part.

Nothing here is about finite width. All of Section 8.2 concerns $\mu \in \mathcal{P}(\mathbb{R}^d)$. A network with m units corresponds to an atomic μ^m, which never has full support, so Proposition 8.9 does not apply to it at any m. The finite-width error is Table 8.1 again, with its own rate and its own metric, and the two must be composed rather than conflated. In particular there is no statement in this volume that almost all critical points of the finite-width risk are global minima, and the mean-field picture does not supply one.

8.3 Mean-Field Langevin Dynamics

The obstruction identified above is removed — not evaded — by adding an entropy. What results is a genuine convergence theorem whose hypothesis can be stated.

Definition 8.11 (Regularised risk and mean-field Langevin dynamics). Fix a *confining reference measure* $\gamma(\mathrm{d}\theta) \propto e^{-V_0(\theta)}\mathrm{d}\theta$ on $\mathbb{R}^d$ with $V_0 \in C^2$, e^{-V_0} integrable, V_0 coercive, and

$$\int e^{c|\theta|^2}\,\gamma(\mathrm{d}\theta) < \infty \qquad \text{for some } c > 0. \tag{8.10}$$

The last requirement — a Gaussian-type moment, satisfied by $V_0 = \frac{1}{2}|\theta|^2$ and by any V_0 growing at least quadratically — is what keeps the variational problem inside $\mathcal{P}_2(\mathbb{R}^d)$; Remark 8.14 says where it is used. For $\lambda > 0$ set

$$F_\lambda(\mu) = R(\mu) + \lambda\,\mathrm{KL}(\mu\,\|\,\gamma)\,, \tag{8.11}$$

and let *mean-field Langevin dynamics* be

$$\mathrm{d}\theta_t = -\nabla_\theta\Big(\frac{\delta R}{\delta\mu}[\mu_t] + \lambda V_0\Big)(\theta_t)\,\mathrm{d}t + \sqrt{2\lambda}\,\mathrm{d}W_t, \qquad \mu_t = \mathrm{Law}(\theta_t), \tag{8.12}$$

whose marginals solve $\partial_t\mu = \nabla\cdot\big(\mu\nabla\big(\frac{\delta R}{\delta\mu} + \lambda V_0\big)\big) + \lambda\Delta\mu$. The *proximal Gibbs measure* at μ is defined relative to γ:

$$\frac{\mathrm{d}\hat{\mu}}{\mathrm{d}\gamma}(\theta) \propto \exp\Big(-\tfrac{1}{\lambda}\tfrac{\delta R}{\delta\mu}[\mu](\theta)\Big). \tag{8.13}$$

Remark 8.12 (Why the entropy is taken relative to γ). Taking the entropy relative to Lebesgue measure on $\mathbb{R}^d$, as $\lambda\int\mu\log\mu$, does not work, and the failure is not technical. First, $\mu \mapsto \int\mu\log\mu$ is not coercive: with $R \equiv 0$ — which is flat-convex with a bounded first variation — and $\mu_s = \mathcal{N}(0, s^2 I_d)$ one has $\int\mu_s\log\mu_s \to -\infty$ as $s \to \infty$, so F_λ has no minimiser. Second, if $\delta R/\delta\mu$ is bounded, as assumed below, then $\exp(-\lambda^{-1}\delta R/\delta\mu)$ is bounded above and below by positive constants, so it is not Lebesgue-integrable on $\mathbb{R}^d$ and the proximal Gibbs measure of (8.13) would not be a probability measure at all. Both objects on which this section rests would fail to exist. Relative to a confining γ neither difficulty arises: $\mathrm{KL}(\cdot\,\|\,\gamma)$ has compact sublevel sets in the weak topology, and the density in (8.13) is bounded, hence integrable against γ.

Proposition 8.13 (Free energy, dissipation, and the self-consistency condition) DERIVED. *Assume R is flat-convex with a bounded, Lipschitz first variation; for part (ii) assume in addition that R is bounded below and weakly lower semicontinuous on $\mathcal{P}_2(\mathbb{R}^d)$. Both extra properties hold for the risk (8.8) when ϕ is bounded and continuous and ℓ is convex, continuous and bounded below, which is the case Section 8.2 supplies; they are stated separately because the rest of the proposition is abstract and does not need them. Let γ be as in Definition 8.11. Then:*

(i) (8.12) is the W_2 gradient flow of F_λ, and along it

$$\frac{\mathrm{d}}{\mathrm{d}t}F_\lambda(\mu_t) = -\int\mu_t\Big|\nabla\Big(\tfrac{\delta R}{\delta\mu}[\mu_t] + \lambda\log\tfrac{\mathrm{d}\mu_t}{\mathrm{d}\gamma}\Big)\Big|^2 = -\lambda^2\,\mathcal{I}(\mu_t\,\|\,\hat{\mu}_t)\,; \tag{8.14}$$

(ii) *F_λ is strictly convex along linear interpolations, so it has at most one minimiser; and since $\mathrm{KL}(\cdot \,\|\, \gamma)$ is weakly lower semicontinuous with weakly compact sublevel sets and R is bounded below and weakly lower semicontinuous, F_λ has at least one. The unique minimiser μ^* is characterised by the self-consistency condition $\mu^* = \hat\mu^*$.*

Convexity gives uniqueness; coercivity of $\mathrm{KL}(\cdot \,\|\, \gamma)$, not convexity, gives existence.

Remark 8.14 (Why the reference needs a Gaussian moment). Weak compactness of the sublevel sets $\{\mathrm{KL}(\mu \,\|\, \gamma) \le C\}$ is Prokhorov's theorem and needs nothing beyond tightness. But this section works in $\mathcal{P}_2(\mathbb{R}^d)$ — the W_2 geometry of (8.12) is defined there — and a weak limit of measures with finite second moment need not have one. The Donsker–Varadhan entropy inequality closes the gap: for any $c > 0$,

$$c \int |\theta|^2 \, \mathrm{d}\mu \;\le\; \mathrm{KL}(\mu \,\|\, \gamma) + \log \int e^{c|\theta|^2} \, \mathrm{d}\gamma,$$

so (8.10) converts an entropy bound into a *uniform* second-moment bound, second moments are weakly lower semicontinuous, and the minimising sequence's limit lies in $\mathcal{P}_2(\mathbb{R}^d)$. Coercivity of V_0 alone does not give this — a reference with heavy tails has KL-sublevel sets containing measures of unbounded second moment — and neither does $\gamma \in \mathcal{P}_2(\mathbb{R}^d)$.

Proof. (i) The first variation of F_λ is $\delta R/\delta\mu + \lambda \log(\mathrm{d}\mu/\mathrm{d}\gamma) + \lambda = \delta R/\delta\mu + \lambda \log \mu + \lambda V_0 + \lambda$ up to a constant, so the flow with velocity $-\nabla\, \delta F_\lambda/\delta\mu$ is exactly (8.12), and the dissipation identity is Proposition 2.12 with $\mathcal{F}$ replaced by F_λ — the proof there used only the flux form (2.18), which holds here with the same computation. For the last equality, $\nabla(\delta R/\delta\mu + \lambda \log(\mathrm{d}\mu/\mathrm{d}\gamma)) = \lambda \nabla \log(\mathrm{d}\mu/\mathrm{d}\hat\mu)$ by (8.13), so the integral is $\lambda^2 \int \mu \,|\nabla \log(\mathrm{d}\mu/\mathrm{d}\hat\mu)|^2 = \lambda^2 \mathcal{I}(\mu \,\|\, \hat\mu)$. (ii) R is convex and $\mu \mapsto \mathrm{KL}(\mu \,\|\, \gamma)$ is strictly convex along linear interpolations, so F_λ is strictly convex there and has at most one minimiser. For existence, $\mathrm{KL}(\cdot \,\|\, \gamma)$ is weakly lower semicontinuous and its sublevel sets $\{\mathrm{KL}(\mu \,\|\, \gamma) \le c\}$ are tight, hence weakly compact by Prokhorov; by (8.10) and the entropy inequality those sublevel sets also have uniformly bounded second moments, so the limit stays in $\mathcal{P}_2(\mathbb{R}^d)$ (Remark 8.14); R is bounded below and weakly lower semicontinuous by hypothesis — for the risk (8.8) this follows from bounded continuous ϕ, which makes $\mu \mapsto f_\mu(x)$ weakly continuous, together with Fatou; so F_λ attains its infimum on any sublevel set, which is non-empty since $F_\lambda(\gamma) < \infty$. The first-order condition $\delta R/\delta\mu + \lambda \log(\mathrm{d}\mu^*/\mathrm{d}\gamma) = \text{const}$ is $\mathrm{d}\mu^*/\mathrm{d}\gamma \propto \exp(-\lambda^{-1} \delta R/\delta\mu[\mu^*])$, which is $\mu^* = \hat\mu^*$. ■

Proposition 8.15 (Exponential convergence under a uniform logarithmic Sobolev inequality) ESTABLISHED. *In the setting of Proposition 8.13, suppose the proximal Gibbs measures satisfy a logarithmic Sobolev inequality with a constant $\alpha > 0$ uniform along the trajectory: $\hat\mu_t$ satisfies $\mathrm{LSI}(\alpha)$ for every t. Then*

$$F_\lambda(\mu_t) - F_\lambda(\mu^*) \;\le\; e^{-2\alpha\lambda t} \left(F_\lambda(\mu_0) - F_\lambda(\mu^*)\right). \tag{8.15}$$

The convergence analysis through the proximal Gibbs measure is due to Nitanda, Wu and Suzuki [20]. Two conventions must travel with the rate: the normalisation of LSI — this volume's is (2.14), and a source using $\mathrm{KL}(\,\|\,) \le \alpha^{-1} I$ shifts the exponent by a factor of two — and the placement of λ, which depends on the time normalisation of (8.12).

Scope of Proposition 8.13 and Proposition 8.15

Assumed, and stated first. $\lambda > 0$; a confining reference γ; flat convexity of R; a bounded Lipschitz first variation, which for a bounded feature map ϕ and a Lipschitz loss holds; and, for the rate, a logarithmic Sobolev inequality for $\hat{\mu}_t$ *uniform in* t. The last is a hypothesis about the whole trajectory and is not implied by the others.

The reference measure is not a technical device. By Remark 8.12, without it neither μ^* nor $\hat{\mu}$ exists under these hypotheses. It also enters the dynamics, as the confinement term $-\lambda\nabla V_0$ in (8.12), and it enters the answer: μ^* depends on γ, so the regularised problem is a different problem for each choice, and γ is a modelling decision on the same footing as λ.

Why this is a repair and not a weakening. The claim it replaces — global exponential convergence from flat convexity — is false, by Proposition 8.8(ii). What is true is (8.15), and it is a genuine theorem about a genuine algorithm: the entropy term is exactly what makes F_λ strictly convex and gives the unique self-consistent minimiser of Proposition 8.13(ii). The price is the regularisation, and the rate $2a\lambda$ vanishes with λ.

Where the uniform inequality fails. $\hat{\mu}_t$ is a bounded perturbation of γ, with $d\hat{\mu}_t/d\gamma \propto \exp(-\lambda^{-1}\delta R/\delta\mu[\mu_t])$; so if γ itself satisfies $\mathrm{LSI}(a_\gamma)$ — for a Gaussian γ it does, by Corollary 2.23 — then Proposition 2.28 supplies a uniform constant $a \geq a_\gamma e^{-\operatorname{osc}(\delta R/\delta\mu)/\lambda}$. That is a genuine uniform inequality and it is exponentially bad in $1/\lambda$ — so the hypothesis holds with a constant that is uniform and tiny, and (8.15) becomes vacuous before λ is small. Whether a better constant is available for a natural class of risks is the substance of the problem, and this volume does not settle it.

What it does not say. Nothing about $\lambda = 0$; nothing about finite width, which is Table 8.1; nothing about the discretisation in time, which is Proposition 4.3 with a state-dependent drift and is not treated; and nothing about generalisation.

What would falsify them. For Proposition 8.13(ii), a flat-convex R with two distinct minimisers of F_λ at $\lambda > 0$.

8.4 Uniform-in-Time and Entropic Propagation of Chaos

Proposition 8.2 is a finite-horizon statement with an exponential constant. Two strengthenings are available and are often quoted together, which obscures that they are different: a uniform-in-time bound, and a bound in relative entropy on k-particle marginals.

One name, used consistently before either statement: the property in play is *block superadditivity of the relative entropy of an exchangeable measure* — the N-particle entropy is at least N/k times the k-marginal entropy — which is the same statement as *monotonicity of the normalised sequence* $k \mapsto \frac{1}{k}\mathrm{KL}(\rho^{N,k} \,\|\, \rho^{\otimes k})$. It is not subadditivity of the entropy.

The divisible case is proved here; the general case is not, and the two are therefore stated separately rather than merged under one badge.

Proposition 8.16 (Block inequality for divisible marginals) PROVED. *Let ρ^N be exchangeable on $(\mathbb{R}^d)^N$ with k-particle marginals $\rho^{N,k}$, and let $\rho \in \mathcal{P}_2(\mathbb{R}^d)$. If k divides N, then*

$$\mathrm{KL}\left(\rho^{N,k} \,\|\, \rho^{\otimes k}\right) \leq \frac{k}{N}\mathrm{KL}\left(\rho^N \,\|\, \rho^{\otimes N}\right). \tag{8.16}$$

Proof. Partition $\{1, \ldots, N\}$ into N/k blocks of size k, apply the chain rule for relative entropy along the blocks, and use exchangeability to identify each block's contribution with

$\mathrm{KL}(\rho^{N,k} \,\|\, \rho^{\otimes k})$ after dropping the conditioning — which cannot increase it. Rearranging gives (8.16). ■

Proposition 8.17 (Normalised marginal-entropy monotonicity, arbitrary k) ESTABLISHED. *For every exchangeable ρ^N and every $1 \le k \le N$, the sequence $k \mapsto \frac{1}{k}\mathrm{KL}(\rho^{N,k} \,\|\, \rho^{\otimes k})$ is non-decreasing. This is the Han-type inequality for the averaged k-marginal entropies of an exchangeable measure, quoted here for arbitrary $k \le N$ and not only for $k \mid N$.*

Corollary 8.18 (The bound at arbitrary k) DERIVED. *Under Proposition 8.17, (8.16) holds for every $1 \le k \le N$, not only for $k \mid N$. In particular a bound $\mathrm{KL}(\rho^N \,\|\, \rho^{\otimes N}) = O(1)$ gives only $O(k/N)$ for the k-marginal, and says nothing per particle beyond that.*

Proof. Monotonicity of $j \mapsto \frac{1}{j}\mathrm{KL}(\rho^{N,j} \,\|\, \rho^{\otimes j})$ applied between $j = k$ and $j = N$ gives $\frac{1}{k}\mathrm{KL}(\rho^{N,k} \,\|\, \rho^{\otimes k}) \le \frac{1}{N}\mathrm{KL}(\rho^N \,\|\, \rho^{\otimes N})$. ■

Remark 8.19 (Why this is split rather than stated once). It is tempting to state the arbitrary-k conclusion as a single proposition and let its proof handle $k \nmid N$ directly, appealing for general k to an external inequality. That would misplace the proof obligation: DERIVED means proved here from stated hypotheses, and an external theorem invoked inside a proof is not that. The split above keeps the two apart — Proposition 8.16 is proved, Proposition 8.17 is quoted and carries the attribution, and Corollary 8.18 is the two-line consequence.

Proposition 8.20 (Sharp entropic propagation of chaos, specialised) ESTABLISHED. *The following is a specialisation of Lacker and Le Flem [14], stated in one of the two regimes their theorem covers. Consider the interacting diffusion (8.2) in either of the following settings:*

(a) Convex regime on $\mathbb{R}^d$. *The drift derives from a confining potential V and a pairwise interaction potential W, both convex and sufficiently regular for the particle system and its McKean–Vlasov limit to be well posed, with V uniformly convex and W even. The initial law is required to have the moment and entropy regularity the source imposes.*

(b) Small-interaction regime on the torus. *The state space is $\mathbb{T}^d$; the interaction kernel is bounded and satisfies an explicit smallness condition relative to the diffusivity — it is a quantitative bound involving $\nabla \cdot K$, not the qualitative phrase "small compared with the noise". The initial density is smooth and bounded above and below, $\lambda^{-1} \le \mu_0 \le \lambda$ for some $\lambda < \infty$.*

The precise constants in the smallness condition and the exact regularity demanded of V, W and μ_0 are those of the source's corollaries and are not reproduced here; they are what decides whether the exponent below is attained, and a reader carrying this statement to a new model should take them from [14] rather than from this paraphrase. Assume moreover that the parameters are in the subregime of the source's corollaries in which the optimal exponent is attained — in the convex case a sufficiently strong ratio of confinement to interaction, in the torus case a threshold on the interaction strictly beyond the one securing uniformity in time. Then, for every $k \le N$,

$$\mathrm{KL}\left(\rho_t^{N,k} \,\|\, \rho_t^{\otimes k}\right) \lesssim \left(\frac{k}{N}\right)^2 \tag{8.17}$$

uniformly in $t \ge 0$, and the same $O((k/N)^2)$ rate holds with the left-hand side replaced by W_2^2 or by squared total variation. The exponent 2 is optimal, and it is not available throughout the two *regimes: in the weaker parameter ranges the source's corollaries give uniform-in-time chaos with a suboptimal exponent. Uniformity in time and optimality of the exponent are two thresholds, not one.*

Two things travel with the statement and are lost if it is quoted loosely. The hypothesis is not "Lipschitz interaction plus a log-Sobolev inequality somewhere": the uniform-in-time conclusion is obtained through time-uniform log-Sobolev inequalities along the hierarchy, and it is the convexity in (a) or the smallness in (b) that supplies them. And the rate is in k/N, not in N alone; the k-marginal is the object, and the bound degrades as k approaches N, as it must.

Scope of Proposition 8.16, Proposition 8.17 and Proposition 8.20

Assumed. Proposition 8.16: exchangeability and $k \mid N$, and nothing else — it is proved here. Proposition 8.17: exchangeability, and an external inequality that this volume quotes; Corollary 8.18 inherits that dependence, so the arbitrary-k bound is not derived here even though the divisible case is. Proposition 8.20: regularity of the interaction and a uniform functional inequality for the limit, which is the hypothesis that makes the bound uniform in time and which is exactly the row of Table 2.2 that fails for multimodal limits.

Three normalisations, three statements. Total relative entropy $\mathrm{KL}(\rho^N \,\|\, \rho^{\otimes N})$; per-particle $\frac{1}{N}\mathrm{KL}(\rho^N \,\|\, \rho^{\otimes N})$; and k-marginal $\mathrm{KL}\left(\rho^{N,k} \,\|\, \rho^{\otimes k}\right)$. Equation (8.16) relates the first and third at rate k/N; (8.17) improves the third to $(k/N)^2$ by a genuinely different argument. A claim of a "dimension-free entropy bound" is meaningless until it says which of the three it is about, whether it is uniform in time, and by which of the two routes it was obtained.

What does not follow from either. Neither bound gives (8.6): relative entropy of marginals does not control the Wasserstein distance of the empirical measure, and the two rows of Table 8.1 they occupy are separate. Pinsker converts entropy into total variation of the k-marginal, not into a statement about μ^N.

What would falsify them. For Proposition 8.16, an exchangeable ρ^N with $k \mid N$ violating (8.16); for Proposition 8.17, an exchangeable ρ^N and a $k \nmid N$ at which the normalised sequence decreases.

8.5 Chapter Summary

Chapter 8 asks in which metric a mean-field limit holds, and the answer differs by object. Proposition 8.3 splits the empirical error into a coupling term and a sampling term; the first is $N^{-1/2}$ with a dimension-free constant, the second is the empirical-measure rate and is dimension-cursed above dimension four, and by Corollary 8.5 the second dominates. Figure 8.1 shows the two limits and the two errors together. A finite-horizon propagation-of-chaos estimate does not justify exchanging $N \to \infty$ with $t \to \infty$; that exchange needs the uniform-in-time hypotheses of Section 8.3.

The chapter's second theme is that flat convexity is not displacement convexity. The risk of a two-layer network is convex in the measure of neurons and this implies nothing about the gradient flow; what is available instead is the entropy-regularised problem of Section 8.3, whose convergence theorem names its hypothesis — a uniform logarithmic Sobolev inequality for the proximal Gibbs law, and for that law and not for an atomic empirical measure.

Exercises

Tier I Direct application.

8.1 *First variation of a two-layer risk*

For $\ell(z,y) = \frac{1}{2}(z-y)^2$ compute $\delta R/\delta\mu$ from Proposition 8.8(iii) explicitly, and write out the McKean–Vlasov equation it generates. Identify the interaction kernel and check that it is Lipschitz when ϕ is bounded with bounded derivative.

8.2 *The two rates, numerically*

Sample N points from $\mathcal{N}(0, \mathrm{I}_d)$ for $d \in \{2, 10, 50\}$ and N up to 10^4, and estimate $W_2(\hat{\rho}^N, \rho)$ by an optimal-transport solver. Fit the exponent and compare with (8.6). State which row of Table 8.1 your measurement is in and which it is not.

Tier II Completing the derivation.

8.3 *Flat against geodesic convexity*

Prove Proposition 8.8(ii) in both directions, and then verify that the quadratic interaction energy

$$W(\mu) = \frac{1}{2} \iint |x-y|^2 \, \mu(\mathrm{d}x)\, \mu(\mathrm{d}y)$$

is displacement convex and *not* flat convex. For the first, write W along a displacement interpolant as $\frac{1}{2}\mathbb{E}\,|X_s - Y_s|^2$ with $X_s = (1-s)X_0 + sT(X_0)$ and Y an independent copy, and observe that this is a quadratic in s with non-negative leading coefficient. For the second, show $W(\mu) = \int |x|^2 \,\mathrm{d}\mu - \left|\int x \,\mathrm{d}\mu\right|^2$ and deduce that along a linear mixture it is the sum of a linear term and a concave one, strictly concave whenever the two means differ. Conclude that neither property implies the other, and say which one Theorem 2.13 and Table 2.2 use.

Hint. Do not use the entropy: $\int \rho \log \rho$ is convex in both senses, and its negative is convex in neither.

8.4 *The dissipation identity for a nonlinear functional*

Carry out the proof of Proposition 8.13(i) in full, checking that every step of the proof of Proposition 2.12 survives the dependence of the drift on μ_t. Identify the one place where the nonlinearity would matter if the functional were not of the form (8.11).

Tier III Probing the boundary.

8.5 *Where the uniform constant collapses*

Take $q = 1$, $\phi(x;\theta) = \tanh(\theta x)$ and a data distribution making $\delta R/\delta\mu$ a double well of depth Δ. Estimate the logarithmic Sobolev constant of the proximal Gibbs measure (8.13) using Example 2.29, and hence the rate in (8.15), as a function of λ. Determine the λ maximising the rate, and report the resulting bias $F_\lambda - F_0$.

8.6 *Atoms never have full support*

Let μ^m be atomic with m atoms. Show directly that the optimality condition of Proposition 8.8(iii) can hold at μ^m while $R(\mu^m) > \inf R$, by exhibiting a ϕ and a target for which a two-atom configuration is stationary and suboptimal. State which hypothesis of Proposition 8.9 your example violates.

8.7 *Composing the two errors*

Combine (8.5) with (8.15) to bound $W_2(\mu_t^N, \mu^*)$ for the N-particle mean-field Langevin dynamics, as a function of N, t, d, α and λ. Determine the t at which the two contributions are equal, and say which one a run with $N = 10^3$ particles in $d = 100$ is limited by.

Tier IV Computational.

8.8 *Mean-field training, measured*

Take a one-dimensional regression problem and a two-layer network with $\phi(x;\theta) = a\tanh(wx + b)$, $\theta = (a, w, b)$.

Read this first. Any finite-m run tests a *composition* — the mean-field approximation together with the flow — and not Proposition 8.15 on its own. That proposition is a statement about the $m = \infty$ dynamics; nothing below isolates it, and part (d) asks you to say so precisely.

(a) Train with the mean-field scaling of Definition 8.7 at $m \in \{10, 10^2, 10^3\}$ and measure $R(\mu^m)$ at convergence. Report the gap to the best value you can obtain, against m.

(b) Add the entropy regulariser by injecting noise as in (8.12) with $\lambda \in \{10^{-1}, 10^{-2}, 10^{-3}\}$ and measure the convergence of F_λ. Fit the exponential rate and compare with (8.15), estimating α from the proximal Gibbs measure numerically.

(c) At several times t, construct the proximal Gibbs *density* $\hat{\rho}_t$ of (8.13) on a one-dimensional grid and estimate a lower bound for its log-Sobolev constant in the convention (2.14). Report whether that bound is approximately uniform in t. *Do not* substitute the atomic empirical particle measure for $\hat{\rho}_t$: the hypothesis of Proposition 8.15 is an inequality for the proximal Gibbs law, and the Dirichlet form of Definition 2.18 is not defined on an atomic measure without a separate discrete generator — measuring one and quoting the other is the object substitution this chapter is written to prevent.

(d) State which of (a)–(c) tests Proposition 8.15, which tests its hypothesis, and which tests neither because it varies m and λ together.

Acceptance criteria: (c) reports the variation of the constant over time as a number, and names the measure it was computed for; (d) classifies all three.

CHAPTER 9

Mean-Field Control and Mean-Field Games

Chapter 8 let N particles interact through their empirical measure and asked where the system goes. This chapter lets them *choose*, and the answer depends on who is choosing.

If one planner chooses the behaviour of all the particles, the problem is optimal control on the space of measures, and it has a value, a minimiser, and — under conditions — a variational structure. If each particle chooses for itself, taking the behaviour of the others as given, the problem is a game, its solution is an equilibrium rather than an optimum, and there is in general no functional whose minimisation produces it. The two are routinely written with the same pair of coupled equations and are not the same problem.

A word on what this chapter is. It is a *bridge chapter*, not a course in mean-field games: it takes existence, uniqueness and regularity for the coupled system as given, proves nothing about the master equation, and stops where a treatment of the subject would begin. What it does supply is the comparison — which object is being solved for, which equation carries the difference, and which structural properties transfer between the control and the game. A reader who needs the theory itself should go to a dedicated treatment; a reader who needs to know whether a mean-field-control argument applies to a game, or the reverse, is in the right place.

Sections 9.1 and 9.2 set the two up so that the difference is visible in one term of one equation. Section 9.3 gives the exact criterion under which they coincide, and it has the same shape as Proposition 6.6: an object is a first variation if and only if a symmetry holds. Section 9.4 gives what replaces convexity when the variational structure is absent, and Section 9.5 says what the limit means for a game with finitely many players.

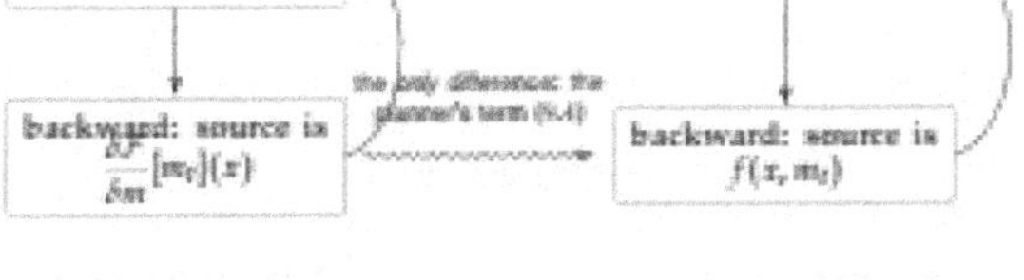

Figure 9.1: The two systems differ in one term of one equation. The forward equations are identical; the backward equations carry different sources, and the difference is the planner's term that a single optimiser sees and an individual agent does not. What this figure does not imply: that the two systems have solutions of the same kind. One has a value and a minimiser, the other an equilibrium; they coincide only under (9.8) (Proposition 9.5), and the planner's term does not vanish merely because the interaction is weak.

9.1 Mean-Field Control

Throughout, agents have state in $\mathbb{R}^d$, dynamics $dX_t = a_t\,dt + dW_t$ with control a, and the running control cost is $\frac{1}{2}|a|^2$; the diffusion coefficient is the identity and the cost quadratic so that the Hamiltonian is explicit. Write m_t for the law of X_t.

Definition 9.1 (Mean-field control problem). Given $f : \mathbb{R}^d \times \mathcal{P}_2(\mathbb{R}^d) \to \mathbb{R}$, $g : \mathbb{R}^d \times \mathcal{P}_2(\mathbb{R}^d) \to \mathbb{R}$ and $m_0 \in \mathcal{P}_2(\mathbb{R}^d)$, the *mean-field control* problem is

$$\inf_a \mathbb{E}\Big[\int_0^T \big(\tfrac{1}{2}|a_t|^2 + f(X_t, m_t)\big)dt + g(X_T, m_T)\Big], \qquad m_t = \mathrm{Law}(X_t), \tag{9.1}$$

the infimum being over controls for which the McKean–Vlasov equation is well posed. Equivalently, writing $\mathcal{F}(m) = \int f(x,m)\,m(dx)$ and $\mathcal{G}(m) = \int g(x,m)\,m(dx)$, it is

$$\inf_{(m,v)} \int_0^T \Big(\tfrac{1}{2}\int |v_t|^2\,dm_t + \mathcal{F}(m_t)\Big)dt + \mathcal{G}(m_T) \tag{9.2}$$

over curves solving $\partial_t m_t + \nabla\cdot(m_t v_t) = \frac{1}{2}\Delta m_t$ with m_0 given.

Proposition 9.2 (Optimality system for mean-field control) DERIVED. *Suppose $\mathcal{F}, \mathcal{G}$ admit first variations $\delta\mathcal{F}/\delta m$, $\delta\mathcal{G}/\delta m$ that are C^1 in x. Then a smooth minimiser of (9.2) satisfies, with $v_t = -\nabla u_t$,*

$$\begin{cases} -\partial_t u - \frac{1}{2}\Delta u + \frac{1}{2}|\nabla u|^2 = \dfrac{\delta\mathcal{F}}{\delta m}[m_t](x), & u(T,\cdot) = \dfrac{\delta\mathcal{G}}{\delta m}[m_T], \\ \partial_t m - \frac{1}{2}\Delta m - \nabla\cdot(m\nabla u) = 0, & m(0,\cdot) = m_0. \end{cases} \tag{9.3}$$

Expanding the first variation of $\mathcal{F}(m) = \int f(x,m)m(dx)$ gives

$$\frac{\delta\mathcal{F}}{\delta m}[m](x) = f(x,m) + \underbrace{\int \frac{\delta f}{\delta m}[m](y)(x)\,m(dy)}_{\text{the planner's term}}, \tag{9.4}$$

and the second summand is present because the planner accounts for the effect of moving one agent on the cost of all the others.

Proof. Introduce a multiplier φ for the constraint and consider

$$\mathcal{J} = \int_0^T \Big(\tfrac{1}{2}\int |v_t|^2\,dm_t + \mathcal{F}(m_t)\Big)dt + \mathcal{G}(m_T) + \int_0^T\!\!\int \varphi_t\Big(\partial_t m_t + \nabla\cdot(m_t v_t) - \tfrac{1}{2}\Delta m_t\Big).$$

Varying v gives $m_t v_t - m_t\nabla\varphi_t = 0$, hence $v_t = \nabla\varphi_t$. Varying m and integrating by parts in t and x gives $-\partial_t\varphi - \frac{1}{2}\Delta\varphi + \frac{1}{2}|\nabla\varphi|^2 - \langle\nabla\varphi, v\rangle + \delta\mathcal{F}/\delta m = 0$, which with $v = \nabla\varphi$ reduces to $-\partial_t\varphi - \frac{1}{2}\Delta\varphi - \frac{1}{2}|\nabla\varphi|^2 + \delta\mathcal{F}/\delta m = 0$, and the terminal condition $\varphi(T) = -\delta\mathcal{G}/\delta m$. Setting $u = -\varphi$ gives (9.3). For (9.4), perturb $m \mapsto m + \epsilon\chi$ with $\int\chi = 0$: $\frac{d}{d\epsilon}\big|_0 \int f(x, m+\epsilon\chi)(m+\epsilon\chi)(dx) = \int f(x,m)\chi(dx) + \iint \frac{\delta f}{\delta m}[m](y)(x)\chi(dx)m(dy)$, which identifies the first variation as (9.4). ■

9.2 Mean-Field Games

Definition 9.3 (Mean-field Nash equilibrium). A flow $(m_t)_{t\in[0,T]}$ and a control a^* form a *mean-field Nash equilibrium* if

(i) given the flow (m_t) held *fixed*, a^* minimises $\mathbb{E}\left[\int_0^T (\frac{1}{2}|a_t|^2 + f(X_t, m_t))dt + g(X_T, m_T)\right]$ for a single agent;

(ii) the law of the state controlled by a^* and started from m_0 is (m_t).

Condition (i) is optimality against the crowd; condition (ii) is the fixed point that makes the crowd consistent.

Proposition 9.4 (The Lasry–Lions system) DERIVED. *Under smoothness, a mean-field Nash equilibrium satisfies*

$$\begin{cases} -\partial_t u - \frac{1}{2}\Delta u + \frac{1}{2}|\nabla u|^2 = f(x, m_t), & u(T,\cdot) = g(\cdot, m_T), \\ \partial_t m - \frac{1}{2}\Delta m - \nabla\cdot(m\nabla u) = 0, & m(0,\cdot) = m_0, \end{cases} \tag{9.5}$$

with optimal control $a^* = -\nabla u$. The system differs from (9.3) in exactly one place: the right-hand side of the backward equation is $f(x, m_t)$ rather than $\delta\mathcal{F}/\delta m[m_t](x)$, that is, it omits the planner's term of (9.4).

Proof. For a fixed flow (m_t), condition (i) of Definition 9.3 is a standard stochastic control problem with running cost $\frac{1}{2}|a|^2 + f(x, m_t)$ and terminal cost $g(\cdot, m_T)$; by Proposition 7.2 its value function solves the Hamilton–Jacobi–Bellman equation (7.3), whose Hamiltonian for a quadratic cost is $\inf_a\{\langle a, \nabla u\rangle + \frac{1}{2}|a|^2\} = -\frac{1}{2}|\nabla u|^2$ attained at $a^* = -\nabla u$; this is the first line of (9.5). Condition (ii) is the Fokker–Planck equation (2.4) for the closed-loop dynamics $dX = -\nabla u\,dt + dW$, which is the second line. The comparison with (9.3) is Proposition 9.2. ■

An equilibrium is not an optimum. In Definition 9.3(i) the flow is held fixed while a single agent optimises; the agent does not account for the effect of its own deviation on the crowd, because in the limit one agent has no such effect. The planner of Definition 9.1 does account for it, and (9.4) is exactly that accounting. The two systems have the same forward equation, the same Hamiltonian and different sources, so they have different solutions whenever the planner's term is non-zero.

Neither system is a gradient flow. Both are forward–backward: the forward equation carries an initial condition and the backward one a terminal condition, and they are coupled. There is no functional of m_t alone whose decrease characterises a solution, and no initial-value problem to integrate. This is the fourth row of Table 11.1 and it is why Chapter 2's language does not transfer.

The N-player game is not the N-particle system. Chapter 8 studied a prescribed interacting dynamics; here the dynamics is the output of an equilibrium computation. A propagation-of-chaos statement for the former is not a statement about the latter, and Section 9.5 needs different tools.

9.3 When the Two Coincide

The planner's term in (9.4) is in general non-zero, and it vanishes identically only in trivial cases. But the mean-field game system (9.5) can still be the optimality system of *some* control problem — a different one, with a different running cost — and the criterion for that is exact.

Proposition 9.5 (Potential criterion) DERIVED. *Suppose there exist functionals* Φ, Ψ *on* $\mathcal{P}_2(\mathbb{R}^d)$ *with*

$$f(x,m) = \frac{\delta \Phi}{\delta m}[m](x), \qquad g(x,m) = \frac{\delta \Psi}{\delta m}[m](x). \tag{9.6}$$

Then the mean-field game system (9.5) is exactly the optimality system (9.3) of the control problem

$$\inf_{(m,v)} \int_0^T \Big(\tfrac{1}{2} \int |v_t|^2 \, \mathrm{d}m_t + \Phi(m_t) \Big) \mathrm{d}t + \Psi(m_T), \tag{9.7}$$

so every equilibrium is a critical point of (9.7), and a minimiser of (9.7) is an equilibrium. Conversely, if f *is* C^1 *in* m *and (9.6) holds for some* Φ*, then*

$$\frac{\delta f}{\delta m}[m](y)(x) = \frac{\delta f}{\delta m}[m](x)(y) \qquad \text{for all } x, y, m, \tag{9.8}$$

and, on a class on which the construction below is legitimate, (9.8) is also sufficient. That class must be fixed before the converse means anything, and this volume fixes it as follows: m *ranges over an affine, convex set* $\mathcal{D}$ *of probability measures closed under the linear interpolation* $m_s = (1-s)m_0 + sm$ *— for instance the smooth positive densities on a compact manifold such as the torus — and* f *is flat-*C^2 *on* $\mathcal{D}$*, meaning both variations exist, are jointly continuous, and are integrable against differences of elements of* $\mathcal{D}$*. The functional* Φ *so produced is unique up to an additive constant, which is the gauge freedom of a first variation and is invisible to (9.5).*

Proof. If $f = \delta\Phi/\delta m$ then the source of the backward equation in (9.3), applied to the problem (9.7) whose running cost is $\Phi(m)$ rather than $\int f \mathrm{d}m$, is exactly $\delta\Phi/\delta m = f$; likewise for the terminal condition. So (9.3) for (9.7) is (9.5), which is the first assertion.

For (9.8), differentiate (9.6) once more: $\delta f/\delta m = \delta^2\Phi/\delta m^2$, and the second variation is symmetric in its two arguments by equality of mixed variations, which holds whenever Φ is twice continuously differentiable in the flat sense. Sufficiency is a Poincaré lemma, and it is a Poincaré lemma *on* $\mathcal{D}$, not on $\mathcal{P}_2(\mathbb{R}^d)$: given (9.8), define $\Phi(m) = \int_0^1 \int f(x, m_s)(m - m_0)(\mathrm{d}x)\,\mathrm{d}s$ along the linear path $m_s = (1-s)m_0 + sm$, which stays in $\mathcal{D}$ by convexity, and verify by differentiating under the integral that $\delta\Phi/\delta m = f$; the verification uses (9.8) exactly once, to exchange the two variations, and uses flat-C^2 regularity to differentiate under the integral. Neither step is available on all of $\mathcal{P}_2(\mathbb{R}^d)$, where the path integral need not be defined and $\delta f/\delta m$ need not be a function. ■

Example 9.6 (A potential and a non-potential coupling). Local congestion, $f(x,m) = \varsigma(m(x))$ for a scalar function ς, is defined only for m with a density, so take $\mathcal{D}$ to be the smooth positive densities on the torus — on $\mathcal{P}_2(\mathbb{R}^d)$ the expression has no meaning at a Dirac, and its flat derivative is distribution valued rather than a function. On $\mathcal{D}$ it satisfies (9.8): here $\delta f/\delta m[m](y)(x) = \varsigma'(m(x))\delta_x(y)$, symmetric as a distribution in (x,y).

and pairing it against smooth densities is legitimate. So it is potential, with $\Phi(m) = \int \varsigma_0(m(x))\mathrm{d}x$ and $\varsigma_0' = \varsigma$.

Now take $d = 1$ and the antisymmetric coupling $f(x,m) = \int K(x-y)\,m(\mathrm{d}y)$ with K odd and non-zero — agents are attracted to those on one side and repelled by those on the other. Then $\delta f/\delta m[m](y)(x) = K(x-y)$, and (9.8) would require $K(x-y) = K(y-x) = -K(x-y)$, hence $K \equiv 0$. So no potential exists, and by Proposition 9.5 the system (9.5) is not the optimality system of any control problem of the form (9.7).

What the criterion does and does not settle. Equation (9.8) decides whether the game system is the optimality system of *some* control problem. It does not make the game equal to the mean-field control problem (9.1) with the same f: those two differ by (9.4) regardless, and the potential control problem (9.7) has running cost $\Phi(m)$, not $\int f\,\mathrm{d}m$. Three problems, not two.

Where it fails. By Example 9.6 an antisymmetric interaction admits no potential at all. More generally (9.8) is a codimension-large condition: the space of couplings is not close to the space of potential ones, so a coupling chosen for modelling reasons will not satisfy it unless it was built to.

> What does not follow
>
> **Mean-field games are not mean-field control.** Proposition 9.4 and Proposition 9.2 differ by (9.4), which is non-zero in general; it vanishes when f does not depend on m — no interaction and no game — and, more precisely, it is invisible to (9.5) whenever it is constant in x, since the backward equation determines u only up to the additive gauge that a time-dependent constant supplies. What is *not* available is a converse reading the vanishing of the planner's term off the couplings: "the two systems agree if and only if there is no interaction" is stronger than (9.4) supports, and this volume does not assert it. Every statement obtained by minimising a functional over flows of measures is a statement about Definition 9.1 or about (9.7), and applies to Definition 9.3 only under (9.8).

A potential is not a Lyapunov functional. Even under (9.8), Φ is the running cost of a control problem whose optimality system is forward–backward. It does not decrease along the equilibrium flow, and there is no dissipation identity of the kind (2.19). The variational structure buys existence and a selection principle among equilibria; it does not buy a descent argument, and the fourth row of Table 11.1 stands.

Equilibria need not be unique. Without further conditions (9.5) can have several solutions, and under (9.8) these correspond to several critical points of (9.7), of which at most one is the global minimiser. Selecting among them is Section 9.4.

9.4 Monotonicity and Uniqueness

What convexity does for control, monotonicity does for games.

Definition 9.7 (Lasry–Lions monotonicity). The coupling f is *monotone* if for all $m, m' \in \mathcal{P}_2(\mathbb{R}^d)$

$$\int_{\mathbb{R}^d} \big(f(x,m) - f(x,m')\big)\,\mathrm{d}(m-m')(x) \;\geq\; 0, \tag{9.9}$$

and *strictly monotone* if equality forces $m = m'$. The same definition is applied to g.

Monotonicity says agents prefer to avoid crowding: moving mass into a region raises the cost there. Example 9.6's congestion coupling with ς increasing is monotone; a coupling rewarding imitation is not.

Proposition 9.8 (Uniqueness under strict monotonicity) DERIVED. *Suppose f and g are monotone and f is strictly monotone, and assume in addition, for the final identification of the value functions, either*

(a) m_t has full support for every $t > 0$; or

(b) once m is fixed, the backward equation of (9.5) has at most one solution in the smooth class under consideration, given its terminal datum.

Then (9.5) has at most one smooth solution.

Both (a) and (b) are analytic facts about the equations rather than about the game, and both are assumed here rather than proved: (a) holds for the non-degenerate diffusion of (9.5) by standard parabolic regularity, and (b) is classical uniqueness for a Hamilton–Jacobi–Bellman equation in the stated class. Naming them in the statement is what keeps the DERIVED *badge honest — the argument below owns its hypotheses instead of borrowing an unnumbered theorem at the last step.*

Proof sketch. Let (u^1, m^1) and (u^2, m^2) be two solutions and put $\bar{u} = u^1 - u^2$, $\bar{m} = m^1 - m^2$. Differentiate $I(t) = \int \bar{u}_t \, d\bar{m}_t$ and substitute both equations of (9.5). The two Laplacian contributions cancel after one integration by parts, and what remains is

$$I'(t) = -\int \big(f(x, m_t^1) - f(x, m_t^2)\big) d\bar{m}_t - \tfrac{1}{2} \int |\nabla \bar{u}_t|^2 \, d(m_t^1 + m_t^2). \tag{9.10}$$

Both signs are negative, and this is the whole content of the argument. To see the second term, collect the Hamiltonian contributions as

$$\int \Big(\tfrac{1}{2} \big|\nabla u^1\big|^2 - \tfrac{1}{2} \big|\nabla u^2\big|^2 - \big\langle \nabla \bar{u}, \nabla u^1 \big\rangle \Big) dm^1 + (1 \leftrightarrow 2),$$

and note that each bracket equals $-\frac{1}{2} |\nabla \bar{u}|^2$ by expanding the square — this is where convexity of the Hamiltonian $\frac{1}{2}|p|^2$ enters, and for a general convex H the same step produces a sum of Bregman divergences of H, again non-negative.

Now integrate (9.10) over $[0, T]$. On the left, $I(0) = 0$ because $\bar{m}_0 = 0$, and $I(T) = \int \big(g(x, m_T^1) - g(x, m_T^2)\big) d\bar{m}_T \geq 0$ by monotonicity of g. Hence

$$0 \leq I(T) = -\int_0^T \int \big(f(x, m^1) - f(x, m^2)\big) d\bar{m} \, dt - \tfrac{1}{2} \int_0^T \int |\nabla \bar{u}|^2 \, d(m^1 + m^2) \, dt \leq 0,$$

the last inequality because both integrals are non-negative, by (9.9) and by the previous paragraph. Every term therefore vanishes; strict monotonicity forces $m_t^1 = m_t^2$ for almost every t, and then $\nabla \bar{u} = 0$ $(m^1 + m^2)$-almost everywhere.

It remains to pass from "$\nabla \bar{u} = 0$ where the mass is" to $\bar{u} \equiv 0$. This is where hypothesis (a) or (b) of the statement is used, and the choice is a hypothesis rather than a detail.

Route 1, under (a). With $\operatorname{supp} m_t = \mathbb{R}^d$ for $t > 0$ — which the uniformly parabolic forward equation supplies — one has $\nabla \bar{u}(t,\cdot) = 0$ everywhere for $t > 0$, so $\bar{u}(t,x) = c(t)$ is constant in x. Substituting u^i = const + common part into the two backward equations, the Hamiltonian terms $\frac{1}{2}\left|\nabla u^i\right|^2$ and the Laplacians cancel in the difference, leaving $-\dot{c}(t) = f(x,m^1) - f(x,m^2) = 0$; with the terminal condition $\bar{u}(T,\cdot) = g(\cdot,m_T^1) - g(\cdot,m_T^2) = 0$ this gives $c \equiv 0$.

Route 2, under (b). Avoiding the positivity statement: once $m^1 = m^2 =: m$, the two functions u^1, u^2 solve the *same* backward Hamilton–Jacobi–Bellman equation with the same terminal datum $g(\cdot, m_T)$, since the coupling term $f(\cdot, m)$ is now identical. Classical uniqueness for that equation in the regularity class assumed gives $u^1 = u^2$. Note that the additive gauge which makes u non-unique in general is fixed here by the terminal condition, so no gauge argument is needed.

Route 1 is the more informative one — it says *why* the support condition is harmless for a non-degenerate diffusion — and Route 2 is the one that survives if the diffusion is degenerate but backward uniqueness is still available. What remains a sketch is the analytic bookkeeping: the differentiation under the integral sign and the boundary terms require regularity that is assumed rather than established here. ∎

Assumed. Smooth solutions, monotone g, strictly monotone f, and a convex Hamiltonian — here $\frac{1}{2}|p|^2$, which is where the quadratic control cost is used. The last step of the proof additionally uses either positivity of m_t for $t > 0$, which the non-degenerate diffusion supplies, or classical backward uniqueness for the Hamilton–Jacobi–Bellman equation; the two routes are given, and a degenerate model must choose the second.

What monotonicity is and is not. It is a condition on the *interaction*, playing the role that convexity plays for a control problem, and it is unrelated to (9.8): there are potential games that are not monotone and monotone games that are not potential. Uniqueness therefore does not follow from a variational structure, and a variational structure does not follow from uniqueness.

Where it fails. Any coupling rewarding agreement — imitation, herding, coordination — is anti-monotone, and such games routinely have multiple equilibria; that multiplicity is the modelling content rather than a defect. Nothing in this volume selects among them.

9.5 The Finite-Player Approximation

Two routes to a convergence statement exist, they have different hypotheses and different conclusions, and they are stated separately because merging them would produce a theorem with no single source.

Neither is stated here as a theorem. Both are recorded as *literature notes*: they carry no status badge, no proof obligation and no downstream licence. The reason is mathematical rather than editorial — the available results use materially different equilibrium notions, compare different objects, measure in different metrics and quantify differently, so no single statement covers them, and a numbered proposition merging them would assert a theorem that none of them proves. The two routes are recorded for orientation, and nothing later in the volume rests on either.

Remark 9.9 (Literature note: convergence of values under monotonicity). Under regularity of the couplings and monotonicity of f and g in the sense of Definition 9.7, the equilibrium value of a representative player in the N-player game is known to converge to the mean-field value as $N \to \infty$. To become a claim in this volume the statement must fix: whether the N-player equilibrium is open- or closed-loop, and in which information structure; whether the value compared is the equilibrium value or an approximate one; the regularity class of f, g and the horizon; and the exponent in $1/N$ together with the constant's dependence on T, d and that class. The exponent is *not* determined by monotonicity alone. The mean-field game formulation and the monotonicity condition are due to Lasry and Lions [15]; for the finite-player convergence problem see Cardaliaguet et al. [5].

Remark 9.10 (Literature note: convergence of states via the master equation). If the master equation associated with (9.5) has a sufficiently regular classical solution on $[0, T]$, the empirical measure of the N-player equilibrium states is known to converge to m_t. To become a claim here the statement must fix the metric — the four columns of Table 8.1 are four different answers — the quantifier in t, and the rate. Those rates are *not* inherited from Table 8.1: the N-player dynamics is the output of a fixed-point computation rather than a prescribed interacting system, so Proposition 8.2 does not apply to it and the table supplies vocabulary only. For the finite-player convergence programme see Cardaliaguet et al. [5].

Connection

There are two routes from a finite game to its mean-field limit, and they answer different questions. Monotonicity controls a *scalar* — the value a representative player obtains — and does so by the same convexity-like mechanism that gave uniqueness in Proposition 9.8. The master equation controls a *measure*, the distribution of the players' states, and does so through a regularity hypothesis on a partial differential equation whose unknown is a function of a measure. The hypothesis sets overlap and neither contains the other, so a rate obtained on one route cannot be quoted for the other.

Neither route says anything about *computing* an equilibrium, or about whether a learning dynamics converges to one; that is a question about a third object again. The source-dependent rates belong in the literature notes above rather than in a statement of this volume.

What does not follow

Convergence of equilibria is not propagation of chaos. In Chapter 8 the dynamics was prescribed and the limit was a law of large numbers. Here the dynamics is the output of a fixed-point computation at each N, and there is no coupling of the N-player and mean-field systems on a common probability space to which Proposition 8.2 could be applied. The rates of Table 8.1 are therefore *not* inherited by either statement above: what Table 8.1 supplies is the vocabulary — which object is being compared, in which metric — and any rate for Remark 9.10 must come from its own source together with its own hypotheses.

The master equation is where the difficulty is. The state-space route of Remark 9.10 passes through a partial differential equation on $[0, T] \times \mathbb{R}^d \times \mathcal{P}_2(\mathbb{R}^d)$ — an equation whose unknown is a function of a measure. Its well-posedness is the substantive hypothesis, it holds under monotonicity or under short horizons, and it is not available in general. This volume states no result about it.

Finitely many players is a different game. At finite N a player's deviation moves the empirical measure, so the equilibrium condition is not Definition 9.3(i). The mean-field problem is the $N \to \infty$ idealisation in which that feedback vanishes, and its solution is an approximate equilibrium for large N — with an error that Remark 9.9 quantifies and that is not zero.

9.6 Chapter Summary

Chapter 9 is a bridge chapter. It takes existence, uniqueness and regularity for the coupled forward–backward system as given, proves nothing about the master equation, and supplies instead the comparison: mean-field control and mean-field games have identical forward equations and differ in one term of the backward equation, the planner's term of (9.4). Proposition 9.5 gives the exact condition under which a game is nevertheless the optimality system of some control problem, on a stated class of measures rather than on all of $\mathcal{P}_2(\mathbb{R}^d)$; Proposition 9.8 gives uniqueness under strict monotonicity, with the analytic fact needed to close its last step named in the statement rather than borrowed in the proof.

Convergence of finite-player equilibria is known and is recorded here as two literature notes rather than as claims: the available results use different equilibrium notions and different metrics, so no single statement covers them. Nothing later depends on either.

Exercises

Tier I Direct application.

9.1 *The planner's term, computed*

For $f(x,m) = \int K(x-y)m(dy)$ with K even, compute both sides of (9.4) explicitly and show that the planner's term equals $f(x,m)$, so that $\delta F/\delta m = 2f$. Interpret the factor of two, and say what it does to (9.3) compared with (9.5).

9.2 *A linear-quadratic mean-field game*

Take $d = 1$, $f(x,m) = \frac{\gamma}{2}(x - \bar{m})^2$ with $\bar{m} = \int y\, m(dy)$, and $g \equiv 0$. Solve (9.5) with a quadratic ansatz $u(t,x) = \frac{1}{2}a_t x^2 + b_t x + c_t$, and identify the equilibrium mean flow. Then solve (9.3) for the same f and report the difference.

Tier II Completing the derivation.

9.3 *Sufficiency in the potential criterion*

Complete the Poincaré-lemma argument in the proof of Proposition 9.5: define Φ along the linear path and verify $\delta\Phi/\delta m = f$, identifying the step at which (9.8) is used. Then explain why the linear path may be replaced by any path joining m_0 to m without changing Φ.

9.4 *Monotone and potential are independent*

Exhibit a coupling that is potential and not monotone, and one that is monotone and not potential. Conclude that neither of Proposition 9.5 and Proposition 9.8 implies the other.

Hint. For the first, take ς decreasing in Example 9.6.

Tier III Probing the boundary.

9.5 *Anti-monotonicity, and where a bifurcation would come from*

Take $f(x,m) = -\lambda \int K(x-y)m(dy)$ on the circle, with K a positive even kernel, $\lambda > 0$, $g \equiv 0$ and m_0 uniform.

(a) Show that f is anti-monotone, so that Proposition 9.8 does not apply, and identify precisely which hypothesis fails.

(b) Show that the uniform flow $m_t \equiv \mathrm{Unif}$, u_t constant in x, solves (9.5) for every λ.

(c) Linearise (9.5) about that solution in a single Fourier mode e^{ikx} and write down the resulting two-dimensional linear forward–backward system for the mode's coefficients. Show that its solvability on $[0,T]$ with the given initial and terminal data fails for λ beyond a value depending on $\hat{K}(k)$, on the diffusion constant, *and* on the horizon T — and say why the threshold cannot be a function of $\hat{K}$ alone.

Part (c) is the linearised, and only the linearised, statement: exhibiting an actual second equilibrium is a bifurcation theorem for the nonlinear system and is not asked for here. Treat it as a computational investigation of where multiplicity would begin.

9.6 *A potential without a descent*

For a potential game, verify by direct computation that $\Phi(m_t)$ is not monotone along the equilibrium flow. Then say precisely which property of Proposition 2.12 fails and why the forward–backward coupling is responsible.

9.7 *Three problems, one coupling*

Fix a coupling f satisfying (9.8). Write down the three distinct problems: the mean-field game (9.5), the mean-field control problem (9.1), and the potential control problem (9.7). Show that the first and third share an optimality system and that the second does not, and identify for a concrete f the size of the discrepancy between the second and the other two.

Tier IV Computational.

9.8 *Solving a mean-field game numerically*

Take $d = 1$ on a torus, $f(x,m) = \varsigma(m(x))$ with ς increasing, and g a smooth target penalty.

(a) Solve (9.5) by fixed-point iteration on the flow (m_t): solve the backward equation given m, then the forward equation given u. Report whether the iteration converges and at what rate.

(b) Compute Φ from Example 9.6 and evaluate (9.7) at the solution and at perturbations of it, confirming that the solution is a critical point.

(c) Plot $\Phi(m_t)$ along the equilibrium flow and confirm that it is not monotone.

(d) Replace ς by a decreasing function and repeat (a). Report what changes, and state which hypothesis of Proposition 9.8 you have removed.

Acceptance criteria: (a) reports the observed convergence rate; (c) shows non-monotonicity explicitly; (d) names the hypothesis.

CHAPTER 10

Path Measures, Schrödinger Bridges, and Steering

Most of the probabilistic comparisons so far have been marginal in time: a law at a fixed instant, compared with another law at the same instant. This chapter changes the principal object to a measure on whole trajectories, and the change buys one thing: a notion of distance between two dynamics. It also changes the absolute-continuity conditions that every such comparison must carry, which is the theme of Proposition 10.2. Two diffusions with the same covariance characteristic induce two measures on the same path space and, *under the absolute-continuity conditions of Theorem 10.1* — an invertible diffusion coefficient, a square-integrable drift difference, and a martingale criterion for the stochastic exponential — their relative entropy is finite and equals an integral of the squared difference of the drifts. That single identity — Theorem 10.1 and Proposition 10.2 below — underlies the training objective of Chapter 3, the control problems of Chapter 7, and everything in this chapter.

The chapter's second concern is a distinction. There are three ways to make a diffusion end up somewhere: reweight its paths by a function of the endpoint, condition it on where it lands, or constrain both of its marginals. These are three different problems with three different solutions, and they are frequently written as one. Section 10.3 separates them, states the conditions under which each pair coincides, and locates score-based generation among them.

10.1 Girsanov, Path-Space Relative Entropy, and Entropy Production

10.1.1 Change of drift

Theorem 10.1 (Girsanov) CLASSICAL. *Let W be a d-dimensional Brownian motion on $(\Omega, \mathcal{F}, (\mathcal{F}_t), \mathbb{P})$ and let θ be progressively measurable with $\mathbb{E}\exp(\frac{1}{2}\int_0^T |\theta_s|^2\, ds) < \infty$ — Novikov's condition. Then the Doléans–Dade exponential Z of Corollary 1.34 is a true martingale with $\mathbb{E}Z_T = 1$, the measure $d\mathbb{Q} = Z_T\, d\mathbb{P}$ is a probability measure, and*

$$\tilde{W}_t := W_t - \int_0^t \theta_s\, ds \tag{10.1}$$

is a Brownian motion under $\mathbb{Q}$.

Proof roadmap. Novikov's condition makes Z uniformly integrable, hence a true martingale by Theorem 1.21; this is the only place the condition is used. It is a *sufficient* criterion and nothing follows from its failure alone: the exponential may still be a true martingale under a weaker criterion such as Kazamaki's, or it may be a strict local martingale, and which of the two holds is decided separately in each application. That $\tilde{W}$ is a $\mathbb{Q}$-Brownian motion is

Lévy's characterisation: $\widetilde{W}$ is continuous, and $Z\widetilde{W}$ and $Z(\widetilde{W}^i\widetilde{W}^j - \delta_{ij}t)$ are $\mathbb{P}$-martingales by Theorem 1.33 and Corollary 1.34, which is exactly the statement that $\widetilde{W}$ and $\widetilde{W}^i\widetilde{W}^j - \delta_{ij}t$ are $\mathbb{Q}$-martingales. The details are standard and quoted; what this volume uses is the statement and Proposition 10.2. □

Proposition 10.2 (Relative entropy on path space) DERIVED. *Let $\mathbb{Q}^1, \mathbb{Q}^2$ be the laws on $C([0,T];\mathbb{R}^d)$ of*

$$\mathrm{d}X_t = b^k(t, X_t)\,\mathrm{d}t + \sigma(t)\,\mathrm{d}W_t, \qquad X_0 \sim \nu^k, \qquad k = 1, 2, \tag{10.2}$$

with the same invertible σ. Assume

(a) $\nu^1 \ll \nu^2$;

(b) conditionally on $X_0 = x$, the stochastic exponential of $\theta = \sigma^{-1}(b^1 - b^2)$ is a true $\mathbb{Q}^2(\cdot \mid X_0 = x)$-martingale — Novikov's condition for θ is one sufficient criterion;

(c) $\mathbb{E}_{\mathbb{Q}^1} \int_0^T \|\theta\|^2\,\mathrm{d}t < \infty$;

(d) the martingale problem for (b^1, a) is well posed, so that the law constructed below is $\mathbb{Q}^1$ and not merely some law with drift b^1.

Then the path-space density factors into an initial part and a drift part,

$$\frac{\mathrm{d}\mathbb{Q}^1}{\mathrm{d}\mathbb{Q}^2} = \frac{\mathrm{d}\nu^1}{\mathrm{d}\nu^2}(X_0)\cdot Z_T, \qquad Z_T = \exp\Big(\int_0^T \langle\theta, \mathrm{d}W^2\rangle - \tfrac{1}{2}\int_0^T \|\theta\|^2\,\mathrm{d}t\Big), \tag{10.3}$$

and consequently

$$\mathrm{KL}\big(\mathbb{Q}^1 \,\|\, \mathbb{Q}^2\big) = \mathrm{KL}\big(\nu^1 \,\|\, \nu^2\big) + \frac{1}{2}\mathbb{E}_{\mathbb{Q}^1}\int_0^T \big\|\sigma(t)^{-1}(b^1 - b^2)(t, X_t)\big\|^2\,\mathrm{d}t. \tag{10.4}$$

Hypothesis (a) is not implied by the others and cannot be dropped: a Girsanov change of measure reweights paths by a functional of the driving noise and leaves the initial law where it was, so without $\nu^1 \ll \nu^2$ the two path measures are not absolutely continuous and (10.4) degenerates to $\infty = \infty$. The alternative of defining $\mathbb{Q}^1$ as the transform of $\mathbb{Q}^2$, which dispenses with (d), produces a law with initial marginal ν^2; obtaining ν^1 requires the further change $\nu^2 \to \nu^1$, and that step is exactly where (a) is consumed. The requirement that the two diffusion coefficients agree is not a convenience. Finiteness of $\mathrm{KL}(\mathbb{Q}^1 \,\|\, \mathbb{Q}^2)$ requires the quadratic-variation characteristics to agree $\mathrm{d}t \otimes \mathbb{Q}^1$-almost surely; if $a^1(t, X_t) \neq a^2(t, X_t)$ on a set of positive $\mathrm{d}t \otimes \mathbb{Q}^1$ occupation measure, then $\mathbb{Q}^1 \not\ll \mathbb{Q}^2$ and $\mathrm{KL}(\mathbb{Q}^1 \,\|\, \mathbb{Q}^2) = +\infty$. This is a one-sided criterion, oriented by the first argument of the divergence. Mutual singularity is a stronger conclusion and needs a separating condition visible to both measures — two deterministic covariance schedules differing on a set of positive time measure, as in Exercise 10.6, is the standard case — and it does not follow from a mismatch charged only by $\mathbb{Q}^2$. Changing the square root while keeping a fixed — replacing σ by $-\sigma$, or by σO with O orthogonal — changes neither the law nor the relative entropy.

Proof. Disintegrate each path measure into its initial law and its conditional law given X_0. By (a) the initial parts are absolutely continuous with density $\mathrm{d}\nu^1/\mathrm{d}\nu^2$, and by (b) the conditional parts are, for ν^2-almost every x, related by the Girsanov density Z_T below; the product of the two is (10.3), and the chain rule for relative entropy gives $\mathrm{KL}(\mathbb{Q}^1 \,\|\, \mathbb{Q}^2) = \mathrm{KL}(\nu^1 \,\|\, \nu^2) + \mathbb{E}_{\nu^1}\left[\mathrm{KL}\left(\mathbb{Q}^1_{|x} \,\|\, \mathbb{Q}^2_{|x}\right)\right]$. It therefore suffices to treat a common deterministic initial point, which is where (b) and (c) are used. There, fix $\mathbb{Q}^2$ as the base measure and let W^2 denote a $\mathbb{Q}^2$-Brownian motion, so that $\mathrm{d}X = b^2\mathrm{d}t + \sigma\mathrm{d}W^2$ under $\mathbb{Q}^2$ and $\mathrm{d}X = b^1\mathrm{d}t + \sigma\mathrm{d}W^1$ under $\mathbb{Q}^1$. With $\theta = \sigma^{-1}(b^1 - b^2)$ the two Brownian motions are related by

$$\mathrm{d}W^2 = \mathrm{d}W^1 + \theta\,\mathrm{d}t, \tag{10.5}$$

and Theorem 10.1, with (b) guaranteeing that the stochastic exponential is a true martingale under $\mathbb{Q}^2(\cdot \mid X_0 = x)$, gives

$$\frac{\mathrm{d}\mathbb{Q}^1}{\mathrm{d}\mathbb{Q}^2} = \exp\Big(\int_0^T \langle \theta_s, \mathrm{d}W_s^2 \rangle - \tfrac{1}{2}\int_0^T |\theta_s|^2\,\mathrm{d}s\Big). \tag{10.6}$$

The minus sign belongs with the $\mathbb{Q}^2$-Brownian motion; substituting (10.5) converts (10.6) into $\exp\big(\int \langle \theta, \mathrm{d}W^1 \rangle + \frac{1}{2}\int |\theta|^2\,\mathrm{d}t\big)$, and it is *that* expression which carries a plus. Writing a $\mathbb{Q}^2$-Brownian motion together with a plus sign, or (10.5) with the opposite sign, are the two errors that cancel each other and leave the final identity looking correct. Taking $\mathbb{E}_{\mathbb{Q}^1}$ of the logarithm in the W^1 form,

$$\mathrm{KL}\left(\mathbb{Q}^1 \,\|\, \mathbb{Q}^2\right) = \mathbb{E}_{\mathbb{Q}^1}\Big[\int_0^T \left\langle \theta, \mathrm{d}W^1 \right\rangle + \tfrac{1}{2}\int_0^T |\theta|^2\,\mathrm{d}t\Big] = \tfrac{1}{2}\,\mathbb{E}_{\mathbb{Q}^1}\int_0^T |\theta|^2\,\mathrm{d}t,$$

the stochastic integral having zero mean by the integrability hypothesis. For the covariance criterion, the quadratic variation $\langle X \rangle_t = \int_0^t a(s, X_s)\mathrm{d}s$ is, by Theorem 1.25, a functional of the path determined almost surely under each measure. Let $A = \{(t, x) : a^1(t, x) \neq a^2(t, x)\}$ and suppose it has positive $\mathrm{d}t \otimes \mathbb{Q}^1$ occupation measure, that is $\mathbb{E}_{\mathbb{Q}^1} Y > 0$ for $Y(\omega) = \int_0^T 1_A(t, X_t(\omega))\mathrm{d}t$. Then $\mathbb{Q}^1(Y > 0) > 0$ — *positive* probability, not full: a positive expectation gives no more, and the conclusion needs no more. Let E^1 be the $\mathbb{Q}^1$-full set on which $\mathrm{d}\langle X \rangle_t/\mathrm{d}t = a^1(t, X_t)$ for almost every t, and put $B = E^1 \cap \{Y > 0\}$, so $\mathbb{Q}^1(B) > 0$. Under $\mathbb{Q}^2$, $\mathrm{d}\langle X \rangle_t/\mathrm{d}t = a^2(t, X_t)$ for almost every t, almost surely; every path in B violates this on a set of times of positive measure, so $\mathbb{Q}^2(B) = 0$. A set of positive $\mathbb{Q}^1$-measure and zero $\mathbb{Q}^2$-measure is exactly the failure of absolute continuity, so $\mathbb{Q}^1 \not\ll \mathbb{Q}^2$ and the relative entropy is infinite. Note what this argument uses and what it does not: it needs A to be charged by $\mathbb{Q}^1$, the measure in the first slot. If A is charged only by $\mathbb{Q}^2$ — for instance if the coefficients differ only on a region $\mathbb{Q}^1$ never visits — the argument gives nothing in this direction, and indeed $\mathbb{Q}^1 \ll \mathbb{Q}^2$ can hold. Mutual singularity requires the separation to be visible from both sides. Since a and not σ enters, two square roots of the same a give the same law. ■

Scope of Theorem 10.1 and Proposition 10.2

Assumed. A common, invertible diffusion coefficient; Novikov's condition; square integrability of the drift difference under $\mathbb{Q}^1$; and well-posedness of the martingale problem for (b^1, a), without which the Girsanov transform of $\mathbb{Q}^2$ is a law with drift b^1 but not necessarily the one named in the statement. The alternative — defining $\mathbb{Q}^1$ as the transform — costs nothing and is what Proposition 7.5 does.

Why the common covariance is not a technicality. The quadratic variation of a path is measurable with respect to the path, and by Theorem 1.25 it is determined almost surely. If the characteristics differ where the first process goes, the two disagree on a set of positive Q^1-measure and zero Q^2-measure — positive is what a positive occupation measure delivers, and positive is all the conclusion needs — so $Q^1 \not\ll Q^2$ and the relative entropy is $+\infty$, not large. The statement is oriented, and the symmetric conclusion — mutual singularity — needs a condition seen by both measures; quoting the symmetric version from a one-sided hypothesis is itself a substitution of the kind Table 6.2 catalogues. Every path-space divergence used in this volume — in Proposition 3.13, in Proposition 3.24, in Proposition 7.5 — silently requires this, and it is why the reverse process of Chapter 3 is given the same g_t as the forward one.

Where Novikov fails. For a drift growing faster than linearly the exponential moment can be infinite, and Novikov's criterion is then unavailable. That is all that follows: Z may still be a true martingale, for instance by Kazamaki's criterion, and it may instead be a strict local martingale, in which case $\mathbb{E}Z_T < 1$ and Q is a sub-probability measure whose defect corresponds to explosion. The two possibilities are distinguished by a separate argument, and in this volume the drifts to which Theorem 10.1 is applied are either bounded or verified case by case.

What would falsify them. Two diffusions with a common σ satisfying the integrability hypotheses whose path-space relative entropy differs from (10.4); it is an identity.

Proposition 10.2 assumes an invertible σ, which is what the generative constructions of Chapter 3 need and what keeps $\theta = \sigma^{-1}(b^1 - b^2)$ unambiguous. Control problems are different: there the drift increment is manufactured *from* σ, and the natural setting allows σ to have a kernel, so that several controls produce the same process. The following is the version Chapter 7 calls, and it is stated here so that the volume maintains one set of hypotheses rather than two.

Proposition 10.3 (Path-space entropy under a possibly singular covariance) DERIVED. *Let $\mathbb{R}$ be the law of $dX = b\,dt + \sigma(t, X)dW$ and let v be a measurable field with*

$$v(t,x) \in \operatorname{Ran}\sigma(t,x) \qquad \text{for } dt \otimes Q^v\text{-almost every } (t,x), \tag{10.7}$$

where σ^{+} denotes the Moore–Penrose pseudoinverse and $\theta := \sigma^{+}v$. Assume the stochastic exponential of θ is a true $\mathbb{R}$-martingale, and define Q^v by that Girsanov transform — so that no uniqueness-in-law hypothesis is needed and Q^v is the law of $dX = (b + v)dt + \sigma dW$ from the same initial law whenever that martingale problem is well posed. If $\mathbb{E}_{Q^v} \int_0^T |\theta|^2\,dt < \infty$, then

$$\mathrm{KL}(Q^v \,\|\, \mathbb{R}) = \tfrac{1}{2}\,\mathbb{E}_{Q^v} \int_0^T \|\sigma^{+} v\|^2\,dt. \tag{10.8}$$

In particular, for $v = \sigma u$ the range condition holds automatically and $\theta = \sigma^{\dagger}\sigma u = \Pi u$ with Π the orthogonal projection onto $(\ker\sigma)^{\perp}$, so (10.8) reads $\frac{1}{2}\mathbb{E}\int \|\Pi u\|^2$.

Proof. The two path measures share the covariance characteristic $a = \sigma\sigma^{\top}$ by construction, so the obstruction of Proposition 10.2 does not arise. Write the exponent in terms of the state path rather than the driving noise. Put $\beta := a^{\dagger}v$, where $a^{\dagger}$ is the pseudoinverse of $a = \sigma\sigma^{\top}$. Under $\mathbb{R}$ one has $dX - b\,dt = \sigma dW$, so

$$\int_0^T \left\langle \sigma^{\dagger} v, dW \right\rangle = \int_0^T \langle \beta, dX - b\,dt \rangle, \qquad \left|\sigma^{\dagger} v\right|^2 = \left\langle v, a^{\dagger} v \right\rangle = \langle \beta, a\beta \rangle. \tag{10.9}$$

both identities being the standard pseudoinverse relations $\sigma^\dagger = \sigma^\top a^\dagger$ and $\sigma^\top a^\dagger \sigma = \Pi$ on $\operatorname{Ran}\sigma^\top$, together with the range condition (10.7) which places v where a^+ inverts a. The right-hand side of (10.9) is measurable with respect to the path of X alone, so the density is a functional on $C([0,T];\mathbb{R}^d)$ and defines a change of measure there — this is the substantive point, and it is why the ambiguity $\theta \in \sigma^\dagger v + \ker\sigma$ in the *Brownian-space* representation does not propagate: different θ give different functionals of W but the same functional of X. Repeating the calculation of Proposition 10.2 with this exponent, and with $\mathbb{Q}^v$ defined as the transform so that the identification step is vacuous, gives (10.8), whose integrand may equivalently be written $\langle v, a^+ v\rangle$. That the *minimal* representative is forced, rather than chosen for convenience, is the content of Remark 7.6: a larger representative leaves the left-hand side unchanged while inflating the right. ■

Remark 10.4 (What the range condition excludes). Condition (10.7) is not decoration, and its failure is oriented in exactly the way Proposition 10.2 is. Suppose v has a component outside $\operatorname{Ran}\sigma$ — a drift perturbation in a direction the noise cannot reach — on a set of positive $dt \otimes \mathbb{Q}^v$ occupation measure. The two laws then share the covariance characteristic a, so the quadratic-variation obstruction does not apply; what obstructs instead is the unreachable component, which appears along a $\mathbb{Q}^v$-typical path as a drift in a direction in which $\mathbb{R}$-typical paths have no martingale part, and is therefore detectable on a set of positive $\mathbb{Q}^v$-measure and zero $\mathbb{R}$-measure. Hence $\mathbb{Q}^v \not\ll \mathbb{R}$ and $\mathrm{KL}(\mathbb{Q}^v \,\|\, \mathbb{R}) = \infty$.

Mutual singularity does not follow, and is not asserted: the argument is run under $\mathbb{Q}^v$ and says nothing about a component living where $\mathbb{Q}^v$ never goes. A two-sided statement needs a separating condition visible from both laws — for instance a constant-rank σ together with an outside-range drift on a time set of positive measure under both — and is left to the reader who needs it. This is the "control acting through a channel other than σ" case flagged in the notes on Proposition 7.5, and it is why the actuation constraint of Section 7.4 is a hypothesis about geometry and not about cost.

10.1.2 Entropy production and the arrow of time

Proposition 2.5 decomposed the drift of a stationary diffusion as $b = -\nabla V + c$ and showed that reversibility is $c = 0$. On path space that dichotomy becomes quantitative.

Proposition 10.5 (The time reversal of a stationary diffusion, and its cost) [illegible]. *Let $\pi \propto e^{-V}$ be stationary for $dX = b(X)dt + \sqrt{2}\,dW$ with $b = -\nabla V + c$ and c admissible in the sense of Definition 2.15, and let $X_0 \sim \pi$. Assume both the forward equation and the equation with drift $-\nabla V - c$ are well posed, and that the ordered pair $(\mathbb{Q}, \mathbb{Q}^-)$ satisfies the hypotheses of Proposition 10.2: the initial laws agree, so absolute continuity there is automatic, but the true-martingale criterion for the drift change and the finite drift energy*

$$\int_{\mathbb{R}^d} |c|^2 \, d\pi < \infty \tag{10.10}$$

are extra assumptions, not consequences of Definition 2.15. Then:

(i) *the time-reversed process $Y_s = X_{T-s}$ is the stationary diffusion with drift $-\nabla V - c$, that is the same equation with c negated;*

(ii) *writing* Q *for the law of* X *and* Q^- *for the law of the diffusion with drift* $-\nabla V - c$ *started from* π,

$$\mathrm{KL}(Q \,\|\, Q^-) = T \int_{\mathbb{R}^d} |c|^2 \, d\pi, \tag{10.11}$$

the entropy production over $[0, T]$;

(iii) *consequently the process is indistinguishable from its time reversal* — $\mathrm{KL}(Q \,\|\, Q^-) = 0$ — *if and only if* $c = 0$, *that is if and only if it is reversible in the sense of Definition 2.4.*

Proof. (i) By Theorem 3.1 with $a = 2I$ and $p_t \equiv \pi$, the reversed drift is $-b + 2\nabla \log \pi = -(-\nabla V + c) - 2\nabla V = -\nabla V - c$. (ii) The two diffusions have the same $\sigma = \sqrt{2}I$, the same initial law π, and drift difference $2c$. Proposition 10.2 gives $\mathrm{KL}(Q \,\|\, Q^-) = \frac{1}{2}\mathbb{E}\int_0^T \|\frac{1}{\sqrt{2}} 2c(X_t)\|^2 dt = \mathbb{E}\int_0^T |c(X_t)|^2 \, dt$, and stationarity makes the integrand's law π at every t, giving (10.11). (iii) The right-hand side of (10.11) vanishes if and only if $c = 0$ π-almost everywhere, which by Proposition 2.5(iv) is reversibility. ∎

Where stationarity is used. Twice, and differently: to identify the reversed drift in (i), where $p_t \equiv \pi$ makes the score term constant in t; and to evaluate the expectation in (ii), where the law of X_t is π at every time. Away from stationarity neither step survives and the reversed process has a time-dependent drift, which is Theorem 3.1 in its general form.

What it does not say. It compares the forward law with the law of the *adjoint* equation, not with the forward law composed with path reversal — those two are equal, by (i), which is the content of the statement rather than a triviality to be checked. And it says nothing about a non-stationary process, for which no such closed form is available here.

> What does not follow
>
> **The arrow of time is the non-gradient part of the drift, and nothing else.** Equation (10.11) says that the statistical distinguishability of a stationary diffusion from its own time reversal is exactly $T\int |c|^2 \, d\pi$. A reversible diffusion is at equilibrium in the strong sense that a film of it played backwards has the same law; a non-reversible one is not, and the discrepancy is measured by the circulating flux $j[\pi] = \pi c$ of Proposition 2.5(ii).
>
> **This is not a derivation of the reverse-time formula of Chapter 3.** Proposition 10.5 concerns a *stationary* diffusion and computes a relative entropy. The reverse-time equation of Theorem 3.1 concerns a time-inhomogeneous process that is at no time stationary, and it is proved by time-reversal theory under its own hypotheses. Fluctuation identities of this kind are a conceptual connection between the two and not a proof of either; presenting one as the physical origin of the other substitutes an analogy for an argument.

10.2 The Schrödinger Problem

Definition 10.6 (Schrödinger problem)**.** Let $\mathbb{R}$ be a reference probability measure on $C([0, T]; \mathbb{R}^d)$ and let $\mu_0, \mu_T \in \mathcal{P}_2(\mathbb{R}^d)$. The *Schrödinger problem* is

$$\inf\left\{\mathrm{KL}(Q \,\|\, \mathbb{R}) \;:\; Q_0 = \mu_0, \; Q_T = \mu_T\right\}, \tag{10.12}$$

the infimum over path measures with the two prescribed marginals.

Proposition 10.7 (Static reduction) DERIVED. *Write Q_{0T} for the joint law of (X_0, X_T) under Q and $Q_{|0T}$ for the conditional law of the path given its endpoints. Then for every $Q \ll R$,*

$$\mathrm{KL}(Q \,\|\, R) = \mathrm{KL}(Q_{0T} \,\|\, R_{0T}) + \mathbb{E}_{Q_{0T}}\Big[\mathrm{KL}(Q_{|0T} \,\|\, R_{|0T})\Big], \tag{10.13}$$

so any minimiser of (10.12) has $Q^*_{|0T} = R_{|0T}$ — it shares the bridges of the reference — and (10.12) reduces to the static problem

$$\inf\Big\{\mathrm{KL}(\gamma \,\|\, R_{0T}) \;:\; \gamma \in \Pi(\mu_0, \mu_T)\Big\} \tag{10.14}$$

over couplings of the two marginals, which is an entropic optimal transport problem.

Proof. Equation (10.13) is the chain rule for relative entropy under the disintegration by (X_0, X_T). The second term is non-negative and depends on Q only through $Q_{|0T}$, which is unconstrained by (10.12); setting it to $R_{|0T}$ makes it zero without changing Q_{0T}, hence without affecting feasibility. What remains is (10.14). ■

The two things usually stated together here have different hypothesis sets, and this volume separates them so that a downstream proposition cannot inherit the second while citing only the first.

Theorem 10.8 (Existence and uniqueness of the *I*-projection) ESTABLISHED. *Let R be a probability measure on $C([0,T];\mathbb{R}^d)$ with Polish path space, let $\mu_0, \mu_T \in \mathcal{P}(\mathbb{R}^d)$, and let*

$$\Pi(\mu_0, \mu_T) = \{Q : Q_0 = \mu_0,\ Q_T = \mu_T\}$$

be the feasible set, which is convex and closed for the weak topology. Suppose the value of (10.12) is finite, that is $\Pi(\mu_0, \mu_T)$ contains some $Q \ll R$ with $\mathrm{KL}(Q \,\|\, R) < \infty$. Then the minimiser Q^* exists and is unique.

Proof roadmap. The argument is the direct method, and it needs one ingredient beyond closedness and lower semicontinuity — on a non-compact Polish space those two alone do not produce a minimiser, because a minimising sequence may escape.

What supplies compactness is the reference measure. For a fixed probability measure R, the sublevel sets $\{Q : \mathrm{KL}(Q \,\|\, R) \le C\}$ are tight; this is a standard consequence of the variational characterisation of relative entropy, and it is where the finiteness hypothesis is used. A minimising sequence lies in one such sublevel set, so by Prokhorov's theorem it has a weakly convergent subsequence. The endpoint maps $Q \mapsto Q_0$ and $Q \mapsto Q_T$ are weakly continuous on $C([0,T];\mathbb{R}^d)$, so the limit is still feasible; lower semicontinuity of $\mathrm{KL}(\cdot \,\|\, R)$ then makes it a minimiser. Uniqueness is strict convexity of relative entropy on the convex set $\Pi(\mu_0, \mu_T)$.

The hypotheses are stated in the theorem rather than left to a citation note, because a reader should not have to consult a footnote to learn whether the feasible set is closed. □

Remark 10.9 (Endpoint factorisation). Under positivity hypotheses on the reference endpoint law R_{0T} — stronger than finiteness of the value alone — Schrödinger-system theory gives the minimiser a product-form density,

$$\frac{dQ^*}{dR} = \varphi(X_0)\,\psi(X_T), \tag{10.15}$$

with $\varphi, \psi \geq 0$ measurable and determined, up to the scaling $(\varphi, \psi) \mapsto (\lambda\varphi, \lambda^{-1}\psi)$, by the two marginal constraints. Léonard's survey [16] is the entry point.

This is stated as a remark rather than as a theorem of this volume, and deliberately so: the admissible class of endpoint pairs, the value range and integrability of the potentials, and the gauge in which uniqueness holds vary between formulations, and quoting the conclusion without fixing them would assert more than any one source gives. Nothing below uses the *necessity* of (10.15) except the identification of the IPF limit with the bridge, which says so where it occurs. The *sufficiency* direction — a feasible product-form measure is the bridge — is what Proposition 10.13(iv) and Proposition 10.15(iii) actually use, and it is proved here as Lemma 10.10.

The sufficiency direction — that a feasible product-form measure *is* the bridge — needs none of that, and it is worth proving here, because it is what Proposition 10.13(iv) actually uses. The argument is the Pythagorean identity for relative entropy, and it is short.

Lemma 10.10 (Product-form sufficiency) DERIVED. *Let* Q *be feasible for* (10.12) *— that is* $Q_0 = \mu_0$ *and* $Q_T = \mu_T$ *— with* $\mathrm{KL}(Q \,\|\, R) < \infty$ *and*

$$\frac{dQ}{dR} = \varphi(X_0)\,\psi(X_T), \tag{10.16}$$

where $\log\varphi$ *is* μ_0*-integrable and* $\log\psi$ *is* μ_T*-integrable. Then for every feasible* Q',

$$\mathrm{KL}(Q' \,\|\, R) = \mathrm{KL}(Q' \,\|\, Q) + \mathrm{KL}(Q \,\|\, R), \tag{10.17}$$

and consequently Q *is the unique minimiser of* (10.12).

Proof. If $\mathrm{KL}(Q' \,\|\, R) = \infty$ there is nothing to prove, so assume it finite; then $Q' \ll R$.

The chain rule below needs $Q' \ll Q$, and this follows from the integrability hypotheses rather than being automatic. Since $\log\varphi \in L^1(\mu_0)$ and $\log\psi \in L^1(\mu_T)$, we have $\varphi > 0$ μ_0-almost everywhere and $\psi > 0$ μ_T-almost everywhere. Every feasible Q' has marginals μ_0 and μ_T, so $\varphi(X_0)\psi(X_T) > 0$ holds Q'-almost surely. Now let A be a Q-null set. Then $\int_A \varphi(X_0)\psi(X_T)\,dR = 0$ with a non-negative integrand, so $\varphi(X_0)\psi(X_T) = 0$ R-almost everywhere on A; since $Q' \ll R$, the same holds Q'-almost everywhere on A, and combining this with the positivity just established gives $Q'(A) = 0$. Hence $Q' \ll Q$, and we may write, Q'-almost surely,

$$\log\frac{dQ'}{dR} = \log\frac{dQ'}{dQ} + \log\frac{dQ}{dR}.$$

Take $\mathbb{E}_{Q'}$. The last term is $\mathbb{E}_{Q'}[\log\varphi(X_0) + \log\psi(X_T)]$ by (10.16), and this depends on Q' only through its two marginals, which are μ_0 and μ_T by feasibility — the same marginals Q has. The integrability hypotheses make both expectations finite, so $\mathbb{E}_{Q'}\log(dQ/dR) = \mathbb{E}_{Q}\log(dQ/dR) = \mathrm{KL}(Q \,\|\, R)$, which is (10.17). Since $\mathrm{KL}(Q' \,\|\, Q) \geq 0$ with equality only for $Q' = Q$, the claim follows. ■

Remark 10.11 (Which direction is proved here, and which is quoted). Lemma 10.10 is a *sufficiency* statement: product form plus feasibility plus integrability implies optimality. What Remark 10.9 supplies is the *necessity* — that the minimiser must have product form

— and that is the clause this volume quotes rather than proves. The distinction matters because the two are used in different places: a construction that produces a product-form measure and checks its marginals, as in Proposition 10.13(iv), needs only the lemma; a claim that *every* solution of a Schrödinger problem factorises needs the theorem. Conflating them would let an external hypothesis leak into a statement that does not require it.

> Notes on Proposition 10.7 and Theorem 10.8
>
> **Assumed.** $Q \ll R$ and finiteness of the value in (10.12). Finiteness is a genuine restriction: it fails whenever μ_0 or μ_T is not absolutely continuous with respect to the corresponding marginal of the reference, and in particular it fails for a reference with compactly supported marginals and a target that is not. It is also, on its own, not enough for the product form, which is why Remark 10.9 is stated separately and carries its own hypotheses. Only the necessity direction is quoted: the identification of the IPF limit with the bridge after Proposition 10.17 cites the factorisation result of Remark 10.9 and inherits the conditions stated there. Proposition 10.13(iv) and Proposition 10.15(iii) need only the sufficiency direction, which is Lemma 10.10 and is proved here; and Proposition 10.17(ii) needs neither, since that the iterates have product form is proved by induction from the half-step formula.
>
> **What the reduction does and does not remove.** It removes the path variable: by (10.13) the solution shares the reference's bridges, so (10.12) is the finite-dimensional problem (10.14) in the endpoint variables. It does not make the problem easy: (10.14) is an entropic optimal transport problem whose solution is available in closed form only for Gaussian data.
>
> **Reference dependence.** The bridges, and hence the whole solution, belong to R. Two references with the same marginals and different bridges give different answers, so there is no reference-free Schrödinger problem and no sense in which a bridge is canonical.

10.3 Three Ways to Steer a Diffusion

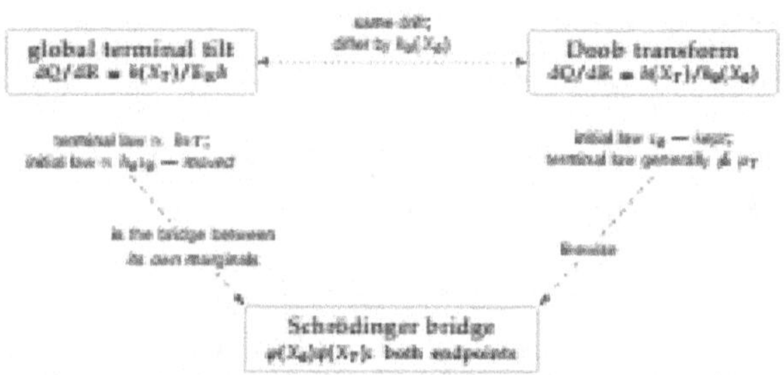

Figure 10.1: Three constructions that steer a reference diffusion, and the marginals each controls. All three preserve the bridges of R; they differ in which endpoint they fix. What this figure does not imply: that the three are the same construction under different names. The two upper constructions share a drift and differ in their initial marginal, agreeing as path measures only when h_0 is constant; and each is the bridge between the marginals it happens to produce, which is a prescription only if one accepts those (Proposition 10.13).

Let $\mathbb{R}$ be the law of a reference diffusion with initial law ν_0 and terminal law ν_T. Three constructions produce a new path measure that ends somewhere else, and they are not the same construction.

Definition 10.12 (Tilting, conditioning, constraining). Let $h \geq 0$ be measurable with $0 < \mathbb{E}_{\mathbb{R}} h(X_T) < \infty$ and write

$$h_t(x) := \mathbb{E}_{\mathbb{R}}\big[h(X_T) \mid X_t = x\big], \qquad \text{so that} \qquad \mathbb{E}_{\mathbb{R}} h(X_T) = \int h_0 \, \mathrm{d}\nu_0. \tag{10.18}$$

(a) *Global terminal tilt* (Feynman–Kac tilting by a terminal cost g, with $h = e^{-g}$): a single normalisation over the whole space,

$$\frac{\mathrm{d}\mathbb{Q}^{(a)}}{\mathrm{d}\mathbb{R}} = \frac{h(X_T)}{\mathbb{E}_{\mathbb{R}} h(X_T)}. \tag{10.19}$$

(b) *Doob h-transform*, normalised *conditionally on the starting point*:

$$\frac{\mathrm{d}\mathbb{Q}^{(b)}}{\mathrm{d}\mathbb{R}} = \frac{h(X_T)}{h_0(X_0)}. \tag{10.20}$$

(c) *Schrödinger bridge*: the solution of (10.12) with both marginals prescribed.

Constructions (a) and (b) differ by the factor $h_0(X_0)$, a function of the initial point alone. They therefore have the same conditional law given X_0 — the same dynamics, the same drift — and differ only in their initial marginal. When X_0 is deterministic, or more generally when h_0 is ν_0-almost surely constant, they coincide.

Proposition 10.13 (The three, and the conditions relating them) DERIVED. *With the notation of Definition 10.12, and $\mathbb{R}$ Markov — part (iii) additionally assuming a diffusion reference, as stated there:*

(i) Both one-sided constructions leave the bridges of $\mathbb{R}$ unchanged: *the conditional law given (X_0, X_T) is that of $\mathbb{R}$, so by Proposition 10.7 each is determined by its endpoint law. Their endpoint marginals are*

$$\mathbb{Q}_T^{(a)}(\mathrm{d}x) = \frac{h(x)\,\nu_T(\mathrm{d}x)}{\int h\,\mathrm{d}\nu_T}, \qquad \mathbb{Q}_0^{(a)}(\mathrm{d}x) = \frac{h_0(x)\,\nu_0(\mathrm{d}x)}{\int h_0\,\mathrm{d}\nu_0}, \tag{10.21}$$

$$\mathbb{Q}_0^{(b)} = \nu_0, \qquad \mathbb{Q}_T^{(b)}(\mathrm{d}x_T) = \int \frac{h(x_T)}{h_0(x_0)}\, \mathbb{R}_{0T}(\mathrm{d}x_T \mid x_0)\, \nu_0(\mathrm{d}x_0). \tag{10.22}$$

In particular the global tilt changes the initial marginal unless h_0 is ν_0-a.s. constant, and the Doob transform preserves it always.

(ii) Reaching a prescribed terminal law. *For a given $\mu_T \ll \nu_T$, $\mathbb{Q}_T^{(a)} = \mu_T$ if and only if $h \propto \mathrm{d}\mu_T/\mathrm{d}\nu_T$ — proportionality, since (10.19) normalises h and an overall constant is invisible; and in particular not $h = \mu_T$ and not $g = -\log \mu_T$. For the Doob transform the terminal density is, by (10.22),*

$$\frac{\mathrm{d}\mathbb{Q}_T^{(b)}}{\mathrm{d}\nu_T}(x_T) = h(x_T)\, \mathbb{E}_{\mathbb{R}}\big[h_0(X_0)^{-1} \mid X_T = x_T\big], \tag{10.23}$$

so with h normalised by $\int h \mathrm{d}\nu_T = 1$,

$$Q_T^{(b)} = \mu_T \iff \mathbb{E}_{\mathbb{R}}\left[h_0(X_0)^{-1} \mid X_T = \cdot\right] = 1 \quad \mu_T\text{-almost everywhere.}$$

Constancy of h_0 *is sufficient for this, and is exactly the condition for the two path measures to coincide; it is not necessary for the terminal marginals to coincide, since a non-constant* h_0^{-1} *may average to 1 given* X_T. *It becomes necessary only when* $\varphi \mapsto \mathbb{E}_{\mathbb{R}}[\varphi(X_0) \mid X_T = \cdot]$ *is injective on the relevant class.*

(iii) Both are h-transforms of the dynamics. *Assume in addition, for this part only, that* $\mathbb{R}$ *is the law of an Itô diffusion* $\mathrm{d}X = b\,\mathrm{d}t + \sigma\,\mathrm{d}W$ *with* $a = \sigma\sigma^{\top}$, *that* $h_t > 0$ *is* $C^{1,2}$, *and that the integrability required by Theorem 10.1 holds. Then under either construction the drift acquires the additional term* $a\nabla \log h_t$, *with* h_t *as in (10.18); the two differ only through their initial law, since* $\mathrm{d}Q^{(a)}/\mathrm{d}Q^{(b)} = h_0(X_0)/\mathbb{E}_{\mathbb{R}}h(X_T)$ *is a function of* X_0 *alone. Parts (i), (ii), (iv) and (v) are identities between measures and use only that* $\mathbb{R}$ *is Markov; this is the one clause that needs a diffusion and its regularity, and it should not be quoted for a general Markov reference.*

(iv) Relation to the bridge. *Each of (a) and (b) has a density of the product form* $\varphi(X_0)\psi(X_T)$ *with respect to* $\mathbb{R}$ — $\varphi \equiv \text{const}$ *for (a),* $\varphi = 1/h_0$ *for (b) — and is therefore, by Lemma 10.10, the* Schrödinger bridge between its own marginals — *a conclusion that needs only the sufficiency direction, and therefore does not depend on that clause. Given a prescribed pair* (μ_0, μ_T), *the global tilt with* $h = \mathrm{d}\mu_T/\mathrm{d}\nu_T$ *solves (10.12) if and only if* $\mu_0 = h_0\nu_0/\int h_0 \mathrm{d}\nu_0$; *in particular it solves it with* $\mu_0 = \nu_0$ *if and only if* h_0 *is* ν_0*-a.s. constant. In general two multipliers are needed and neither one-sided construction suffices.*

(v) *Consequently no single statement identifies all three, and the two one-sided constructions are not interchangeable either: they answer "which marginal do you insist on keeping?" differently.*

Proof. (i) Both densities are of the form $\varphi(X_0)\psi(X_T)$, hence measurable with respect to the endpoint pair; by Proposition 10.7 the conditional law given (X_0, X_T) is therefore unchanged, and the endpoint law is reweighted by the same factor. For (a) the joint endpoint law is $h(x_T)\mathbb{R}_{0T}(\mathrm{d}x_0, \mathrm{d}x_T)/\mathbb{E}_{\mathbb{R}}h$; marginalising in x_0 gives the first identity in (10.21), and marginalising in x_T gives $\nu_0(\mathrm{d}x_0)\int h(x_T)\mathbb{R}_{0T}(\mathrm{d}x_T \mid x_0) = h_0(x_0)\nu_0(\mathrm{d}x_0)$, normalised by $\int h_0 \mathrm{d}\nu_0 = \mathbb{E}_{\mathbb{R}}h(X_T)$ from (10.18). For (b) the same computation with the extra factor $1/h_0(x_0)$ gives $Q_0^{(b)} = \nu_0$ exactly, and the stated terminal law.

(ii) The first claim is (10.21), whose right-hand side is unchanged by $h \mapsto ch$ for $c > 0$, so the criterion is proportionality rather than equality. For the second, disintegrate (10.22) over X_T: writing $\mathbb{R}_{0T}(\mathrm{d}x_0, \mathrm{d}x_T) = \mathbb{R}_{0|T}(\mathrm{d}x_0 \mid x_T)\,\nu_T(\mathrm{d}x_T)$,

$$Q_T^{(b)}(\mathrm{d}x_T) = h(x_T)\left(\int h_0(x_0)^{-1}\,\mathbb{R}_{0|T}(\mathrm{d}x_0 \mid x_T)\right)\nu_T(\mathrm{d}x_T),$$

which is (10.23); comparing with $\mu_T = h\,\nu_T$ gives the stated criterion. Constancy of h_0 makes the conditional expectation constant, and equal to 1 because $\int h_0 \mathrm{d}\nu_0 = \int h \mathrm{d}\nu_T = 1$ by (10.18); but a non-constant h_0^{-1} can average to 1 given X_T, so the converse holds only when $\varphi \mapsto \mathbb{E}_{\mathbb{R}}[\varphi(X_0) \mid X_T = \cdot]$ is injective on the relevant class. Absorbing $h_0(x_0)^{-1}$ into a normalising constant is legitimate at the level of *path* measures and not at the level of the terminal marginal, which is where the two statements part.

(iii) $h_t(X_t)$ is a $\mathbb{R}$-martingale by the tower property, so $\mathrm{d}\mathbb{Q}^{(b)}/\mathrm{d}\mathbb{R}\big|_{\mathcal{F}_t} = h_t(X_t)/h_0(X_0)$ and $\mathrm{d}\mathbb{Q}^{(a)}/\mathrm{d}\mathbb{R}\big|_{\mathcal{F}_t} = h_t(X_t)/\mathbb{E}_{\mathbb{R}}h$; in both cases Theorem 10.1 identifies the drift change as $\sigma\sigma^T\nabla \log h_t = a\nabla \log h_t$, and the ratio of the two densities is $h_0(X_0)/\mathbb{E}_{\mathbb{R}}h$.

(iv) The direction used here is the sufficiency one, and it is Lemma 10.10: a measure with a product-form density whose marginals are the prescribed pair, with the stated integrability, is the unique minimiser — no appeal to Remark 10.9 is needed for that. (The converse, that every minimiser factorises, is the quoted clause; it is not used here.) Applying the lemma to (a) with $h = \mathrm{d}\mu_T/\mathrm{d}\nu_T$: its terminal marginal is μ_T by (ii) and its initial marginal is $h_0\nu_0/\int h_0 \mathrm{d}\nu_0$ by (i), so it is the bridge for the prescribed pair exactly under the stated condition.

(v) is the content of (i)–(iv). ∎

Remark 10.14 (Where the old identification goes wrong). It is often said that a terminal tilt "does not touch the initial distribution". That is true of the *conditional* dynamics and false of the *marginal*: reweighting a path by a function of its endpoint favours those starting points from which the favoured endpoint is easy to reach, and $h_0(x_0) = \mathbb{E}_{\mathbb{R}}[h(X_T) \mid X_0 = x_0]$ is exactly that bias. The statement becomes true when X_0 is deterministic — the case of a control problem with a fixed initial state, which is why the issue does not arise in Proposition 7.5 — and, more generally, when h_0 is constant. Whenever ν_0 is genuinely spread out, the choice between (10.19) and (10.20) is a modelling decision that must be made and stated.

Table 10.1: Three constructions, three problems. All three keep the bridges of $\mathbb{R}$; they differ in which endpoint they control. The second column is the one usually asserted without checking, and (10.21) is the check. Throughout, $h_0(x) = \mathbb{E}_{\mathbb{R}}[h(X_T) \mid X_0 = x]$.

Construction	Initial law obtained	Terminal law obtained	Is the bridge for (μ_0, μ_T) when
Global tilt (10.19) by $h = e^{-g}$	$\propto h_0\nu_0$; equal to ν_0 only if h_0 is constant	$\propto e^{-g}\nu_T$, *not* $\propto e^{-g}$	$\mu_T \propto e^{-g}\nu_T$ and $\mu_0 \propto h_0\nu_0$; with $\mu_0 = \nu_0$ this forces h_0 constant
Doob transform (10.20), $h = \mathrm{d}\mu_T/\mathrm{d}\nu_T$	exactly ν_0	(10.22); equal to μ_T if h_0 is constant, and otherwise only under the weaker conditional-expectation condition of Proposition 10.13(iii)	$\mu_0 = \nu_0$ *and* its terminal law happens to be μ_T; h_0 constant is sufficient
Schrödinger bridge	exactly μ_0	exactly μ_T	always — that is its definition; it needs *both* multipliers $\varphi(X_0)\psi(X_T)$

Proposition 10.15 (Where score-based generation sits) [illegible]. *Let $\mathbb{R}$ be the law of the forward noising process of Section 3.2 started from p_{data}, so $\nu_0 = p_{\mathrm{data}}$ and $\nu_T = p_T$, and write $\bar{\mathbb{R}}$ for $\mathbb{R}$ read backwards, a path measure with initial law p_T and terminal law p_{data}. Then:*

(i) *the exact reverse-time process of Theorem 3.1 run from p_T is $\bar{\mathbb{R}}$: it is the trivial case $h \equiv 1$ of Definition 10.12, a construction with one free endpoint;*

(ii) *run instead from a prior* $\varpi \ll p_T$*, it is the reweighting of* $\bar{\mathbb{R}}$ *by the function* $f(X_0) = \mathrm{d}\varpi/\mathrm{d}p_T(X_0)$ *of the initial coordinate of the reversed process. This is the mirror of Definition 10.12(a): it preserves every bridge of* $\bar{\mathbb{R}}$ *and every conditional law given the starting point — the reverse-time drift is unchanged, which is why the sampler is the same algorithm — and it moves the terminal law away from* p_{data} *to* $\int \bar{\mathbb{R}}_{0T}(\cdot \mid x)\varpi(\mathrm{d}x)$*;*

(iii) $\varpi = p_T$ *is sufficient for it to solve the Schrödinger problem (10.12) with reference* $\bar{\mathbb{R}}$ *and prescribed marginals* $(\varpi, p_{\text{data}})$*. It is also necessary provided the endpoint transition operator* $\varpi \mapsto \int \bar{\mathbb{R}}_{0T}(\cdot \mid x)\,\varpi(\mathrm{d}x)$ *is injective on the class of admissible priors — an assumption about the noising semigroup, not a consequence of uniqueness of the reversed reference, since a non-injective kernel can carry two distinct initial laws to the same terminal law. Under that injectivity and for* $\varpi \neq p_T$*, the construction is the bridge between* ϖ *and its own terminal law, and the second multiplier that would restore* p_{data} *is exactly what score-based generation does not compute.*

By Proposition 3.10, $\mathrm{KL}(p_T \,\|\, \varpi)$ *is positive for every finite T when* $\varpi = \mathcal{N}(0,1)$ *unless* p_{data} *is already the reference's invariant law, in which case* $p_T = \varpi$ *exactly. That exception is degenerate — it says the data were Gaussian to begin with — but it is the one case in which the criterion is met, and it is named rather than dismissed.*

Proof. (i) The reverse-time process of Theorem 3.1 has by construction the law of $\mathbb{R}$ under path reversal, so starting it from p_T reproduces $\bar{\mathbb{R}}$.

(ii) Changing only the starting law of a Markov path measure multiplies its density by a function of the initial coordinate: if $\mathbb{Q}$ has the transition kernels of $\bar{\mathbb{R}}$ and initial law ϖ, then $\mathrm{d}\mathbb{Q}/\mathrm{d}\bar{\mathbb{R}} = (\mathrm{d}\varpi/\mathrm{d}p_T)(X_0)$. Such a factor is measurable with respect to the endpoint pair, so Proposition 10.7 applies and the bridges are unchanged; and since the factor depends on X_0 alone, the conditional law given X_0 — hence the drift — is unchanged too. The terminal law is the stated mixture.

(iii) The density $\mathrm{d}\mathbb{Q}/\mathrm{d}\bar{\mathbb{R}}$ has the product form with $\psi \equiv 1$, so by Lemma 10.10 it is the bridge between its own marginals, namely ϖ and the mixture of (ii). It is the bridge for the prescribed pair $(\varpi, p_{\text{data}})$ exactly when that mixture is p_{data}. Taking $\varpi = p_T$ gives that mixture, since $\bar{\mathbb{R}}$ has those two marginals, which is sufficiency. Necessity requires excluding a second prior with the same image under the endpoint kernel, which is precisely the injectivity hypothesis; uniqueness of the reversed reference does not supply it, being a statement about path measures given a fixed initial law. The mismatch is Proposition 3.10, strictly positive for finite T unless p_{data} is already the reference's invariant law. ■

> **Interpretation**
>
> The three constructions differ in one question: which endpoint do you insist on keeping? A terminal tilt fixes where the process ends and lets the starting distribution move. A Doob transform fixes where it starts and lets the ending move. A bridge fixes both, and pays for it with a second multiplier that has to be found rather than written down. All three leave the reference's bridges alone, which is why they look alike on paper and behave differently in use. Figure 10.1 is the picture; Table 10.1 is the same information as a table.

Notes on Proposition 10.13 and Proposition 10.15

The composite claim that does not exist. It is often written that Feynman–Kac tilting, the Doob h-transform and the Schrödinger bridge are the same construction. Proposition 10.13 shows what is true: the first two share a drift and differ in their initial marginal, coinciding exactly when h_0 is constant; and either is the bridge *between its own marginals*, which is a prescription only if one already accepts whichever endpoint law the construction happens to produce. There is no single statement covering all three, and asserting one is asserting a conjunction of conditions without stating any of them.

Where the tilting error bites, at both ends. By (10.21) a terminal cost g produces $\propto e^{-g}\nu_T$. Setting $g = -\log p_{\text{data}}$ therefore produces $\propto p_{\text{data}}\,\nu_T$, which is not p_{data} unless ν_T is constant — and no probability density on $\mathbb{R}^d$ is constant. This is Remark 7.8 again, now with the correction: use $h = \mathrm{d}\mu_T/\mathrm{d}\nu_T$. The second, less visible error is at the other end: that same reweighting multiplies the initial law by h_0, so a construction advertised as "steering the endpoint" has silently moved the starting distribution as well. Remark 10.14 says when it has not.

What Proposition 10.15 does not say. It does not say that score-based generation is wrong, or that a bridge is better. It says the two solve different problems: one endpoint free against two endpoints fixed. Their solutions agree when the prior equals the reference's terminal law, and Proposition 3.10 bounds the disagreement. Note also which side is being reweighted: replacing p_T by a prior is an *initial* reweighting of the reversed process, so it leaves the reverse drift alone and moves the generated law; a terminal reweighting would change the drift instead. The vocabulary of "h-transform" covers both and distinguishes neither, which is why Definition 10.12 writes the density down.

10.4 Iterative Proportional Fitting

Definition 10.16 (Iterative proportional fitting). Let $\mathcal{C}_0 = \{\mathrm{Q} : \mathrm{Q}_0 = \mu_0\}$ and $\mathcal{C}_T = \{\mathrm{Q} : \mathrm{Q}_T = \mu_T\}$, both convex. Starting from $\mathrm{Q}^0 = \mathbb{R}$, define

$$\mathrm{Q}^{2n+1} = \operatorname*{arg\,min}_{\mathrm{Q}\in\mathcal{C}_T} \mathrm{KL}(\mathrm{Q}\,\|\,\mathrm{Q}^{2n}), \qquad \mathrm{Q}^{2n+2} = \operatorname*{arg\,min}_{\mathrm{Q}\in\mathcal{C}_0} \mathrm{KL}\left(\mathrm{Q}\,\|\,\mathrm{Q}^{2n+1}\right). \tag{10.24}$$

Proposition 10.17 (Structure of the iteration) DERIVED. *Each half-step of (10.24) is explicit: the projection onto $\mathcal{C}_T$ multiplies the density by a function of X_T alone,*

$$\frac{\mathrm{d}\mathrm{Q}^{2n+1}}{\mathrm{d}\mathrm{Q}^{2n}}(X) = \frac{\mathrm{d}\mu_T}{\mathrm{d}\mathrm{Q}^{2n}_T}(X_T), \tag{10.25}$$

and symmetrically for $\mathcal{C}_0$ with a function of X_0. Consequently:

(i) *every iterate has the same bridges as $\mathbb{R}$, so by Proposition 10.7 the iteration is entirely determined by the sequence of joint endpoint laws — it is a Sinkhorn iteration for (10.14);*

(ii) *every iterate has the product form $\mathrm{d}\mathrm{Q}^n/\mathrm{d}\mathbb{R} = \varphi_n(X_0)\psi_n(X_T)$ — by induction from (10.25) starting at $\varphi_0 = \psi_0 \equiv 1$, so this clause is proved here and does not depend on Remark 10.9; what does depend on it is the identification of a product-form limit with the bridge;*

(iii) *the Pythagorean identity holds: for $\mathrm{Q} \in \mathcal{C}_T$,*

$$\mathrm{KL}(\mathrm{Q}\,\|\,\mathrm{Q}^{2n}) = \mathrm{KL}\left(\mathrm{Q}\,\|\,\mathrm{Q}^{2n+1}\right) + \mathrm{KL}\left(\mathrm{Q}^{2n+1}\,\|\,\mathrm{Q}^{2n}\right), \tag{10.26}$$

so $n \mapsto \mathrm{KL}(\mathrm{Q}^*\,\|\,\mathrm{Q}^n)$ is non-increasing.

Proof. The constraint set $\mathcal{C}_T$ fixes only the X_T-marginal, so writing $\mathrm{KL}(Q \,\|\, Q^{2n}) = \mathrm{KL}(Q_T \,\|\, Q^{2n}_T) + \mathbb{E}_{Q_T}\mathrm{KL}\left(Q_{|T} \,\|\, Q^{2n}_{|T}\right)$ by the chain rule, the first term is determined by the constraint and the second is minimised, at zero, by taking the conditional law given X_T equal to that of Q^{2n}. That is exactly (10.25).

(i) and (ii) follow because (10.25) multiplies the density by a function of one endpoint, which by Proposition 10.7 leaves the bridges unchanged and preserves the product form, starting from $\varphi_0 = \psi_0 = 1$. (iii) is the same chain-rule computation applied to a $Q \in \mathcal{C}_T$: the conditional terms combine because Q^{2n+1} and Q^{2n} share their conditional laws given X_T, leaving (10.26). Monotonicity follows by applying it with $Q = Q^*$, which lies in both constraint sets. ∎

Proposition 10.18 (Convergence of $\mathrm{KL}(Q^* \,\|\, Q^n)$ to zero) [illegible]. *If the value of* (10.12) *is finite then* $\mathrm{KL}(Q^* \,\|\, Q^n) \to 0$ *— the same orientation of the divergence that* (10.26) *makes monotone, and the one in which the iteration is a sequence of projections towards* Q^*. *See Léonard [16] for the Schrödinger problem and the iterative scheme.*

The rate at which that convergence happens is a separate statement resting on a separate hypothesis. It is recorded as a note rather than as a clause of the proposition above, so that a reader who needs only convergence does not also inherit the positivity condition.

Remark 10.19 (Literature note: linear rate under positivity). Under a positivity condition on R_{0T} — boundedness of the density away from zero and infinity on the product of the supports — the convergence is reported in the literature to be linear, with a rate given by the Hilbert-metric contraction coefficient of the associated positive integral operator. *Which divergence that rate is stated in, and in which orientation, is not settled here*, and the two orientations are not interchangeable (Remark 10.20). Nothing in this volume is derived from the rate; Table 10.2 records it as vocabulary. See Léonard [16] for the Schrödinger problem and the iterative scheme.

Remark 10.20 (Which divergence, and why it is not a detail). Relative entropy is not symmetric, and "converges in relative entropy" names no quantity until the ordered pair is written down. Here the natural one is $\mathrm{KL}(Q^* \,\|\, Q^n)$: it is what (10.26) bounds, decreasing at every half-step by the amount $\mathrm{KL}\left(Q^{2n+1} \,\|\, Q^{2n}\right)$ moved at that step. A bound on the reverse orientation $\mathrm{KL}(Q^n \,\|\, Q^*)$ does not follow from it, and would require its own argument; where a downstream statement needs one rather than the other, it says which.

What does not follow

Iterative proportional fitting is alternating projection, not a Picard iteration. Proposition 10.17 identifies (10.24) as alternating relative-entropy projection onto two convex sets, with the Pythagorean identity (10.26) as its structural fact. A Picard-iteration description applies to one representation of the same algorithm — the fixed-point iteration on the pair (φ, ψ) of (10.15) — and the two are not interchangeable: the projection description gives monotonicity for free by (10.26), and the fixed-point description gives a linear rate only under the positivity condition of Remark 10.19, which is a hypothesis and not a consequence. Claiming linear convergence from the fixed-point form without that condition is claiming a contraction that has not been established.

Each half-step is exact, and none of them is computable. The projection (10.25) requires the marginal Q_T^{2n}, which is the terminal law of a diffusion with a learned drift. Every implementation replaces it by an estimate, and the resulting iteration is not (10.24); the error of one iterate enters the target of the next. Open Problem 11.4 states the missing error analysis, and this volume does not supply it.

Sharing the bridges is a property of the reference, not a simplification. By Proposition 10.7 the solution and every iterate share $\mathbb{R}$'s bridges. That is what makes the problem finite-dimensional in the endpoint variables, and it is also why changing the reference process changes the answer: the bridge is a bridge *of a reference*, and there is no reference-free Schrödinger problem.

Scope of Proposition 10.17 and Proposition 10.18

Assumed. Proposition 10.17: only that the two projections are well defined, that is that the constrained problems have finite value at every step. Proposition 10.18: finiteness of the value, and nothing more. The positivity condition on R_{0T} — bounded above and below on the product of the supports — belongs to the *rate* and not to the convergence, and the rate is Remark 10.19, a note rather than a claim.

Where the positivity condition fails. It fails for a reference whose endpoint coupling is nearly degenerate, which is exactly the small-noise regime in which (10.14) approaches unregularised optimal transport. The rate therefore degrades in the limit in which the problem becomes most interesting, and the Hilbert-metric coefficient quantifies the degradation.

What the structure buys. Monotonicity, for free, by (10.26); and the reduction to a Sinkhorn iteration on endpoint laws. It does not buy a rate, and it does not survive the replacement of the exact projections by learned approximations, which is what an implementation does.

What would falsify them. A half-step of (10.24) whose density is not a function of one endpoint alone; the proof is the chain rule under the disintegration by that endpoint.

10.5 What Can Be Compared

The constructions of this chapter and of Chapter 3 are often ranked by a single complexity figure. They cannot be. Table 11.4 gives the fields a complexity claim must fill; the following instantiates it and shows that the rows do not align.

Table 10.2: Three families, and why no common expression exists. Each column is a field of Table 11.4. A single bound covering all three rows would have to delete the last column.

Family	Oracle counted	Error metric	What is assumed about the learned object
Score-based generation	Evaluations of a learned score	KL($\cdot\|\cdot$) or TV at time t (Proposition 3.24)	A weighted L^2 score error against the true score, under the true p_t

continued overleaf

Table 10.2: (continued)

Family	Oracle counted	Error metric	What is assumed about the learned object
Bridge methods	Evaluations of a learned drift, times the number of outer iterations	$\mathrm{KL}(Q^{*} \,\|\, Q^{*})$, in that order (Proposition 10.18; the rate is Remark 10.19)	Nothing is assumed: the error of one outer iterate enters the next, and no analysis is available (Open Problem 11.4)
Path-space samplers	Evaluations of $\nabla \log \pi$ along simulated paths	Variance of an importance weight, or $\mathrm{KL}(\cdot \,\|\, \cdot)$ to the tilt	An approximate control; the weight's variance is the operative quantity and is not a norm on the control

What does not follow

No unified complexity bound is stated in this volume. The three rows count different oracles, measure error in different quantities, and make different assumptions about what is learned. A common expression in d and ε can be written only by suppressing the last column, which is where the families differ. Table 11.4 exists so that this is visible rather than absent.

An outer loop is not a constant factor. The second row's cost is the inner cost times the number of outer iterations, and by Remark 10.19 the number of outer iterations is controlled only under a positivity condition, and only in whichever divergence that note's source turns out to control. Reporting the inner cost alone, or treating the outer count as $O(1)$, removes the term that distinguishes the row.

10.6 Chapter Summary

Chapter 10 changes the principal object to a measure on trajectories. Proposition 10.2 is the single source of truth for the path-space relative entropy identity, with its four hypotheses stated separately — absolute continuity of the initial laws is not implied by the others — and its conclusion oriented: a mismatch of covariance characteristics where the first process goes gives $+\infty$ in that order, while mutual singularity needs a condition both measures see.

The chapter's contribution is to separate three constructions usually written as one. A global terminal tilt and a Doob transform share a drift and differ in their initial marginal; each is the Schrödinger bridge between the marginals it happens to produce, which is a prescription only if one accepts those; and the bridge for a prescribed pair needs two multipliers. Lemma 10.10 supplies the direction of the factorisation this volume actually uses and is proved here, while the converse remains quoted rather than proved (Remark 10.9).

Exercises

Tier 1 Direct application.

10.1 *Entropy production of a rotated Ornstein–Uhlenbeck process*

For the process of Proposition 4.14 in $d = 2$ with $\Sigma = I$ and $c = sJx$, compute $\int |c|^2 \, d\pi$

and hence (10.11) explicitly as a function of s. Confirm that it vanishes only at $s = 0$ and compare with the spectral improvement of (4.13).

10.2 *Tilting a Brownian motion*

Let $\mathbb{R}$ be the law of a Brownian motion on $[0,1]$ from δ_0, so $\nu_1 = \mathcal{N}(0,1)$. Compute the terminal law of the tilt by e^{-g} with $g(x) = \frac{1}{2}\lambda x^2$, and the h needed to reach $\mathcal{N}(0,\tau^2)$. Verify (10.21) in both cases and report the g that a naive reading would have used.

Tier II Completing the derivation.

10.3 *The h-transform drift, and the two normalisations*

Prove Proposition 10.13(iii) in full: show $h_t(X_t)$ is a $\mathbb{R}$-martingale, apply Theorem 10.1 on $\mathcal{F}_t$, and identify the drift change. Verify that for h the indicator of a set the construction is conditioning on the event $\{X_T \in A\}$, and compute h_0 in that case; say for which reference and which A the conditioning leaves ν_0 unchanged.

10.4 *Pythagoras for relative entropy*

Prove (10.26) directly from the chain rule, and state the general form: for a convex set $\mathcal{C}$ and the relative-entropy projection $\Pi_{\mathcal{C}}$, $\mathrm{KL}(\mathrm{Q}\,\|\,\mathrm{Q}') \geq \mathrm{KL}(\mathrm{Q}\,\|\,\Pi_{\mathcal{C}}\mathrm{Q}') + \mathrm{KL}(\Pi_{\mathcal{C}}\mathrm{Q}'\,\|\,\mathrm{Q}')$ for $\mathrm{Q} \in \mathcal{C}$. Identify where the equality in (10.26) comes from.

Tier III Probing the boundary.

10.5 *When the bridge is not a tilt*

Take $\mathbb{R}$ the law of a standard Brownian motion on $[0,1]$ from $\nu_0 = \mathcal{N}(0,1)$, and let $\mu_0 = \mathcal{N}(1,1)$, $\mu_T = \mathcal{N}(0,2)$. Show that no h-transform solves (10.12), compute the Schrödinger system in this Gaussian case explicitly, and report φ and $\hat{\varphi}$. Identify which hypothesis of Proposition 10.13(iv) fails, and compute h_0 for the tilt that reaches μ_T to see how far its initial law is from μ_0.

10.6 *Mutual singularity, and the orientation of the criterion*

Let Q^1 and Q^2 be the laws of Brownian motions with constant diffusion coefficients 1 and $1+\epsilon$.

(a) Exhibit a path functional separating them almost surely, conclude that they are *mutually* singular, and hence that both $\mathrm{KL}(\mathrm{Q}^1\,\|\,\mathrm{Q}^2)$ and $\mathrm{KL}(\mathrm{Q}^2\,\|\,\mathrm{Q}^1)$ are infinite for every $\epsilon \neq 0$. Identify where the argument uses that the coefficients are deterministic, so that the mismatch is seen by both measures.

(b) Now let $a^1 \equiv 1$ and let a^2 equal 1 on a region the first process reaches with probability one and 2 on a disjoint open set. Explain why Proposition 10.2 gives nothing about $\mathrm{KL}(\mathrm{Q}^1\,\|\,\mathrm{Q}^2)$ from the mismatch alone, and say which occupation measure the criterion is stated against.

Say what (a) implies for a generative construction whose reverse process is given a different noise schedule from its forward one.

10.7 *Where the outer loop is the cost*

Using (10.26) and a linear rate $\varrho < 1$ *assumed* in the form of Remark 10.19 — state which orientation of $\mathrm{KL}(\,\|\,)$ you are assuming it for — express the total number of drift evaluations needed to reach relative-entropy accuracy ε as a function of ϱ and the inner cost. Then say what happens as the reference process becomes more diffusive, and which entry of Table 10.2 you have been estimating.

Tier IV Computational.

10.8 *Three steerings, one reference*

Take $\mathbb{R}$ the law of an Ornstein–Uhlenbeck process on $[0,1]$ in $d = 1$ from $\nu_0 = \mathcal{N}(0,1)$, and $\mu_T = \mathcal{N}(2, \frac{1}{2})$.

(a) Simulate the tilt by e^{-g} with $g = -\log \mu_T$ and measure the terminal law. Compare with μ_T and with $\propto \mu_T \nu_T$, and report which one it matches.

(b) Repeat with $h = \mathrm{d}\mu_T / \mathrm{d}\nu_T$ and confirm the terminal law is μ_T to Monte-Carlo accuracy.

(c) Now impose $\mu_0 = \mathcal{N}(1,1)$ as well and run (10.24). Report the endpoint marginals after each half-step and the decrease of $\mathrm{KL}(Q^* \,\|\, Q^n)$.

(d) State which of (a)–(c) tests each row of Table 10.1, and which quantity in (c) is guaranteed monotone by (10.26).

Acceptance criteria: (a) identifies the matching law explicitly; (c) reports the monotone quantity.

CHAPTER 11

Continuous-Time Methods for AI: An Equivalence and Design Atlas

11.1 What this chapter is

The preceding ten chapters described learning, sampling, generation, control and collective behaviour as continuous-time probabilistic systems. That single sentence is the volume's claim to be a book rather than a survey, and this chapter is where it is either made precise or given up.

It is made precise by refusing to state it as a theorem. There is no single equation of which the Fokker–Planck equation, the reverse-time stochastic differential equation, the Hamilton–Jacobi–Bellman equation and the Lasry–Lions system are special cases, and a book that asserts otherwise buys its unity by making its strongest claims unfalsifiable. What there is instead is a map: a small number of distinct dynamical structures, a vocabulary for the relations between them, and — for each relation — the exact conditions under which it holds and the failure that occurs outside them. That fourth item is the contribution. Anyone can assert a connection; saying precisely how much weight it carries is the work.

This chapter therefore contains almost no new mathematics. Every claim in the Atlas of Section 11.5 is proved somewhere earlier, and where nothing is proved the row says so. What is new is the sorting.

11.2 Four structures, not one

Principle 11.1 (The three-layer reading) INTERPRETIVE. Every construction in this volume can be located by answering three questions in order. *What moves?* — a state, the law of a state, or a measure on whole trajectories. *In what geometry?* — which norm measures the velocity of that object. *Under what constraint?* — free evolution, a soft penalty, or a hard endpoint condition. Two constructions that answer all three the same way belong to the same *structural class*; two that differ in any answer are related by something weaker than identity, and that something must be named.

Membership of one class is not identity. Two reversible Langevin dynamics with different potentials answer all three questions identically — a law moves, in W_2, freely — and are different equations with different stationary measures and different rates. Deciding identity requires a fourth question, which the Atlas rows answer in their conditions column: *what fixes the vector field?* — which functional is descended, against which reference process, towards which target. Two constructions in the same class with the same answer to the fourth question are the same construction in different notation.

227

Principle 11.1 is interpretive: it organises the volume and is not a theorem. Since this is the final chapter, the usual requirement that nothing later depend on an interpretive claim is satisfied trivially; the substantive version of the requirement is that no *proof* anywhere in the volume depends on it, and that is enforced mechanically.

The first question already separates four structures that behave differently enough that no single variational principle covers them.

Table 11.1: The four dynamical structures of this volume. The last column is what goes wrong if a system in that row is described in the language of the first row, which is the most common failure in this subject and the one the volume was reorganised to prevent.

Structure	Canonical object	What decreases	Cost of calling it a W_2 gradient flow
Dissipative	Reversible diffusion with gradient drift; Fokker–Planck equation (Chapter 2)	A free energy, at a rate equal to its squared metric speed (Proposition 2.12)	Nothing; this is the row for which the description was proved
Hamiltonian / kinetic-hypocoercive	Hamiltonian and kinetic dynamics; underdamped Langevin; symplectic integrators (Section 4.2)	A free energy, but with a *degenerate* dissipation acting only in the velocity variable. The row is named for its conservative core: a purely Hamiltonian flow dissipates nothing, while underdamped Langevin adds friction and noise in the velocity alone and so is neither conservative nor uniformly dissipative	Grönwall does not close (Proposition 4.8); a rate obtained by pretending it does is not a theorem, and hypocoercivity is required
Controlled	Value functions, path measures, Girsanov densities (Section 7.1, Section 10.1)	Nothing on the state space; the variational object lives on path space	The optimum is characterised by a backward equation, not by descent; a forward descent argument answers a different question
Strategic	Forward Fokker–Planck coupled to backward Hamilton–Jacobi–Bellman (Proposition 9.4)	Nothing in general; a potential exists only for special classes	Equilibrium is not optimum (Proposition 9.4); descent on a single functional finds the wrong fixed point unless the game is potential (Proposition 9.5)

Two remarks about the table are worth more than the table.

First, the dissipative row is not defined by "the free energy decreases". That is far weaker than steepest descent, and Proposition 2.16 exhibits a family of diffusions that dissipates the free energy at *exactly* the same instantaneous rate as the W_2 gradient flow, travels strictly faster than it dissipates, and is not a W_2 gradient flow of it. Two different statements are easy to conflate here. The scalar dissipation identity $\frac{d}{dt}\mathcal{F} = -\mathcal{I}(\rho_t \,\|\, \pi)$ holds for the non-reversible flow too, and therefore distinguishes nothing. The *metric* energy dissipation equality (2.21) — in which the functional's decrease is split into the metric speed and the metric slope, each appearing squared — is what fails: the non-reversible curve's speed $|\rho'|$ strictly exceeds its slope, so (2.21) cannot hold with equality even though the scalar identity does. The distinguishing property is the metric equality, or equivalently the evolution variational inequality, and not monotonicity, and not the scalar identity either.

Second, the rows are not a partition of algorithms; they are a partition of *descriptions*. Stein variational gradient descent is in the first row under one norm and nowhere at all under another (Table 5.1); an underdamped sampler is conservative in its velocity variable and dissipative in the same variable at once. The question "which row is this algorithm in?" is badly posed. The question "which row is this *claim* in?" always has an answer.

11.3 The six type errors

Reviewing the volume's own corrections produced a short list. Almost every error that survived a first careful reading was one of six substitutions, and all six are invisible to proofreading because both sides of the substitution are true statements about objects with the same units and often the same symbols. The list is the volume's most transferable output; it is also the rule set against which the manuscript is checked.

Table 11.2: The six substitutions. Each row names the question that detects it and a place in this volume where the distinction is load-bearing — in every case a place where the two sides look identical on the page.

Substitution	The question that detects it	Where it is load-bearing here
Pathwise vs. in law	Is this a statement about sample trajectories or about the law of the process? Girsanov relates measures; time reversal is a statement in law	Proposition 1.35, where only the identity in expectation fails; Proposition 1.44, where the mean and almost every path move in opposite directions; Theorem 1.46, where the Tanaka equation (1.43) has uniqueness in law and no pathwise uniqueness — a discontinuity in σ, not in the drift
Metric substitution	Is the metric in the proof the metric in the statement? No free conversion among KL(‖), TV, W_2 and Fisher-type quantities	Corollary 2.14, where the W_2 rate is half the entropy rate; Proposition 5.14; and Corollary 8.5, where one rate is dimension free and the other is not

continued overleaf

Table 11.2: (continued)

Substitution	The question that detects it	Where it is load-bearing here
Regime substitution	Which of $\eta \to 0$, small η, fixed η; $N \to \infty$ or finite N; finite time or uniform in time?	Table 6.1 and Table 6.2; Proposition 6.4, where the discrepancy diverges at the threshold; Proposition 5.26, where the flow error and the sampling error carry different rates
Geometry substitution	A gradient flow in which metric? The phrase is undefined without one	Table 5.1: the same functional descended in two geometries gives two different flows with two different dissipation rates
Constraint substitution	Free evolution, soft terminal reward, or hard endpoint marginal? These are three different optimisation problems with three different solutions	Proposition 10.13, where a tilt reaches $\propto e^{-f}\nu_T$ and moves ν_0 as well, while a constraint reaches μ_T with μ_0 held; Remark 7.8; Proposition 3.23 and its mini-ledger
Normalisation	Itô or Stratonovich; $\frac{1}{2}g^2$ or g^2; which end of $[0,1]$ carries the data; which log-Sobolev convention; sum or average loss	The convention boxes of Section 1.5.4, Section 2.2.1 and Section 3.4.1, each of which fixes one such choice once for the whole volume

11.4 The vocabulary of relations

A row of the Atlas asserts a relation between two named objects. Five types are used, and only five.

- EXACT An identity. The two sides are equal as mathematical objects under the stated conditions, with no error term and no limit. The word *is* may be used.

- LIMIT One side is the limit of the other along a stated parameter, in a stated mode of convergence, with a stated order where one is known. The word *is* may not be used; *admits as a limit* may.

- VARIATIONAL Both sides solve optimisation problems whose solutions coincide, though the problems themselves differ. The correspondence is between minimisers, not between objectives, unless the objectives are also shown to coincide.

- REPRESENTATION One side can be written in the form of the other after a change of object — a reference process, a coupling, a parameterisation — which must be named in the conditions column.

- ANALOGY A structural resemblance with pedagogical value and no theorem behind it. Nothing may be derived from a row of this type, and the build fails if a proof cites one.

Convention

Five rules govern the Atlas, and they are what keep it from becoming decoration.

(i) Every row has four fields: the relation, its type, the exact conditions, and the failure outside them. A row whose fourth field cannot be written is a row whose relation has not been understood, and it is not admitted.

(ii) The type is not a hedge. A relation is EXACT or it is not; that it is exact only under conditions is what the conditions field is for. A sixth type meaning *exact under conditions* would duplicate the conditions field in the type field, and duplicated information is information that will eventually disagree with itself; the same reasoning excludes an eighth status tag.

(iii) Withdrawn relations stay in the table. A claim that this volume argues against is entered with the badge WITHDRAWN rather than omitted, so that a reader who arrives holding it finds it addressed rather than absent. Each such row is also carried on the list of claims checked for staleness.

(iv) The type never appears in a claim heading. Status tags are properties of claims; relation types are properties of relations. Rendering one where the other belongs is how a two-axis system collapses into a one-axis system with too many labels.

(v) A row may not be more certain than the statement it cites. If the cited claim has an open attribution, or one of its clauses has, the row carries the same marker and its conditions field says which clause. The Atlas is a derived artefact: it summarises the theorems and never strengthens them, and a row that has drifted ahead of its source is a defect of the same kind as a stale cross-reference. Remark 10.9 is the one result quoted rather than proved on which any *row* depends. Relations whose *type* itself is undecided are held outside the canonical table, in Section 11.8. Row A26 does not inherit the pending factorisation clause, because its bridge identification uses only the sufficiency direction proved in Lemma 10.10.

11.5 The Atlas

Rows are grouped by the pair of areas they connect. Cross-references point to the statement that establishes the row, or, for ANALOGY and WITHDRAWN rows, to the discussion that limits it.

Table 11.3: The Equivalence Atlas. Column one carries the relation and its type; column two the conditions under which it holds exactly; column three the failure outside them. The third column is the reason the table is worth reading.

Relation	Exact conditions	What goes wrong outside them
Diffusions, geometry, and rates		
A1. Fokker–Planck flow of a reversible diffusion = W_2 gradient flow of the free energy EXACT Theorem 2.13	Gradient drift $-\nabla V$; constant isotropic diffusion; the functional (2.12); $\nabla^2 V \succeq -\kappa I$; $\mathcal{F}[\rho_0] < \infty$; uniqueness of the JKO step only for $\tau < 1/\kappa$, and proved along generalized geodesics based at the previous iterate	For an admissible non-reversible b the dissipation identity survives unchanged and the gradient structure does not (Proposition 2.16); kinetic dynamics loses both (Proposition 4.8)
A2. Free energy decreases = the flow is steepest descent WITHDRAWN Proposition 2.16	—	Monotonicity is strictly weaker, and so is the scalar dissipation identity, which the non-reversible flow also satisfies. The distinguishing property is the metric energy dissipation equality (2.21) *for the functional in question*, which fails here because the curve's speed exceeds its slope for $\mathcal{F}$. Whether another functional would serve is open (Open Problem 11.2)
A3. π stationary = detailed balance = $\rho_t = \pi$ WITHDRAWN Proposition 2.6	—	Three conditions on different objects: $\nabla \cdot j[\pi] = 0$, $j[\pi] = 0$, and a property of the initial datum. In the reversible case $j[\rho_t] = 0$ exactly when $\rho_t = \pi$; recovering ρ_0 from ρ_t needs injectivity of P_t^{π}, not irreducibility (Remark 2.7)
A4. Reversible = gradient drift EXACT Proposition 2.5	Constant isotropic a; $V \in C^2$ with e^{-V} integrable; decay enough for two integrations by parts	With state-dependent a the criterion is $j[\pi] = 0$, and "gradient drift" is no longer coordinate free

continued overleaf

Table 11.3: (continued)

Relation	Exact conditions	What goes wrong outside them
A5. Spectral gap = a lower bound for any sampling algorithm WITHDRAWN Proposition 4.17	—	The gap belongs to one semigroup. Two generators with the same π have gaps differing by a factor κ; what exists is Proposition 4.18, whose class restriction is in its statement
A6. SVGD mean-field velocity = the pointwise steepest-descent direction of $\mathrm{KL}(\cdot \,\Vert\, \pi)$ in the Stein tangent norm VARIATIONAL Theorem 5.10	The mean-field equation; bounded $C^{1,1}$ kernel; the Stein norm (5.17) and no other. This is a tangent-level statement at a fixed law, and it is proved; the curve-level assertion is U1	With the $L^2(\rho)$ norm one gets Langevin instead; at finite particle number or finite step the identification is uncontrolled (Proposition 5.22)
A8. Variational inference = W_2 gradient flow of $\mathrm{KL}(\cdot \,\Vert\, \pi)$ WITHDRAWN Proposition 5.3	—	The unconstrained flow leaves the family at once. Three flows exist, of which only the constrained one is a W_2 gradient flow and only two are reparameterisation invariant
Generative dynamics		
A9. Marginal velocity field = affine image of the score EXACT Proposition 3.17	Gaussian conditional path; $\sigma_t > 0$ and snr_t strictly monotone at t, equivalently $g_t \neq 0$ (Proposition 3.8)	No affine dictionary for general interpolants (Proposition 3.28), bridges, manifold-valued or discrete-state paths; none where snr_t is stationary
A10. Conditional flow matching objective = denoising score matching objective EXACT Theorem 3.20	Parameterisations tied by (3.28) and weights related by $\tilde{w} = c_t^2 w$ — both — and the time-integrated square-integrability (3.27), or a truncation away from the endpoints	Equality is *not guaranteed*: there exist restricted classes for which the objectives and their minimisers differ, and Example 3.21 is one explicit witness, computing a factor of 3.6. Degenerate classes may still agree by coincidence, so no universal separation is claimed. Without (3.27) both objectives can be $+\infty$ for every θ and the identity is vacuous

continued overleaf

Table 11.3: (continued)

Relation	Exact conditions	What goes wrong outside them
A11. Reverse-time SDE and probability-flow ODE have identical one-time marginals under the exact score EXACT Proposition 3.4	Exact score; the two are $\lambda = 1$ and $\lambda = 0$ of one family, all of whose members share the time marginals	With an approximate score the marginals differ, and differently in λ; couplings and error propagation differ for every λ. The exact-likelihood identity along the $\lambda = 0$ member holds on $[\delta, T]$; carrying it to the data endpoint needs Corollary 3.6 and fails outright for a data law with no density
A12. Diffusion model = infinitely deep variational autoencoder WITHDRAWN Section 3.3.2	—	A hierarchical latent-variable correspondence with the weight $\frac{1}{2}g_t^2$ and a terminal term; the encoder is fixed here and learned there
A13. Score estimation is minimax optimal = the sampler is optimal WITHDRAWN Table 3.1	—	Optimality statements concern the estimation of $\nabla \log p_t$; the sampler needs all three rows of Table 3.1 simultaneously, and their hypotheses are in tension
A14. Discrete state spaces have no time reversal WITHDRAWN Proposition 3.30	—	The reversal is exact and elementary for a finite chain with positive marginals; what continuous time supplies is a differential formalism, not the reversal
Optimisation and control		
A15. Discrete gradient descent = its continuous-time limit LIMIT Table 6.1	$\eta \to 0$ for the flow; for fixed small η, weak approximation of a stated order over a bounded horizon, with ηS bounded away from 2	At $\eta S \geq 2$ there is no valid order estimate, and the stationary discrepancy diverges there (Proposition 6.4)
A16. Learning rate = temperature EXACT under a condition, Proposition 6.5	Constant isotropic noise covariance; then $\beta_{\text{eff}} = 2B/(\eta\sigma_1^2)$ for the modified equation	A *reversible* Gibbs stationary law exists if and only if Σ commutes with $\nabla^2 L$ (Proposition 6.6); otherwise the stationary law circulates, and that it is also not of the form $e^{-\beta L}$ is a separate computation, done for the Gaussian case in Example 6.7

continued overleaf

Table 11.3: **(continued)**

Relation	**Exact conditions**	**What goes wrong outside them**
A17. Large step size = flat minima = better generalisation ANALOGY Conjecture 6.14	None. The middle term is not reparameterisation invariant (Proposition 6.13)	Sharpness at a fixed function can be made arbitrarily large; any conclusion from its value is a conclusion about coordinates
A18a. Static KL-regularised policy optimisation = Gibbs variational tilting EXACT Proposition 7.4	An arbitrary measurable space, a reference distribution and a reward; no dynamics, no time index, no stochastic calculus	The identity is about one optimisation over one measure. It says nothing about a sequential decision process, and it is not a statement in discrete time — it has no time
A18b. Entropy-regularised path-space control = a tilt of a reference process EXACT Proposition 7.5	A deterministic initial state, a control entering through the noise channel, a quadratic cost with identity weight; the hypotheses of Proposition 10.3	A different construction from A18a, on a different object: a measure on trajectories rather than a measure on outcomes
A18c. Sequential discrete-time KL control = continuous-time entropy-regularised control ANALOGY	*None available.* A LIMIT or REPRESENTATION reading would require a scaling theorem relating a sequential decision process to a diffusion, with its step-size regime and its convergence mode	No such theorem is stated in this volume. Writing an autoregressive process as a diffusion is a modelling step whose hypotheses neither A18a nor A18b supplies
A19. Path-integral control = a linear equation EXACT Proposition 7.12	The matrix condition $\lambda B R^{-1} B^{\top} = \sigma\sigma^{\top}$, equivalently $B R^{-1} B^{\top} = a/\lambda$, at every point, with a single scalar λ	Under-actuated systems are excluded; R is not free; and no direction may be scaled independently. A scalar statement of the condition loses all three
A20. Continuous-time reinforcement learning = small-step reinforcement learning WITHDRAWN Proposition 7.11	—	The per-transition signal-to-noise ratio is $O(\sqrt{\Delta})$ and vanishes; the aggregate over a fixed horizon does not. The two readings fail in opposite directions
Interacting systems		

continued overleaf

Table 11.3: **(continued)**

Relation	**Exact conditions**	**What goes wrong outside them**
A21. N-particle system = its mean-field limit LIMIT Proposition 8.2	A named metric and a named horizon; the tagged-particle coupling rate is $N^{-1/2}$ with a dimension-free constant	The empirical measure converges at $N^{-1/d}$ for $d > 4$ over the moment class (8.7) with $q > 4$, and by Corollary 8.5 that term dominates. The weak-error rate is not established here (Remark 8.6)
A22. Convexity of the mean-field risk = global convergence of the flow WITHDRAWN Proposition 8.8	—	Flat convexity holds always and is not displacement convexity, which is what Theorem 2.13 and Table 2.2 use. The repair is Proposition 8.15, under a uniform inequality
A23. Mean-field control = mean-field game WITHDRAWN Proposition 9.4	—	The two systems differ by the planner's term (9.4), which is non-zero in general; it vanishes with the x-dependence of f, and is invisible to (9.5) when it is x-independent, the additive gauge of a first variation. No converse is claimed
A24. Mean-field game = the optimality system of a control problem VARIATIONAL Proposition 9.5	The symmetry (9.8) of the second variation of the coupling, on an affine convex class of measures on which f is flat-C^2 — not on all of $\mathcal{P}_2(\mathbb{R}^d)$, where the construction is unavailable	An antisymmetric interaction admits no potential at all (Example 9.6); and the potential problem's running cost is Φ, not $\int f\,dm$ — those problems, not two
Path space		

continued overleaf

Table 11.3: **(continued)**

Relation	**Exact conditions**	**What goes wrong outside them**
A25. Two diffusions with a common covariance characteristic = $\mathbb{Q}^1 \ll \mathbb{Q}^2$ with an exact path-space KL(‖) identity EXACT Proposition 10.2	A common *covariance characteristic* $a = \sigma\sigma^{\top}$; an invertible σ; $\nu^1 \ll \nu^2$; a true-martingale criterion such as Novikov's, imposed under $\mathbb{Q}^2$ conditionally on X_0; square-integrable drift difference under $\mathbb{Q}^1$ — four hypotheses, not one. The conclusion is *one-sided*: mutual absolute continuity needs the same four in the reverse direction, starting with $\nu^2 \ll \nu^1$	If the covariance characteristics differ where the *first* process goes, KL(‖) in that order is $+\infty$, not large — a one-sided criterion; mutual singularity needs a mismatch both measures see. Changing the square root of a changes nothing
A26. Score diffusion = a Schrödinger bridge REPRESENTATION Proposition 10.15	The generation prior equals the reference's terminal law, $\rho = p_T$ — sufficient always, necessary under injectivity of the endpoint transition operator. The bridge identification in part (iii) uses only the sufficiency direction (Lemma 10.10)	Score-based generation is a one-endpoint construction; by Proposition 3.10 the mismatch is positive for every finite T unless p_{data} is already the invariant law
A27. Terminal reward = terminal marginal constraint WITHDRAWN Proposition 10.13	—	A tilt by e^{-g} produces $\propto e^{-g}\nu_T$ at one end and $\propto h_0\nu_0$ at the other. To reach μ_T one takes $h \propto \mathrm{d}\mu_T/\mathrm{d}\nu_T$, which still moves the initial law; and no probability density on $\mathbb{R}^d$ is constant
A28. Feynman–Kac tilting = Doob h-transform = Schrödinger bridge WITHDRAWN Proposition 10.13	—	The first two share a drift and differ in their initial marginal, agreeing *as path measures* only when $h_0 = \mathbb{E}_{\mathbb{R}}[h(X_T) \mid X_0]$ is constant — their *terminal* laws agree under the weaker condition $\mathbb{E}_{\mathbb{R}}[h_0(X_0)^{-1} \mid X_T] = 1$. Each is the bridge between its *own* marginals, and the bridge for a prescribed pair needs two multipliers. Three statements with three conditions, not one triple identity

continued overleaf

Table 11.3: **(continued)**

Relation	**Exact conditions**	**What goes wrong outside them**
A30. Path-space distance to the time reversal = entropy production: $\mathrm{KL}(Q \,\|\, Q^-) = T \int \lvert c\rvert^2 \, d\tau$, zero iff reversible EXACT Proposition 10.5	A *stationary* diffusion; well-posedness of both equations; the hypotheses of Proposition 10.2 for the ordered pair (Q, Q^-), including a true-martingale criterion and $\int \lvert c\rvert^2 \, d\tau < \infty$	For a time-inhomogeneous process none of it applies, and a fluctuation identity is not a derivation of Theorem 3.1. The thermodynamic reading of the identity as an "arrow of time" is an interpretation, not part of the equality
A31. Three families of method = one complexity bound WITHDRAWN Table 10.2	—	They count different oracles, in different metrics, under different assumptions about what is learned. A common expression exists only by deleting the column in which they differ

The table carries thirty-two numbered rows — counting A18a, A18b and A18c as three, since they are three relations between three different pairs of objects — of which thirteen are withdrawals. Three further relations are held outside the table in Section 11.8 and are not counted here, because a relation whose type is undecided is not yet a row. Identifiers are stable: A29 is absent because that relation is held in Section 11.8 as U2, and the remaining rows are not renumbered, so that a citation to a row remains unambiguous across formats. A row is admitted only where the third column can be written from something proved; a connection that cannot yet be limited belongs in the prose of its own chapter, where it is visibly local, and not here, where it acquires the authority of the table.

11.6 Complexity: an accounting template, not a formula

A recurring temptation is to close a book like this one with a single complexity statement — a bound of the form $\tilde{O}(d/\varepsilon^2)$ said to cover diffusion sampling, bridge methods and policy optimisation at once. No such statement is available, and the reason is not that the constants differ. The three families count different oracles, measure error in different metrics, and depend on hypotheses that are not comparable, so a common expression can be written only by suppressing the terms in which they differ, which is where their difference lives.

What can be written is a template: a fixed set of fields to be filled for each method, so that two methods can be compared field by field and the fields that do not compare are visible rather than absent.

Table 11.4: The accounting template. A complexity claim in this volume fills every field or states that a field is not applicable. The last column is the one a single unified bound would have to delete.

Field	What it must state	Why it blocks a common formula
Oracle counted	Gradient of $\log \pi$; a learned score evaluation; a forward pass of a sequence model; a solve of a proximal subproblem	These have different costs and are not interconvertible; a "function evaluation" means four different things
Error metric	TV, W_2, $\mathrm{KL}(\,\|\,)$, or an expectation of a test class	Rates in different metrics are not comparable without a conversion, and the conversions are lossy and dimension-dependent
Hypotheses that fix the rate	Log-concavity or an isoperimetric constant; smoothness and its constant; condition number; manifold or intrinsic dimension; warm start	Two methods analysed under different hypotheses have not been compared, however similar the exponents look
Approximation error of a learned object	Whether the analysis assumes an exact score, a field with L^2 error $\varepsilon_{\text{score}}$, or a trained network	The first two give bounds of different content, and only the third is a statement about a system anyone runs
Discretisation	Step size, solver order, and whether the bias vanishes with the number of steps	A method with no asymptotic bias and one with an $O(\eta)$ bias are not on the same axis
Regime of validity	Which row of Table 6.1, and finite or infinite particle number	A bound proved in one regime is not a bound in another, and this is the field most often left blank

A common exponent is not a comparison. Two methods reported as $\tilde{O}(d/\varepsilon^2)$ may be counting different oracles, in different metrics, under different smoothness hypotheses, one with an exact score and one with a learned one. The template above exists so that this is visible on the page.

Dimension-free is a statement about a constant, not about a bound. A bound in which d does not appear explicitly is dimension-free only if every constant in it is also dimension-free — including the isoperimetric constant, the initial divergence, and the approximation error of any learned object. Corollary 2.14 carries no explicit d and is not dimension-free, and the notes there say so.

The volume proves no end-to-end bound for any deployed system. Every rate here is for an idealisation identified in the row that carries it. This is a limitation of the field and is recorded as such in the boundary statement below.

11.7 Choosing a continuous-time language

The Atlas is retrospective. The following procedure is the prospective form of the same content: given a method that this volume does not discuss, it determines which chapter's language applies and which claims about it would need an Atlas row. It is a checklist, not an algorithm, and its value is that each question has a wrong answer that is cheap to detect.

Q1. What is the moving object? A single trajectory, the law of a process, or a measure on whole paths. If two claims about the method are about different objects, they are not the same claim, however similar their notation. This is the first row of Table 11.2.

Q2. Is there a fixed target law? If yes, the method is sampling or generative and Chapter 2, Chapter 3 or Chapter 5 apply. If the target is defined by an optimality criterion, the method is control Chapter 7. If it is defined by what other agents do, it is a game (Chapter 9). These are the second, third and fourth rows of Table 11.1.

Q3. Is the drift a gradient, and is the diffusion constant and isotropic? If both, a free energy exists and the W_2 gradient-flow language is available with everything Theorem 2.13 supplies. If either fails, the canonical Theorem 2.13 identification — *this* free energy, *this* metric — is no longer available, and using it anyway is the error Proposition 2.16 was written to exhibit. Whether some other functional or some other geometry would serve is a separate question, open in general (Open Problem 11.2), and a negative answer for one pair is not a negative answer for all.

Q4. Which norm measures a velocity? $L^2(\rho)$, a reproducing kernel Hilbert space, a path-space relative entropy. Naming it fixes the geometry, and hence fixes the dissipation rate — which is then a derived quantity and not a design choice. Table 5.1 is the worked case.

Q5. Are the endpoints free, softly penalised, or constrained? Free gives time reversal; a soft terminal reward gives Feynman–Kac tilting; a hard marginal gives a bridge. Rows A26–A28 of Table 11.3 record what happens when these are merged.

Q6. Which regime is each claim in? Continuous time, small step, fixed step; finite or infinite particle number; finite horizon or uniform in time. Write the answer next to the claim, not in a footnote.

Operational reading

A worked application. Suppose a method fine-tunes a pretrained flow model against a reward, few-step, and is described in its documentation as "solving a Schrödinger bridge". Q1: the moving object is a measure on paths, since the objective compares the fine-tuned process with the pretrained one. Q2: the target is defined by an optimality criterion, so this is control. Q3: not applicable — there is no free energy on the state space. Q4: the natural norm is a path-space relative entropy, so the dissipation language of Chapter 2 does not transfer. Q5: the reward is a soft terminal penalty, not a marginal constraint. That last answer settles it: by row A27 the method is a Feynman–Kac tilting of the pretrained process, and calling it a bridge is a REPRESENTATION claim that requires a hard endpoint the method does not impose. The correct exact statement is the KL-regularised control one, and it is available.

That is one step beyond the Atlas, and this box stops there. Whether the method works, and how its cost compares with alternatives, is Table 11.4, filled field by field, and not a matter this chapter can settle.

11.8 Unresolved relations

Three relations that would otherwise belong to Table 11.3 are held here instead, because for each of them the *strongest justified relation type* is unsettled. The type shown is the one currently warranted; establishing what the second column asks for would promote it, and that promotion is exactly what a canonical row asserts. Each entry says what would have to be shown, and each is a live question rather than a gap in the exposition.

Relation	What would have to be true	What is undecided
U1. SVGD mean-field flow = gradient flow of $\mathrm{KL}(\cdot \,\|\, \pi)$ in the Stein geometry, as a metric statement ANALOGY Interpretive Reading 5.11	Everything in row A6, plus a metric-space structure for $\mathcal{g}^S$ in which the metric derivative is the one used — completeness or a restricted domain, and the identification of the metric derivative. None of this is constructed in this volume, so the reading is carried as INTERPRETIVE	A source establishing the distance, the metric derivative and the domain would promote this to VARIATIONAL and return it to the canonical table; until then nothing may be derived from it (Remark 5.13)
U3. SVGD mean-field flow = the W_2 gradient flow of $\mathrm{KSD}^2(\cdot \,\|\, \pi)$ ANALOGY Section 5.2.4, Open Problem 5.17	A comparison of the $L^2(\rho)$-tangential velocity $\Pi_\rho \phi^*_\rho$ with the Wasserstein gradient of the first variation of KSD^2, at some law where the two can be computed	Proposition 5.15 shows only that a *raw* representative of the velocity need not be a gradient; since only the tangential part moves the measure, that does not refute a curve-level representation. Neither asserted nor refuted here
U2. Iterative proportional fitting = a Picard iteration with a linear rate REPRESENTATION Proposition 10.17, Remark 10.19	The fixed-point form of the Schrödinger system, together with a contraction estimate under the positivity condition	In which orientation of $\mathrm{KL}(\,\|\,)$ the linear rate holds, or whether the contraction is stated for a Hilbert projective metric on the potentials rather than for a divergence between path measures — a third object again (Remark 10.20)

The *representation* half of U2 — that the iteration has the Picard form — is not in doubt and is Proposition 10.17; what is held here is the rate. Similarly, row A6 stands on its own: what is held here is the metric-space upgrade, not the variational statement.

11.9 Open problems

Open Problem 11.2 (A functional for the non-reversible flow) OPEN PROBLEM. Proposition 2.16 shows that the flow (2.25) is not the W_2 gradient flow of $\mathcal{F}$, by exhibiting a curve whose metric speed strictly exceeds its metric slope for that functional. It does not address the weaker question: is there a functional $\mathcal{G}$, or a metric tensor other than W_2, for which (2.25) is a gradient flow? A negative answer would need an invariant of the *curve of laws* that no choice of functional and metric can supply. The non-vanishing stationary probability current is *not* such an invariant: it records the failure of detailed balance at the level of paths, while at $\rho = \pi$ the law itself is stationary, exactly as a gradient flow at a critical point would be. A plausible route instead is to ask whether the linearisation of (2.25) about π can be symmetrised by any inner product on the tangent space — a gradient flow in some metric has self-adjoint linearisation with respect to that metric, and the rotational part obstructs this in the Gaussian computation of Proposition 4.14. A positive answer for some class of admissible b would change how Table 2.1 should be read. Neither is attempted here, and no claim in this volume depends on the answer.

Open Problem 11.3 (A geometry containing all four structures) OPEN PROBLEM. Table 11.1 separates four structures by exhibiting, for each of rows two to four, a property of row one that fails. Is there a single geometry — in the sense of a metric tensor on a space of measures, or a bracket in the style of the conservative–dissipative decompositions of non-equilibrium thermodynamics — in which all four are gradient or gradient-plus-conservative flows, and in which the failures above appear as degeneracies rather than as different categories? A positive answer would replace this table by a theorem. A negative answer would presumably identify an invariant separating the strategic row from the rest. The heuristic reason to expect one is that an equilibrium is not an optimum; but that is a heuristic, and no theorem here excludes a variational representation of the strategic row in a geometry not yet considered.

Open Problem 11.4 (End-to-end error for two-endpoint generative dynamics) OPEN PROBLEM. For score-based generation there are results bounding the error of the sampled law in terms of the L^2 error of a learned score, the terminal mismatch and the solver order (Table 3.1). No comparable statement is available for methods with a hard endpoint constraint, where the learned objects appear inside an iterative projection and the error of one iterate propagates into the target of the next. State and prove the analogue of the three-term decomposition for iterative proportional fitting, or identify the obstruction.

11.10 Chapter Summary

Chapter 11 states no theorem. It is an audit of the volume: four structures that no single variational principle covers (Table 11.1); six substitutions by which a cross-domain claim fails — geometry, regime, normalisation, metric, constraint, and pathwise versus in law (Table 11.2); five relation types and the disposition WITHDRAWN; and Table 11.3, which assigns one of them to each of thirty-two relations and states, in its last column, what goes wrong outside the conditions. Thirteen of those rows are withdrawals, kept rather than deleted so that a reader arriving with the belief finds it addressed.

Three relations are held outside the canonical table in Section 11.8, because for each the strongest justified relation type is unsettled. The chapter's own rule is that a row may never be more certain than the statement it cites; where this volume has had to choose between a clean table and an accurate one, it has chosen the second.

Exercises

Tier I Direct application.

11.1 *Typing five sentences*

For each of the following, state which row of Table 11.1 it belongs to, which relation type of Section 11.4 it asserts, and whether it is admissible as written: (a) "Langevin dynamics minimises the free energy"; (b) "diffusion models are Schrödinger bridges"; (c) "SGD is a stochastic differential equation"; (d) "RLHF is entropy-regularised control"; (e) "mean-field games are optimal transport problems". For each inadmissible one, write the shortest admissible replacement.

Tier II Completing the derivation.

11.2 *Writing a row*

Choose any cross-domain claim from a recent paper in generative modelling. Fill in the four fields of an Atlas row for it. If you cannot write the fourth field, say what you would have to know in order to write it, and whether that is a mathematical question or a matter of reading the paper more carefully.

11.3 *Filling the template*

Take two sampling methods analysed in different papers and fill Table 11.4 for each. Identify every field in which they are not comparable. Then state what a fair comparison would require, and estimate how much of it is available in the two papers.

Tier III Probing the boundary.

11.4 *A fifth relation type?*

The five types of Section 11.4 were chosen so that each excludes the others. Propose a cross-domain claim in this volume that fits none of them. Then either place it in one of the five by sharpening its statement, or argue that a sixth type is needed — and, if so, say what the conventions box of Section 11.4 would have to be changed to, and what would be duplicated.

11.5 *The strategic row*

Open Problem 11.3 asks whether a single geometry contains all four structures. Give the strongest argument you can that the strategic row cannot be included — for instance by exhibiting a two-player mean-field game whose equilibrium is not the minimiser of any functional of the population measure — and identify precisely which step of your argument would fail for a potential game.

11.6 *Auditing a withdrawn row*

Row A28 splits a composite claim — Feynman–Kac tilting = Doob h-transform = Schrödinger bridge — into three. Reconstruct the three separate statements, decide the relation type of each of the three pairs among them, and write the conditions under which each pair coincides. Then explain why the composite, which is false, is more attractive than any of the three, which are true.

Tier IV Computational.

11.7 *Testing a row you did not write*

Pick one EXACT row of Table 11.3 and design a numerical experiment that would detect a violation of it if the conditions column were wrong. Run it in a setting where the conditions hold and confirm agreement to the precision of your solver; then deliberately violate one condition and measure the size of the resulting discrepancy as that condition is relaxed continuously. Report both, and state which of the two runs constitutes evidence about the row.

Notation

Symbols are listed with the convention this volume fixes for them. Where a symbol carries more than one normalisation in the literature, the entry says which one is used here; those are the entries a reader arriving from another source should check first.

Symbol	Meaning, and the convention fixed here
$\mathcal{P}_2(\mathbb{R}^d)$, $\mathcal{P}_2(\mathbb{R}^q)$	Probability measures on $\mathbb{R}^d$, resp. $\mathbb{R}^q$, with finite second moment. Parameter measures in Chapter 8 live on $\mathbb{R}^q$ (Section 8.2).
W_2	Quadratic Wasserstein distance; its Benamou–Brenier form and the tangent space of $\mathcal{P}_2(\mathbb{R}^d)$ are Theorem 2.10.
$\mathrm{KL}(\mu \,\|\, \nu)$	Relative entropy of μ with respect to ν. *Always an ordered pair*; the two orientations are never interchanged, and where a result needs one specifically the statement says which (Remark 10.20).
$\mathcal{I}(\mu \,\|\, \pi)$	Relative Fisher information. Not the Fisher information *matrix* of statistics, which does not appear in this volume.
$\mathcal{E}(f, g)$	Dirichlet form, $-\int f\,\mathcal{L}g\,\mathrm{d}\pi$ (Definition 2.18). Not symmetric for a non-reversible generator; the *energy* $\mathcal{E}(g, g)$ is what the rate estimates use, and Corollary 2.26 concerns the diagonal only.
Γ	Carré du champ (Definition 2.18).
$\mathcal{L}$, $\mathcal{L}^c$	Generator; the superscript denotes an admissible non-reversible perturbation (Definition 2.15).
$\mathrm{LSI}(\alpha)$	Log-Sobolev inequality in the normalisation (2.14), $\mathcal{I}(\cdot \,\|\, \pi) \ge 2\alpha\mathrm{KL}(\cdot \,\|\, \pi)$, giving entropy decay $e^{-2\alpha t}$. A source using $\mathrm{KL}(\|) \le \alpha^{-1}\mathcal{I}(\|)$ shifts every exponent by a factor of two.
$\mathrm{P}(\lambda)$, $\mathrm{T}_2(\beta)$	Poincaré and Talagrand inequalities (Definition 2.24). T_2 is a converter between metrics, not a decay mechanism (Table 2.2).
$a = \sigma\sigma^\top$	Covariance characteristic. The load-bearing object: two square roots of the same a give the same law (Proposition 10.2).
$\mathbb{R}$, $\mathbb{Q}$	Reference and controlled/tilted path measures on $C([0, T]; \mathbb{R}^d)$.
$j[\rho]$	Probability flux, $b\rho - \frac{1}{2}\nabla\cdot(a\rho)$ (2.6).
μ^N, $\bar{\mu}^N$	Empirical measure of an interacting system, resp. of independent copies. The distinction carries the rates of Table 8.1.
s_π	Score $\nabla \log \pi$.
snr_t, c_t	Signal-to-noise ratio of a Gaussian path and its logarithmic derivative (3.28).

Bibliography

[1] L. Ambrosio, N. Gigli and G. Savaré. *Gradient Flows in Metric Spaces and in the Space of Probability Measures*, 2nd ed. Lectures in Mathematics ETH Zürich, Birkhäuser, 2008.

[2] B. D. O. Anderson. Reverse-time diffusion equation models. *Stochastic Processes and their Applications*, 12(3):313–326, 1982.

[3] D. Bakry, I. Gentil and M. Ledoux. *Analysis and Geometry of Markov Diffusion Operators*. Grundlehren der mathematischen Wissenschaften 348, Springer, 2014.

[4] J.-D. Benamou and Y. Brenier. A computational fluid mechanics solution to the Monge–Kantorovich mass transfer problem. *Numerische Mathematik*, 84(3):375–393, 2000.

[5] P. Cardaliaguet, F. Delarue, J.-M. Lasry and P.-L. Lions. *The Master Equation and the Convergence Problem in Mean Field Games*. Annals of Mathematics Studies 201, Princeton University Press, 2019.

[6] R. Carmona and F. Delarue. *Probabilistic Theory of Mean Field Games with Applications*, volumes I and II. Probability Theory and Stochastic Modelling 83–84, Springer, 2018.

[7] X. Cheng, N. S. Chatterji, P. L. Bartlett and M. I. Jordan. Underdamped Langevin MCMC: a non-asymptotic analysis. *Proceedings of the 31st Conference on Learning Theory*, PMLR 75:300–323, 2018.

[8] J. M. Cohen, S. Kaur, Y. Li, J. Z. Kolter and A. Talwalkar. Gradient descent on neural networks typically occurs at the edge of stability. *International Conference on Learning Representations*, 2021. arXiv:2103.00065.

[9] N. Fournier and A. Guillin. On the rate of convergence in Wasserstein distance of the empirical measure. *Probability Theory and Related Fields*, 162(3–4):707–738, 2015.

[10] U. G. Haussmann and E. Pardoux. Time reversal of diffusions. *The Annals of Probability*, 14(4):1188–1205, 1986.

[11] A. Hyvärinen. Estimation of non-normalized statistical models by score matching. *Journal of Machine Learning Research*, 6:695–709, 2005.

[12] R. Jordan, D. Kinderlehrer and F. Otto. The variational formulation of the Fokker–Planck equation. *SIAM Journal on Mathematical Analysis*, 29(1):1–17, 1998.

[13] I. Karatzas and S. E. Shreve. *Brownian Motion and Stochastic Calculus*, 2nd ed. Graduate Texts in Mathematics 113, Springer, 1991.

[14] D. Lacker and L. Le Flem. Sharp uniform-in-time propagation of chaos. *Probability Theory and Related Fields*, 187(1–2):443–480, 2023.

[15] J.-M. Lasry and P.-L. Lions. Mean field games. *Japanese Journal of Mathematics*, 2(1):229–260, 2007.

[16] C. Léonard. A survey of the Schrödinger problem and some of its connections with optimal transport. *Discrete and Continuous Dynamical Systems*, 34(4):1533–1574, 2014.

[17] Y. Lipman, R. T. Q. Chen, H. Ben-Hamu, M. Nickel and M. Le. Flow matching for generative modeling. *International Conference on Learning Representations*, 2023. arXiv:2210.02747.

[18] Q. Liu and D. Wang. Stein variational gradient descent: a general purpose Bayesian inference algorithm. *Advances in Neural Information Processing Systems* 29, 2016.

[19] Q. Liu. Stein variational gradient descent as gradient flow. *Advances in Neural Information Processing Systems* 30, 2017.

[20] A. Nitanda, D. Wu and T. Suzuki. Convex analysis of the mean field Langevin dynamics. *Proceedings of the 25th International Conference on Artificial Intelligence and Statistics*, PMLR 151:9741–9757, 2022.

[21] F. Otto and C. Villani. Generalization of an inequality by Talagrand and links with the logarithmic Sobolev inequality. *Journal of Functional Analysis*, 173(2):361–400, 2000.

[22] G. A. Pavliotis. *Stochastic Processes and Applications: Diffusion Processes, the Fokker–Planck and Langevin Equations*. Texts in Applied Mathematics 60, Springer, 2014.

[23] D. Revuz and M. Yor. *Continuous Martingales and Brownian Motion*, 3rd ed.. Grundlehren der mathematischen Wissenschaften 293, Springer, 1999.

[24] S. Särkkä and A. Solin. *Applied Stochastic Differential Equations*. Institute of Mathematical Statistics Textbooks, Cambridge University Press, 2019.

[25] Y. Song, J. Sohl-Dickstein, D. P. Kingma, A. Kumar, S. Ermon and B. Poole. Score-based generative modeling through stochastic differential equations. *International Conference on Learning Representations*, 2021.

[26] A.-S. Sznitman. Topics in propagation of chaos. In *École d'Été de Probabilités de Saint-Flour XIX—1989*, Lecture Notes in Mathematics 1464, pages 165–251. Springer, 1991.

[27] S. Vempala and A. Wibisono. Rapid convergence of the unadjusted Langevin algorithm: isoperimetry suffices. *Advances in Neural Information Processing Systems* 32, pages 8092–8104, 2019. arXiv:1903.08568.

[28] C. Villani. *Topics in Optimal Transportation*. Graduate Studies in Mathematics 58, American Mathematical Society, 2003.

[29] C. Villani. *Optimal Transport: Old and New*. Grundlehren der mathematischen Wissenschaften 338, Springer, 2009.

[30] P. Vincent. A connection between score matching and denoising autoencoders. *Neural Computation*, 23(7):1661–1674, 2011.

[31] G. Wang. *High-Dimensional Statistics for AI: Probability, Geometry, Spectra, and the Limits of Learning-Theoretic Transfer*. Mathematics for AI 1, 2026.

[32] G. Wang. *Optimal Transport for AI: Theory, Computation, and Applications*. Mathematics for AI 2, 2026.

[33] G. Wang. *Information Theory for AI: Compression, Representation, and the Limits of Machine Knowledge*. Mathematics for AI 6, 2026.

Subject Index